CW00740148

HOW PLANTS WORK

Edited by **Stephen Blackmore**

HOW PLANTS WORK

FORM, DIVERSITY, SURVIVAL

With a foreword by Peter Crane

PRINCETON UNIVERSITY PRESS

PRINCETON AND OXFORD

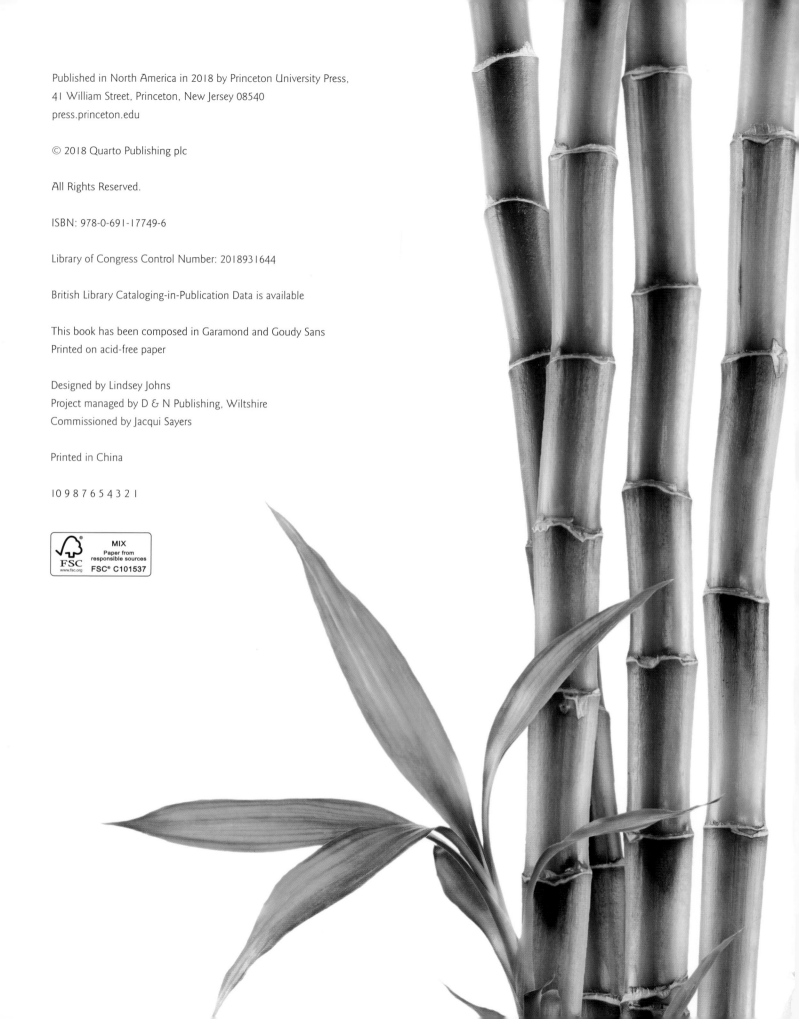

Published in North America in 2018 by Princeton University Press,
41 William Street, Princeton, New Jersey 08540
press.princeton.edu

ISBN: 978-0-691-17749-6

Library of Congress Control Number: 2018931644

British Library Cataloging-in-Publication Data is available

This book has been composed in Garamond and Goudy Sans
Printed on acid-free paper

Designed by Lindsey Johns
Project managed by D & N Publishing, Wiltshire
Commissioned by Jacqui Sayers

Printed in China

10 9 8 7 6 5 4 3 2 1

FSC
www.fsc.org
MIX
Paper from
responsible sources
FSC® C101537

Contributors

Preface, chapter 8 and consultant editor
Stephen Blackmore CBE, VMH, PhD, FRSE
Queen's Botanist and Honorary Fellow, Royal Botanic Garden Edinburgh, United Kingdom
Dr Stephen Blackmore completed his PhD at the University of Reading in the United Kingdom. He then worked on Aldabra Atoll in the Indian Ocean, before being appointed Lecturer in Biology and Head of the National Herbarium and Botanic Gardens at the University of Malawi. In 1980, he was appointed Head of Palynology at the Natural History Museum in London. He was the 15th Regius Keeper of the Royal Botanic Garden Edinburgh from 1999 until 2013, and was appointed Her Majesty's Botanist in Scotland in 2010.

Chapter 1
Andrew Drinnnan PhD
School of BioSciences, University of Melbourne, Australia
Dr Andrew Drinnan is an Australian botanist who completed his MSc and PhD in the School of Botany at the University of Melbourne. After several years at the Field Museum in Chicago, he returned to the University of Melbourne in 1990, where he is now Associate Professor in the School of BioSciences. His research involves palaeobotany and the fossil record of plants, particularly in Australia and Antarctica, and the structure and development of living plants. He teaches a wide variety of plant science subjects at both undergraduate and graduate levels.

Chapter 2
Taryn Bauerle PhD
School of Integrative Plant Science, Cornell University, United States of America
Dr Taryn Bauerle received her doctorate from the Pennsylvania State University and is currently an Associate Professor at Cornell University, an August-Wilhelm Scheer Professor at the Technische Universität München and an Atkinson Center for a Sustainable Future Faculty Fellow. Dr Bauerle's research addresses changes in whole plant water relations resulting from climate change, with a focus on root and rhizosphere responses to water stress.

Chapter 3
Jarmila Pittermann PhD
Department of Ecology and Evolutionary Biology, University of California, United States of America
Dr Jarmila Pittermann received her doctorate at the University of Utah, and is currently an Associate Professor at the University of California, Santa Cruz. Her research focuses on plant water relations, ecophysiology, and the structure and function of vascular tissue through the prisms of ecology, evolution and climate change.

Chapters 4, 6 and 7
Timothy Walker MA, MHort, PGDipLATHE, FHEA, FLS
Somerville College, University of Oxford, United Kingdom
Timothy Walker read botany at the University of Oxford and then worked for 34 years in botanical gardens, including 26 years as Director of the Oxford Botanic Garden. He is now a lecturer and tutor at the University of Oxford, and has a particular interest in plant conservation, evolution, pollination and euphorbias.

Chapter 5
Frederick B. Essig PhD
Associate Professor, Emeritus, University of South Florida, United States of America
Dr Frederick B. Essig received his BS in botany from the University of California at Riverside, and his PhD from Cornell University. He has studied the systematics of palms, *Clematis* and mosses. He was Director of the University of South Florida (USF) Botanical Gardens in Tampa and a member of the USF faculty until his retirement in 2010. He maintains an active research interest in the taxonomy of the mosses of Florida, and blogs regularly on many aspects of botany and wildflowers at botanyprofessor.blogspot.com.

Contents

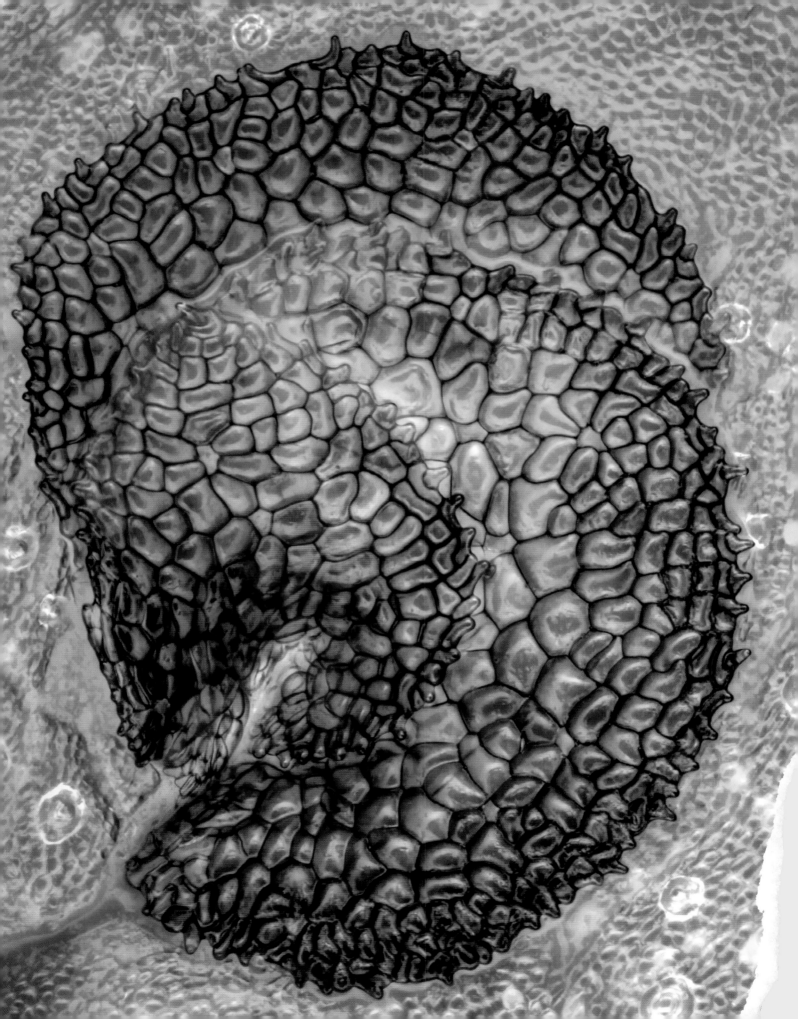

Foreword

Peter Crane FRS

Plants are the foundation of human existence. They are our indispensible companions on this singular planet in the vastness of space. Our species evolved against a changing backdrop of forests and grasslands, and our bodies are shaped by millions of years of interactions with plants. Today, plants remain the ultimate source of all the chemical energy on which our day-to-day existence depends. And make no mistake, the present and future of all of humanity depends on the continuing miracle of photosynthesis, which provides food for us all.

Yet plants, with no voice of their own, are easy to take for granted. The direct connections between people and plants that once shaped all human lives are now obscured by the rise of modernity. Our specialized lifestyles push plants to the background, and with more than half of all people now living in cities, the lives of plants can seem increasingly remote.

This fascinating and richly illustrated book reintroduces us to the world of plants and the intricacies of their existence, including how they live, grow and reproduce. It is an intimate, close-up portrait that deepens our understanding of the commonplace and the exotic. At the same time, it reveals the beauty of plants in new ways. The diversity of plants is brought to life through exemplars that engage, and through insights that enrich. To borrow a phrase from Darwin, there is grandeur in this view of plants. I am sure you will enjoy it.

⟨ Even the smallest plants appear striking when highly magnified. The most widespread form of the common liverwort (*Marchantia polymorpha* ssp. *ruderalis*), found in man-made or disturbed habitats, has colourful scales that provide protection to its growing tips.

Ⓛ The foliage of ferns grows and unfurls in a characteristic shape, called a 'crozier' because of its resemblance to the head of the staff carried as a symbol of office by a bishop. The small scales covering its surface provide protection as the frond grows.

Editor's preface: the world of plants

(A) mid the hustle and bustle of our fast-paced lives, an everyday miracle is happening all around us, often unnoticed. Plants are working their silent and unobtrusive magic – capturing the energy of the sun and powering life on Earth. At first glance, plants rarely seem to be doing anything at all, perhaps just swaying gently in the breeze. But look more closely, as this book does, and we soon discover that plants are hard at work. Absorbing minerals and water from the soil and carbon dioxide from the air, they are creating themselves, fuelling their growth with the products of photosynthesis.

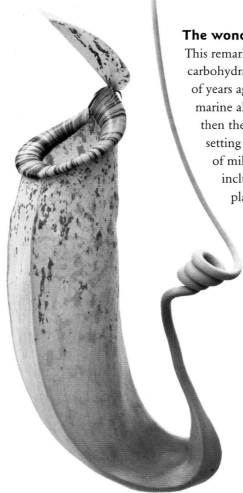

The wonder of photosynthesis

This remarkable process, which creates carbohydrates and oxygen, began billions of years ago in the first single-celled marine algae. It made the oceans and then the atmosphere rich in oxygen, setting the stage for the evolution of millions of species of animals, including our own. Animals and plants face the same challenges: the need for food or a source of energy, and the need for

⊤ Flowerheads of purple salsify (*Tragopogon porrifolius*) open only to expose their bright orange pollen in the morning and then close in the early afternoon.

◁ The leaves of tropical pitcher plants (*Nepenthes* spp.) have highly modified tips, which form liquid-filled traps containing enzymes to capture and digest insects.

a safe place in which to live and reproduce. Being solar powered, plants have solved the energy problem but, unlike animals, they must tackle the other challenges without moving from the spot they are rooted to.

By many ingenious devices, they have clearly succeeded. There are more than 400,000 species of plants, with more being discovered every year, and they have spread everywhere on Earth where water is freely available. Only the driest deserts, and the permanently frozen polar regions and highest mountains, are devoid of vegetation. The diversity of vegetation is enormous – from tropical rainforests to savannah, boreal forests, tundra, mangrove swamps and alpine meadows – and it shapes and defines the landscapes of our planet. Our world is full of life because it is green.

This book is an exploration of how plants have diversified through time, from the dawn of life on land, and how they have become adapted for survival in a thousand and one different forms. It examines the microscopic and the short-lived, right up to the giant redwoods (*Sequoiadendron giganteum*), thousands of years old and the largest living things our planet has ever known. Starting with the story of plant evolution, we explore plants organ by organ to understand the ways in which their internal anatomy and external form are fine-tuned solutions to physical and biological challenges.

The life of roots is revealed as a dynamic interaction between the plant and the physical environment of soil and rock. This goes far beyond simply anchoring the plant and absorbing water. A host of biological interactions are hidden from sight: some roots partner with fungi, which help them to absorb nutrients from the soil; others harness the ability of bacteria to capture nitrogen from the air to enrich the soil they grow in. Moving upwards, stems raise plants higher towards the light, and branch, with complex architecture, to increase the area exposed to light. Specialized cells and tissues allow water to be transported to the tips of the highest trees, and sugars to be moved upwards or downwards between the leaves and the root system. The leaf is the fundamental powerhouse of photosynthesis but it can take many forms. Some float on the surface of water, while others are traps that capture and digest insects and other small animals.

The life cycles and reproduction of plants are explored next. Sexual reproduction is, in many ways, more varied in plants than it is in animals: some have sperm cells, male gametes, that swim freely through films of water, while others deliver sperm to the egg cell in pollen tubes. Often, plants increase their numbers by vegetative propagation, which does not involve sex, so a single individual can establish a new colony. The form of plant reproductive organs – cones and flowers – displays huge diversity, reflecting both evolutionary history and mode of sexual reproduction. Wind, water and animals – from insects and birds to mammals – are employed as pollinating agents. The dazzling diversity of seeds and fruits reveals adaptations for dispersal and many further interactions with animals.

Finally, the ways in which we humans depend on plants in our daily lives are explored. The domestication of plants and adoption of settled agriculture marked the birth of civilizations around the world. On today's crowded planet, many plants are facing extinction.

The contributors to this book hope to heighten your appreciation of plants and inspire you to think more carefully about our relationship with them. We need to see them as our life-support system and treasure them accordingly. This is also a celebration of the beauty and rich complexity of plant life. Plants inspire us in gardens and in art, we like to bring them into our homes and a bouquet of flowers is our warmest gift. So, settle down, take a deep breath, remember where that oxygen came from, and read on.

⌄ Flower of thale cress (*Arabidopsis thaliana*), used as a model organism in developmental studies and the first plant to have its genome sequenced. Artificial colours identify petals (lilac), stamens (lime-green anthers and brown filaments), stigma (blue) and ovary (olive).

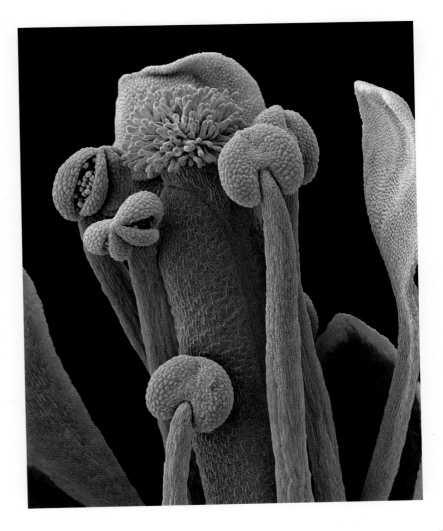

An introduction to plant morphology

Plants are such a familiar part of the landscape that we commonly take them for granted. Of course, we know their importance to us as sources of food, materials and medicines, but we often do not appreciate just how fundamental plants are to life on Earth. Through the process of photosynthesis, they trap energy from sunlight and use that energy to convert carbon dioxide and water into organic molecules. When plants are eaten by other organisms, this carbon and energy is incorporated into their bodies. And it is not just through the production of carbon and energy that plants are useful; with their roots firmly embedded in the soil, they take up virtually all the minerals that enter the biosphere. Through this intimate contact with sunlight, air and soil, plants provide the link between the living and non-living components of the Earth's system.

Plants have not always been part of Earth's terrestrial environment. However, their colonization of the land over the past 400 million years has been the game changer that allowed animals to follow them out of the water, and eventually led to the complex biotas we see today. This chapter introduces plants – what they are, where they came from, how we describe their structure and the different types of plants – and sets the scene for exploring them in greater detail through the rest of the book.

What is a plant?

Plants are multicellular photosynthetic organisms that have adapted to life on land. Their cells are specialized and arranged to form tissues and organs, they have an outer protective cover of cutin, and they have reproductive organs with an outer layer of vegetative cells to protect their sex cells. After the egg cell is fertilized during sexual reproduction, the embryo into which it develops is retained on the parent plant, where it continues to be nourished; this feature gives land plants their formal scientific name, the Embryophyta, or embryophytes.

△ Tree-sized horsetails once dominated the Earth's vegetation, but only herbaceous species like the common horsetail (*Equisetum arvense*) are present today.

▽ *Conocephalum* species are among the largest of the liverworts.

Bryophytes

The most primitive plants are the liverworts, hornworts and mosses, together informally called the bryophytes. These plants lack water-conducting tissue and hence are also often referred to as non-vascular land plants. They are usually found in wet or damp environments, and are small and grow close to the substrate, allowing them to absorb water readily. Bryophytes reproduce by dispersing small spores into the environment.

Some liverworts, like common garden weeds in the genera *Marchantia* and *Lunularia*, are simple flattened plants that are anchored to the soil by a mat of hair-like rhizoids; other liverworts are small and leafy. Hornworts are superficially like the flattened liverworts. In contrast, all mosses are leafy, and although they lack specialized strengthening and supporting tissue and are generally small in stature, some species are quite stiff and robust: the largest species, tall dawsonia (*Dawsonia superba*), reaches up to 40 cm (16 in) in height in southeastern Australian eucalypt forests.

Fern allies

Another group of plants that disperse their spores freely, but do conduct water throughout their tissues and so can grow to be larger and more complex, are known as the fern allies. The lycopods, which include clubmosses, quillworts and spikemosses, have a fossil record extending back about 400 million years, making them the most ancient lineage of living plants. Whisk or fork ferns (Psilotales) and adder's-tongue ferns (Ophioglossales) are also ancient lineages,

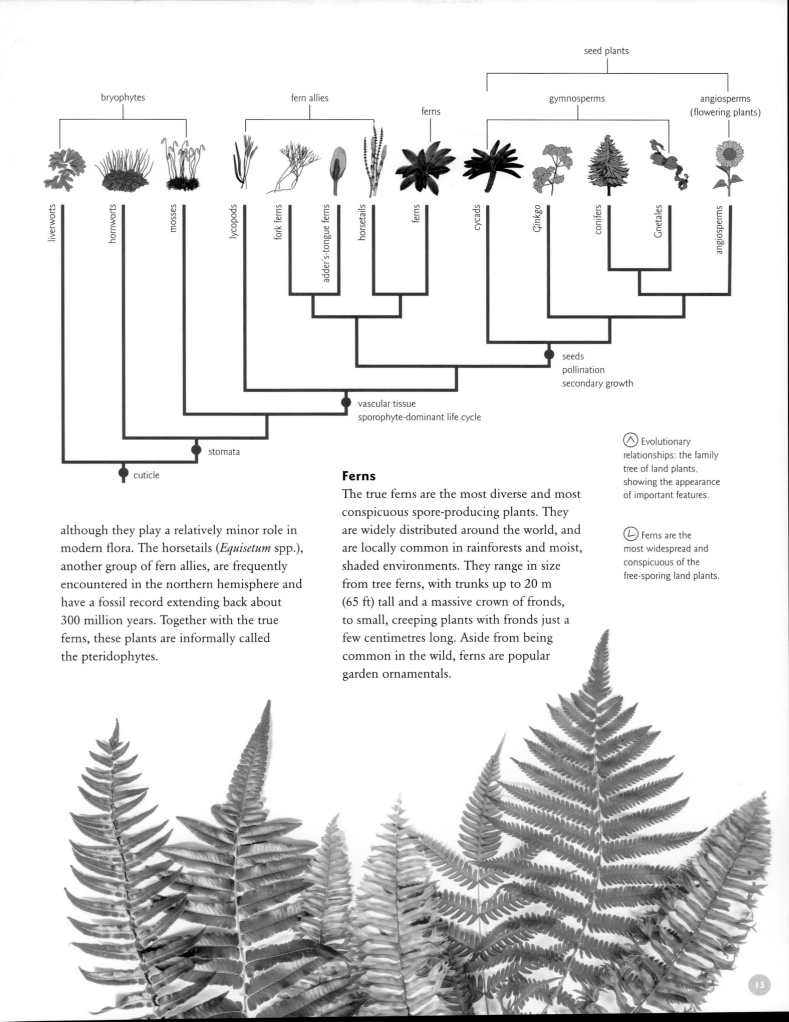

seed plants

bryophytes

fern allies

ferns

gymnosperms

angiosperms
(flowering plants)

liverworts

hornworts

mosses

lycopods

fork ferns

adder's-tongue ferns

horsetails

ferns

cycads

Ginkgo

conifers

Gnetales

angiosperms

seeds
pollination
secondary growth

vascular tissue
sporophyte-dominant life cycle

stomata

cuticle

Evolutionary
relationships: the family
tree of land plants,
showing the appearance
of important features.

Ferns are the
most widespread and
conspicuous of the
free-sporing land plants.

Ferns

The true ferns are the most diverse and most
conspicuous spore-producing plants. They
are widely distributed around the world, and
are locally common in rainforests and moist,
shaded environments. They range in size
from tree ferns, with trunks up to 20 m
(65 ft) tall and a massive crown of fronds,
to small, creeping plants with fronds just a
few centimetres long. Aside from being
common in the wild, ferns are popular
garden ornamentals.

although they play a relatively minor role in
modern flora. The horsetails (*Equisetum* spp.),
another group of fern allies, are frequently
encountered in the northern hemisphere and
have a fossil record extending back about
300 million years. Together with the true
ferns, these plants are informally called
the pteridophytes.

Gymnosperms

Plants that produce seeds are called spermatophytes and are split into two main groups, the gymnosperms (meaning 'naked seeds') and the angiosperms, or flowering plants. The gymnopserms include the cycads, ginkgo (*Ginkgo biloba*), the gnetophytes and the conifers.

Cycads, numbering about 70 species in three extant families, look superficially like small palm trees, but they represent the remnants of a group that reached its greatest importance in the Mesozoic period, 252–66 million years ago – the age of the dinosaurs. The genus *Ginkgo*, with just one surviving species, is another relict – its relatives were also prominent in vegetation of the Mesozoic, but it is now restricted to a small region of China. Its alternative common name, maidenhair tree, refers to its delicate and ornamental foliage, and it has long been prized as a horticultural specimen tree. The gnetophytes also arose in the Mesozoic, and today they comprise three genera: *Gnetum*, *Welwitschia* (with just one species) and *Ephedra*.

The conifers, numbering around 630 species, are the most familiar of the non-flowering seed plants and mainly produce their seeds in woody cones. Pines (*Pinus* spp.), spruces (*Picea* spp.), firs (*Abies* spp.) and larches (*Larix* spp.) dominate large areas of boreal forests, and other conifers are represented in most vegetation types of the world, from rainforests to arid areas. The giant redwood (*Sequoiadendron giganteum*) of California is the tallest of all plants, some individuals comfortably exceeding 100 m (330 ft) in height. The Great Basin bristlecone pine (*Pinus longaeva*), also from California, and additionally Nevada and Utah, is the longest-lived of all plants – tree ring counts have determined some individuals to be around 5,000 years old. Conifers are an important forestry resource, providing softwoods for timber and paper.

(V) Pine trees (*Pinus* spp.) are grown extensively for softwood timber and paper production.

Angiosperms

Flowering plants, or angiosperms, are the most diverse and abundant group of land plants. Including more than 250,000 species, the group comprises the most prominent and most dominant components of the world's vegetation. They range from tiny herbs through to woody shrubs and the tallest trees – a mountain ash (*Eucalyptus regnans*) in Tasmania, Australia, is 99.8 m (327.4 ft) in height, just shy of the giant redwood conifers. With such an enormous diversity, the angiosperms have managed to occupy most habitats worldwide, including the seagrasses in marine environments. Their reproductive strategies include pollination and seed dispersal by wind, water and a range of interactions with animals, the latter including bizarre examples of reward and deception.

Ⓛ *Amborella trichopoda*, a small rainforest tree from New Caledonia, is the most primitive flowering plant.

⊘ Mountain ash (*Eucalyptus regnans*), the tallest flowering plant, dominates extensive montane forests in southeastern Australia.

REASSESSING RELATIONSHIPS

The traditional method of determining the relationships between plants, and classifying them, has relied on morphological characters that can be readily observed, such as stomata, vascular tissue, woody tissues and flowers. However, since the 1990s there has been an ever-increasing emphasis on DNA sequence data, to the extent that most systematic studies now rely on this. DNA (deoxyribonucleic acid) is the genetic material in cells, and in plant cells there are three sets of DNA, one each in the nucleus, chloroplast and mitochondria; all three can be used to assess the degree of similarity between different plant groups. DNA data has led to some significant reassessments of plant relationships over the past two decades, including the recognition of lycophytes as the most primitive vascular plants, the grouping of Psilotales with ferns, the affinities of gnetophytes with conifers, and the clarification that *Amborella trichopoda* from New Caledonia is the most primitive flowering plant.

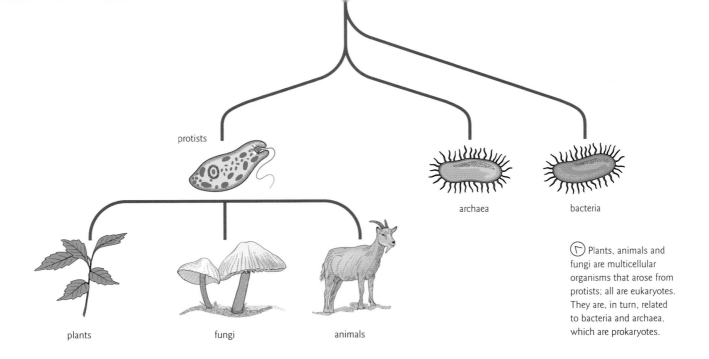

protists

archaea

bacteria

plants

fungi

animals

ⓘ Plants, animals and fungi are multicellular organisms that arose from protists; all are eukaryotes. They are, in turn, related to bacteria and archaea, which are prokaryotes.

Where did plants come from?

Ⓟlants are such a ubiquitous part of our environment that it is difficult to imagine a world without them. But there was a time when there were no plants. Prior to the end of the Cambrian period, about 485 million years ago, life on Earth was much simpler, and it was mostly found in the sea, or at least in water. Animals had already begun their evolutionary journey, and were rapidly expanding the variety of their body forms in the sea. But plants were late starters, lagging several hundred million years behind their animal cousins.

Six kingdoms

All living organisms are arranged into a classification of six kingdoms. A major division is between the two prokaryote kingdoms (bacteria and archaea) and the eukaryote kingdoms (protists, plants, fungi and animals). The prokaryotes are simple cells lacking a well-defined nucleus, whereas the eukaryotes have more complex cells with a membrane-bound nucleus and intricate intracellular structures. A significant feature of all eukaryote cells is the presence of mitochondria. These specialized organelles harvest energy from food molecules, and were themselves once bacteria that became incorporated into eukaryotic cells in a process known as endosymbiosis.

The protists

The simplest of the eukaryotes are the protists, a heterogeneous assemblage of mostly single-celled organisms. They are primarily organisms of aquatic or damp environments, but they also include some of our most significant disease-causing parasites, and they may be either photosynthetic and non-photosynthetic (colourless). Diatoms and dinoflagellates are photosynthetic protists, and between them account for a major proportion of the sun's energy captured in the biosphere. Dinoflagellates are responsible for toxic red tides, and their close relatives the apicomplexans are a group of parasites that cause diseases such as malaria and toxoplasma.

The colourless protists include amoebae, ciliates and flagellates, some of which are also responsible for human disease. Because most protists are single-celled organisms, and microscopic, they are usually observed only as slime in ponds or in damp places. Only the large multicellular seaweeds, species of brown, red and green algae, can be readily observed with the naked eye.

Higher organisms

The remaining three eukaryote kingdoms, the fungi, animals and plants, arose independently from within the protists, and are notable for their evolution as complex, multicellular organisms. Animals and fungi originated close to one another, which is evinced by some similar features such as cell walls made of chitin.

Plants arose from the lineage of protists that comprises the red and green algae. This lineage is referred to as the primary photosynthetic group, and it represents the descendants of the organisms that captured the original chloroplast around 900 million years ago

(see page 145). Green algae share a number of features with land plants, such as the photosynthetic pigments chlorophyll *a* and *b*, the same storage carbohydrate (starch), and cell walls made of cellulose.

⌃ Green seaweeds are protists on the evolutionary lineage leading to plants.

DISTANT RELATIVES

The freshwater green algae of the orders Charales and Coleochaetales (in particular members of the genus *Coleochaete*), are thought to be the closest modern-day relatives of land plants. They share a number of important characters, including a particular type of cell division that is not seen in most other algae, similarly shaped sperm cells, and some similar enzymes. The study of *Coleochaete* DNA sequences confirms this close relationship.

Because Charales and Coleochaetales species are aquatic, it is thought that their divergence from plants occurred well before colonization of the land.

⌄ Stonewort algae, such as this *Chara globularis*, are among the green algae most closely related to plants. They are common in freshwater environments.

Why are plants green?

Perhaps the most obvious feature of plants, and the vegetation they compose, is their consistent colour. While there is some variation in the hue, from the deep green of forests to the bright green of grassy fields and the grey-green of desert vegetation, the overwhelming colour of plants is green. Curiously, this greenness we perceive as so important is visible to us because the colour green is of no importance to plants.

Reflecting on colour

Plants are fundamentally different to animals because they make their own food molecules – the sugars that all cells use as an energy source. Animals, including humans, get their food molecules by eating plants, or eating things that have in turn eaten plants.

The process plants use to make these food molecules is called photosynthesis, whereby they capture the energy from sunlight and use that energy to convert carbon dioxide from the air, along with water, into sugars.

Plants contain a number of different pigment molecules that capture the sunlight and harvest its energy. The most important of these pigments, and the one that captures the vast majority of light, is chlorophyll.

The overall colour of sunlight is white, but this visible spectrum is actually a combination of many different wavelengths, each with its own colour. These different wavelengths can be separated, as when sunlight shines through raindrops to create a rainbow. Red, orange and yellow are at the long wavelength end; blue, indigo and violet are at the short wavelength end; and green is in between. Chlorophyll pigments are choosy when it comes to light. They absorb light at the reddish and bluish regions of the spectrum, but not in the middle. As green light is not collected by the plant, it is reflected back, which is why plants appear green.

$\vee$ Chloroplasts are small organelles in plant cells that contain membranes with the photosynthetic pigments.

CONTRASTING CHLOROPLASTS

The chloroplasts of red algae have a similar structure and pigments to those in a cyanobacterial cell. The knobbly appearance of the membrane system, which is the location of specialized red and blue light-capturing pigments called phycobilins, is indicative of both the link between cyanobacteria and chloroplasts and the primitive condition of the chloroplast in red algae. Chloroplasts of green algae and plants differ from those found red algae, in that their membranes are folded into pancake-like stacks and are dominated by the green light-capturing pigment chlorophyll.

DIFFERENT GREENS

There are two types of chlorophyll in plants, designated chlorophyll *a* and chlorophyll *b*, which absorb slightly different wavelengths of light. Plants in low light environments have more chlorophyll *b* than *a*, and appear dark green; plants in brighter environments are just the opposite – they have more chlorophyll *a* and appear brighter green. Plants of drier environments appear greyish green because their thick, waxy cuticle affects the way light is reflected.

Chlorophylls dominate the pigment composition of healthy leaves, so these structures almost invariably appear some shade of green. But there are other pigments in leaves that reflect different colours – the yellow xanthophylls, orange carotenoids and purple anthocyanins. As leaves of deciduous trees cease function in autumn, the chlorophyll breaks down and the colours of the other pigments are revealed.

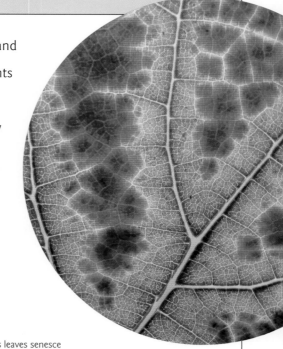

⑦ As leaves senesce in autumn, the green chlorophylls break down to reveal the underlying colours of other pigments.

The origins of photosynthesis

Plants did not invent photosynthesis; the process occurred in cyanobacteria more than 2.5 billion years ago. In one of the most important events in life's history, which took place almost a billion years ago, a cyanobacterium was engulfed by another cell and incorporated as a specialized organelle, a chloroplast. This created a new lineage of photosynthetic organisms that has led to red algae, green algae and plants, all of which inherited their chloroplasts from the same common ancestor.

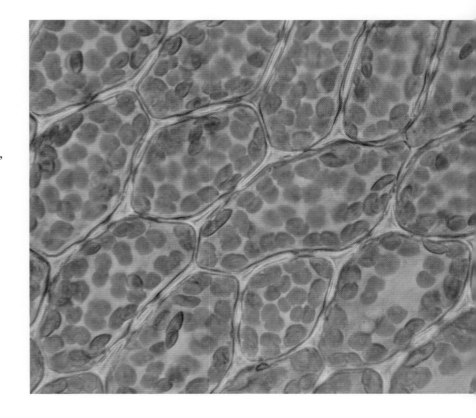

▷ Cells of the moss *Plagiomnium affine* contain numerous chloroplasts, which are pressed against the cell walls so as to be exposed to sunlight.

Living on land versus living in water

P lants are terrestrial organisms, living on land. However, we know from their evolutionary origins among the green algae that they started off in the water. Living on land is very different to living in water, and the move to a terrestrial way of life presented plants with a number of challenges. These essentially relate to four major functions: water balance and transport, structural support, gas exchange, and reproduction.

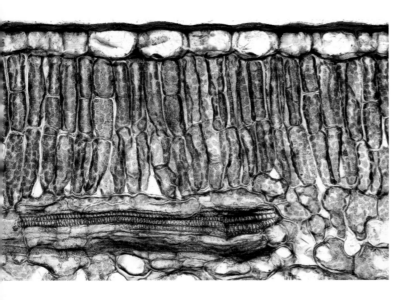

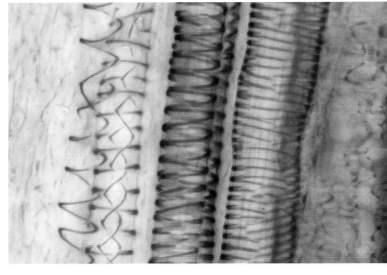

∧ The cuticle is a waxy layer secreted onto the surface of the leaf to prevent the underlying cells from losing water to the environment.

⌐ Xylem forms a plumbing system to transport water throughout the plant. The walls of individual cells are reinforced with a spiral lignin skeleton to prevent them from collapsing.

Water balance and transport

Aquatic seaweeds, such as green algae, are surrounded by water. Consequently, they have no need for special water-absorbing structures, tissues to move water internally or mechanisms to prevent them drying out. In contrast, land plants inhabit an environment where water is scarce, and usually confined to the soil in which they grow. They require specialized organs – roots – to extract water from the soil, they need complex tissue to transport this water to above-ground parts of the plant, and they need a protective, water-resistant coating to minimize water loss to the atmosphere. The degree to which these features are developed in a plant largely determines its growth form and ecological tolerance.

Xylem is the conducting tissue that transports water throughout the plant, forming a continuous plumbing system of elongated, hollow cells. The walls of these cells are conspicuously reinforced by a rigid framework of lignin, which prevents them from collapsing as water is sucked through them. Because of the need to supply water to every part of the plant, the size of a plant is limited by its vascular system.

Structural support

Aquatic plants get their structural support from their buoyancy. For example, the giant kelp (*Macrocystis pyrifera*) forests of the American Pacific coast can grow up to 50 m (160 ft) in height, but if the water were removed, they would collapse onto the sea

⊙ The intricate system of veins in a leaf contains xylem to supply water to all the leaf cells.

floor. In a terrestrial environment, if plants want to gain height to compete for sunlight with their neighbours, they must provide their own support. Lignin, the rigid compound that reinforces the walls of water-conducting cells, is also used to strengthen other tissues in the stem. In shrubs and trees, the old wood that is no longer needed for water transport provides the structural support for trunks and branches.

Reproduction

Algae reproduce in the water, many releasing their male sex cells into the water, where they must swim around to locate an egg cell. In a terrestrial environment, relying on free water for reproduction is risky. Primitive land plants such as liverworts, mosses and ferns do just this, and as a consequence they are restricted to environments where moisture is regularly available. Seed plants have overcome this need for watery reproduction by developing the process of pollination. The sperm cells are produced by germinated pollen grains, but only after these have been transported to the location of the egg cell. The pollen tube grows directly to the egg cell and deposits the sperm cells in precisely the right place.

AQUATIC ABOUT-TURN

A number of flowering plants are aquatic, but all of these have evolved from terrestrial ancestors that have subsequently returned to the water. Most have colonized freshwater habitats, but a few, such as the seagrasses, are found in the sea. Aquatic plants have had to revisit the challenges of living in water with a suite of characters that were adapted for life on land. For the large part, they have simply reverted: submerged plants do not produce a cuticle or stomata, their water-conducting tissue is reduced, and structural support is provided by buoyancy rather than strengthening tissue. However, one area in which they have had to innovate is reproduction. Constrained by flowers and pollination, which are apparently too hard (or too good) to be undone, they bloom at or above the surface of the water and achieve pollination in a conventional manner.

⊙ Pollination in the seagrass *Enhalus acoroides*. Female flowers, with three stigmas, lie on the surface of the water, where they contact the detached and floating pollen-bearing male flowers.

The evolution of plants on land

T he increase in the complexity and specialization of plant vegetative and reproductive features that occurred with their colonization of the land is the key to understanding the remarkable diversity of plant life on Earth. This is best explored in the context of the plant life cycle.

⊙ Gametophyte of the liverwort *Fossombronia*, with male sex organs. These antheridia will burst to release swimming sperm cells.

⊙ Liverwort sporophytes grow from the gametophyte and are totally dependent on them. They consist of a clear stalk and a terminal spore-producing capsule, as in these *Fossombronia* sporophytes.

The life cycle of a land plant

The land plant life cycle describes the process of sexual reproduction. This involves a specialized cell division that halves the chromosome number (meiosis) on one side of the cycle, followed by the production and fusion of sex cells (syngamy) to return to the original chromosome number on the other side.

There are two discrete stages in the life cycle of land plants, which are referred to as separate generations. This is a uniform feature from liverworts and mosses to flowering plants, and is the foundation for any comparative study of plant diversity. The gametophyte

generation is haploid (one set of chromosomes per cell) and produces two types of sex cells (gametes). One type of gamete (conventionally referred to as the male gamete, or sperm cell) is shed, while the other gamete (the female gamete, or egg cell) is retained. Male and female gametes then fuse, doubling the chromosome number. This new cell is called the zygote, and it divides and grows into the multicellular sporophyte generation, which is diploid (two sets of chromosomes per cell).

Sporophytes undergo the reduction division of meiosis to produce spores (now haploid again), which in turn germinate into new gametophytes. Because sporophyte and gametophyte generations in turn give rise to the other, they are said to alternate in the life cycle – called the alternation of generations (see box on page 185). Taking a liverwort, the most primitive land plant, as an example, the persistent green plants are the gametophyte generation in its life cycle, while the sporophyte is the white stalk with a spherical black capsule on the end. Spores are produced by meiosis in the capsule, which splits open for spore dispersal. The gametes and the organs that produce them can be observed only with a microscope.

Evolution of the plant life cycle

To explore the evolution of the life cycle of land plants, we must take a comparative approach, and the obvious comparison is with the closest relative within the green algae, *Coleochaete*. *Coleochaete* is a small disc-shaped, multicellular green alga that grows in fresh water. It is haploid, and it produces egg cells that are retained and sperm cells that are dispersed to locate and fertilize an egg cell. The fusion product of this fertilization event is a diploid zygote. So far, so good. But in *Coleochaete*, the zygote does not develop into a multicellular sporophyte. Instead, it immediately undergoes meiosis to form four haploid spores, which in turn regenerate the gametophyte generation.

Assuming that the *Coleochaete* life cycle is a primitive condition from which the land plant life cycle has evolved, we interpret the multicellular sporophyte generation as unique to land plants; a developmental stage that has been inserted into the life cycle between the zygote and meiosis. So why is this significant? At the most advanced levels of land plant evolution, for example the giant redwood conifers or the mighty mountain ash eucalypts, the plants that we observe are the sporophyte generation. These massive trees are the direct life-cycle equivalent of the tiny stalk and capsule of a liverwort. All of the structure and complexity associated with the success of plants on land, and which is encompassed in the diversity of conifers and flowering plants, has occurred in the sporophyte generation, and that sporophyte generation is unique to land plants.

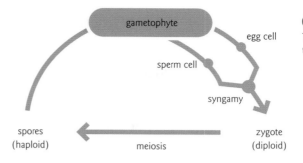

⊲ *Coleochaete* life cycle. The haploid gametophyte is the only multicellular stage.

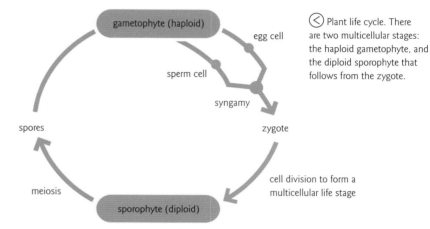

⊲ Plant life cycle. There are two multicellular stages: the haploid gametophyte, and the diploid sporophyte that follows from the zygote.

⊳ Giant sequoias (*Sequoiadendron giganteum*) are the world's tallest plant. The tree that we see is the sporophyte generation.

Evolutionary adaptations to life on land

There are number of major plant features that were critical to the success of plants on land. Some of these, such as a water-resistant cuticle and specialized water-conducting cells, have already been mentioned. These features did not all appear at once – they evolved gradually over several hundred million years, each building on the success of those that had come before them. Curiously, some of these features were tried out by different plants several times, only for them to stagnate or disappear, while others survived and succeeded to the present day.

Successful specializations

Cuticle is a feature of all plants, and was the first of the key features to appear. In liverworts and mosses the cuticle is very thin, but plants cannot survive exposed to the air without it. Stomata, the specialized pores that allow carbon dioxide to enter the plant, followed a little later; they are absent in liverworts, and first occurred in moss and hornwort sporophytes. Specialized conducting or vascular tissue first appears in lycophytes. This is also the stage of plant evolution at which we see the sporophyte becoming the prominent generation in the life cycle, and where it becomes branched and more complex in its anatomy and morphology. It is also in the lycophytes, and more noticeably in the ferns, that we see the differentiation of the plant into specialized organs such as stems, roots and leaves.

With increased complexity and an accompanying increase in size came a demand for more conducting tissues than were available. This problem was solved by the evolution of a special growing region in the stem called the vascular cambium, which produces wood, leading to the formation of shrubs and trees.

Reproductive remodelling

All plants from liverworts to ferns shed their spores directly, leaving them to fend for themselves. But in plants from the cycads up – in other words, all seed plants – the spores are surrounded by a protective covering and are retained on the parent plant. Here, the spore develops into the gametophyte, produces an egg cell, is fertilized, and forms the embryo that will become the seedling of the next sporophyte generation, all the while being nourished by its parent sporophyte – until it is shed as a seed. Seed plants also developed pollination, which delivers the sperm cells directly to the egg, removing the requirement of free water for reproduction.

Running clubmoss (*Lycopodium clavatum*). Lycophytes were the first plants to develop vascular tissue and a sporophyte-dominant life cycle.

KINGS OF THE CARBONIFEROUS

∧ This fossilized bole of *Sigillaria* demonstrates the enormous size of these now extinct arborescent lycophytes.

Lycophytes were the first vascular plants, and although they now form only a minor part of the world's flora, they were once one of the most successful and certainly one of the most innovative groups. Just as history is written by the victors, the seed plants get most of the credit as the modern pinnacle of plant evolution. But lycophytes had done much of it much earlier.

The lycophyte sporophyte is branched and complex, and the shoot is differentiated into stems and leaves. These are not the same as the true leaves of other plants, which developed independently, but they have successfully served the function of photosynthetic organs for more than 400 million years.

Lycophytes were also the first plants to have roots. Around 300 million years ago, during the Carboniferous period, they even developed a type of woody tissue from their own specialized vascular cambium, which allowed them to grow as tall trees with thick woody trunks. Forests of these lycophyte trees dominated swampy environments, and they contributed to the formation of extensive coal deposits across Europe and North America.

Sadly for lycophytes, and despite their early success, they were still constrained by a free-sporing reproductive biology that needed water for fertilization. In an unforgiving terrestrial realm, these kings of the Carboniferous were ultimately displaced by newcomers that had solved that problem.

∨ Reconstruction of a swamp forest of arborescent lycophytes, showing just how bizarre they would have looked.

Palaeobotany and the plant fossil record

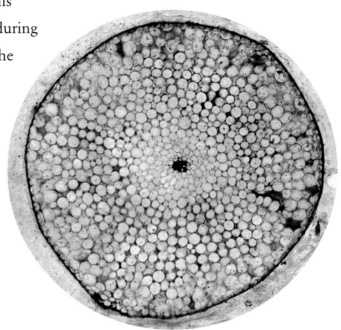

⌄ This cross section of a *Rhynia gwynne-vaughanii* stem reveals the remarkable cell detail preserved in the Rhynie Chert.

⊤he fossil record of plants is extensive, and tells a complex story from the time plants appeared during the Silurian more than 400 million years go to the present day. Although we can infer much about the evolution of plants by looking at those that are living today, the fossil record provides a unique window into the past. It tells us about groups of plants that are now extinct, about the ages of different plant groups and when key features appeared, and about geographic distributions that are very different to those we see today.

First fossils

The first fossil evidence of land plants is dispersed spores, but we do not know anything about the plants that produced them. Owing to the delicate nature of primitive plants at the bryophyte level, their preservation as fossils is unlikely, and indeed the earliest fossils we find are of vascular plants, which are more robust.

One of the most impressive fossil beds is the Rhynie Chert, located near the village of Rhynie in Scotland. Here, about 410 million years ago, plants that were growing in the margins of a spring were embedded in rock that crystallized around them from the mineral-rich water. When these rocks are

⌄ A polished piece of Rhynie Chert, showing numerous stems of a *Rhynia* plant that were preserved in their living position when they were entombed in the rock.

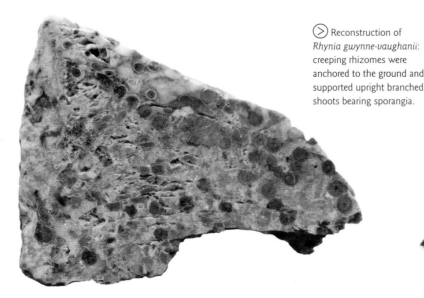

⟩ Reconstruction of *Rhynia gwynne-vaughanii*: creeping rhizomes were anchored to the ground and supported upright branched shoots bearing sporangia.

ONE AND THE SAME

In many locations worldwide, plants in the genus *Archaeopteris* are found as fossils in Late Devonian deposits dating back around 380 million years. They have large frond-like leaves, and were originally thought to be ferns because they have unprotected sporangia that split open and shed their spores. Also found in deposits of the same era are pieces of fossilized wood known as *Callixylon*. This wood is superficially conifer-like, and was clearly produced by a stem that had a vascular cambium. In the 1960s, some particularly well-preserved *Archaeopteris* fronds were discovered, and the stalks of these fronds contained *Callixylon* wood. It turned out that the two different fossils were indeed parts of the same plant. All living plants that produce wood also have seeds, and vice versa; yet here was a fossil with wood, but not seeds. We now have a number of different *Archaeopteris/ Callixylon* fossil plants, which are classified in an extinct group called the progymnosperms. These finds tell us that woody growth and the development of trees preceded the evolution of seed plant biology – important knowledge that we could not have discovered from living plants alone.

(∧) *Archaeopteris* was one of the first trees. It had a conifer-like trunk and wood, but frond-like foliage that bore sporangia.

sliced very thinly and observed with a microscope, all of the cell details are visible, providing a wealth of information about their structure and biology. We have learnt two new and particularly important things from the Rhynie fossils. The first is that sporophytes were branched before the evolution of vascular tissue; in living plants, these features always occur together. The second is that in early land plant evolution, the gametophytes were branched and complex like the sporophytes, and had stomata and possibly conducting tissues; these features are absent in gametophytes of living vascular plants.

Glossopteris and Gondwana

Between 250 and 300 million years ago, during the geological age known as the Permian period, the world was a very different place from today. In the southern hemisphere continents, sedimentary rocks deposited during this time now contain massive coal deposits – the fossilized remains of swamp vegetation. These deposits are dominated by a now extinct group of plants in the order Glossopteridales, which includes the genera *Glossopteris* and *Vertebraria*. These were large trees with many distinctive features that make their fossils easily recognizable. Foremost, their leaves were broad

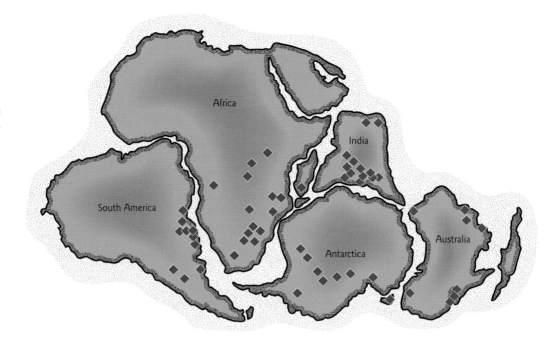

> Arrangement of the southern continents in the single landmass of Gondwana, showing the distribution of *Glossopteris* fossils in Permian deposits.

> Robert Falcon Scott and his party taking a 'selfie' at the South Pole. During their trek they discovered *Glossopteris* leaf fossils in the Transantarctic Mountains.

and tongue-shaped (hence the genus name *Glossopteris* – *glosso* is Greek for 'tongue' and *pteris* means 'fern'), with a very characteristic reticulate or net-like vein pattern. Their roots contained numerous air spaces to prevent waterlogging in their swampy home, and they appear jointed when fossilized (hence the genus name *Vertebraria*, because of the superficial resemblance of the roots to a backbone). It was this distribution of glossopterid fossils in South America,

Africa, India and Australia that led geologists in the late nineteenth century to first propose that these now separate continents were once connected long ago as the super-continent Gondwana.

GREAT SCOTT!

During British explorer Robert Falcon Scott's ill-fated trek to the South Pole in 1912, *Glossopteris* fossils were discovered in the Transantarctic Mountains. Edward Wilson, the expedition doctor and a keen naturalist, realized their importance to the debate on continental connections, and so the fossils were retained by the group to the very end; the rescue party recovered the specimens the following summer, and they now reside in the Natural History Museum in London. The presence in Antarctica of Permian *Glossopteris* and coal seams, and indeed other plant fossils from as recent as several million years ago, reveals that the continent has only recently become the frozen landscape that we recognize today.

The tongue-shaped leaves of *Glossopteris* are widespread in Gondwana fossil deposits from the Permian period.

Cross section through a *Vertebraria* root. Large wedge-shaped spaces between the spokes of wood allowed air to circulate in the roots.

Fossil pollen and flowers

Flowering plants, with more than 250,000 species, dominate the Earth's modern-day vegetation in terms of sheer diversity and variety. The English naturalist Charles Darwin, whose seminal work *On the Origin of Species*, published in 1859, gave us the first formulation of evolutionary theory, was at a loss for an explanation of flowering plants, describing them as an 'abominable mystery' in a letter to his colleague Joseph Hooker in 1879. The ensuing 150 years has seen a vast amount of research piecing together the history and relationships of these important plants, and palaeontology has played a critical role in this.

The earliest definitive evidence of flowering plants in the fossil record are pollen grains from the early Cretaceous period, approximately 145 million years ago, although rare pollen finds from Triassic sediments dating back almost 200 million years are very intriguing. Angiosperm pollen can be recognized by its distinctive cell wall structure, but nothing is known about the plants that produced this early pollen. By the mid-Cretaceous, about 100 million years ago, angiosperms were common enough that their leaves and pollen are well represented in fossil floras. But plant relationships are based largely on the features of their flowers, not their leaves, and since it is the flowers that also produce pollen, fossil flowers provide the most definitive information. Unfortunately, fossil flowers are not often discovered.

Mid-Cretaceous sediments from the eastern seaboard of North America have just the right combination of factors for the preservation of early fossil flowers – they were very fine, silty clays deposited in still water, and have remained unaltered by geological processes such as compression and extreme temperatures. These beds have yielded fragments of flowers at different developmental stages, such as small intact flower buds, stamens that still contain pollen, young fruits with pollen grains on the stigmas, and twigs showing the arrangement of flowers and their parts. This has allowed not only the reconstruction of the flowers and determination of their relationships, but also their link to pollen that is widely distributed in the fossil record but whose affinities were previously unknown.

Plant structures and their functions

The plant kingdom includes such a wide diversity of form, from the most primitive liverworts to the most complex of flowering plants, that it is impossible to generalize a common structure for all. However, in our ordinary everyday interaction with plants, which for the most part involves the flowering plants and conifers, the enormous diversity of form can be encapsulated into a generalized ground plan, with each component or organ assigned a specific function.

Below- and above-ground distinctions

Most plants grow in the ground, and the first obvious distinction is between the root system, which anchors the plant and absorbs water and nutrients from the soil, and the shoot system, the above-ground part of the plant that interacts with the atmosphere. The shoot system is primarily composed of stems and leaves, and intermittently produces the reproductive structures such as flowers in angiosperms and cones in conifers. Despite the enormous diversity of plants, roots and shoots play consistent fundamental roles, and their essential features such as growth, structure and shape are relatively uniform. Indeed, most of the variations we find in plants reflect adaptations of the different organs to allow them to function better in a particular environment.

▷ This seedling's root has a fine covering of hairs, which greatly increase the surface area for water absorption from the soil.

▽ The partially exposed root system of this katsura tree (*Cercidiphyllum japonicum*) shows the extent to which it occupies the soil in its search of water and nutrients.

Rooted to the spot

Roots live in a tough physical environment. They have soft, delicate tips, yet have to penetrate and grow through firm, abrasive soil. Their growing tips are protected by a cap of cells embedded in mucilage that are continually replaced as they are damaged. Just a few millimetres behind the tip, a zone of root hairs marks the region where water and nutrients are absorbed. Roots absorb water only at their tips; the rest of the root stem develops a thick, protective outer layer, and serves to transport the water from the tips back to the shoot system.

Taking a leaf

Leaves are predominantly sites of photosynthesis. Their cells are rich in the light-harvesting chlorophyll pigment, and as a consequence they are predominantly some shade of green. They are usually flat and thin, which maximizes their surface-to-volume ratio and provides the largest area for capturing sunlight. Leaves account for most of a plant's surface area and are covered with a waxy cuticle to prevent them from drying out. They have stomata pores on their surface to absorb carbon dioxide from the air, and they have a spongy internal structure to allow air to circulate. Leaves vary enormously in size, shape and texture between species, but these differences usually relate to optimizing the efficiency of photosynthesis while preventing unnecessary water loss.

Stemming the tide

Stems are transport networks that connect the leaves and the roots. They move the energy-containing sugars made by photosynthesis from the leaves, where they are produced, to other parts of the plant, where they are either stored or used for growth. In addition, they transport water from the roots, where it is absorbed, to the leaves to replace moisture that is unavoidably lost through the stomata. Older stems become woody to form trunks and branches, providing structural support in trees and shrubs, but they also continue to play their primary transport role.

Reproductive organs

Reproduction occurs in cones or flowers. In conifers, pollen and seeds are produced in separate cones, and pollen is dispersed by wind. In flowering plants, the pollen organs (stamens) and seed organs (carpels) can be aggregated into a single flower, or born on separate flowers, and are commonly surrounded by an attractive whorl of colourful petals and protective sepals. In general, complex, colourful, perfumed and nectar-producing flowers are pollinated by insects and birds, while simple, drab flowers are wind pollinated.

⌃ Seeds of conifers such as this spruce (*Picea* sp.) are borne in woody cones.

⌃ Pedunculate oak (*Quercus robur*) have pendulous catkins containing numerous small flowers that release copious amounts of wind-dispersed pollen.

⌃ Sturt's desert pea (*Swainsona formosa*) has brightly coloured flowers to attract pollinators.

A short history of plant morphology

We are so familiar with plants and their appearance – firmly rooted in the ground, and with stems that bear leaves and flowers – that we are often unaware that the way we describe their form is just a convention. We are also often unaware that this convention dates back millennia, and that it is not the only perspective.

Plato (in red) and Aristotle (in blue) in deep discussion. The early Greek philosophers were also well versed in matters of science, including the nature of life itself.

Early thinkers

We can trace our ideas on plant morphology back to antiquity and the time of the great Greek philosophers Plato and Aristotle in the fourth century BCE. These great early thinkers were absorbed with unifying concepts, and they attempted to reconcile plants and animals (including humans) using a common theme. The most basic of these was the concept of psyche (or soul) – plants were said to have a nutritional psyche, animals a sentient psyche, and humans a reasoning psyche. Indeed, the roots of plants were thought to correspond to the mouths of animals.

It was Aristotle's student Theophrastus who first cautioned against an overly rigid plant–animal comparison, and he gave us the first notions of a framework based predominantly on plants. The philosopher divided the plant into permanent parts (root and stem) and temporary parts (leaves, flowers and fruits). He recognized the indeterminate and modular growth of plants, which continually produce new shoots and leaves every year, in contrast to the unitary growth of animals, in which the organs are of fixed and definite number. In keeping with the other ancients, he considered the root to be the primary organ, possibly due to its importance in medicines and the fact that the shoots of many plants die back in winter, only to resprout from the rootstock again the following spring. The leaves attracted little attention; their function was then unknown, and they were considered little more than an accessory to fruit production.

COMMON TERMS

The direct comparison of plants and animals made by the ancient Greek philosophers is the source of some of our current terms and concepts. Examples include the division of both into tissues and organs, and the use of the words 'nerve' for the leaf veins (left) and 'heartwood' for the middle of the trunk. We even refer to the gametes of plants as sperm (motile) and eggs (retained), and convey gender categories of maleness and femaleness on the structures that produce them.

'Everything is leaf'

This notion of plant morphology lasted for 2,000 years. Then in 1790, Johann Wolfgang von Goethe, the famous German statesman, wrote an essay on plant morphology that underpinned a new paradigm. Goethe is best known as a philosopher and poet, and as the author of the great play in which Faust sells his soul to the devil, but he was especially interested in the natural and physical world, and was responsible for a number of important scientific works ranging from natural history to colour theory. During a trip to Italy, he visited the Botanical Garden of Padua, where he observed the European fan palm (*Chamaerops humilis*). Fascinated by the gradation of different-shaped leaves along the stem and their transition into the flowering region, he postulated that they were all variations of the same structure – that they were all types of leaf. Goethe used the phrase '*Alles ist Blatt*', or 'Everything is leaf'. What he really meant was that there was an overarching concept of 'leaf', and that all the variable organs were different manifestations of that archetypal leaf; this reflected a philosophy that had been dominant since Plato, and underscored Goethe's deep metaphysical approach.

Johann Wolfgang von Goethe was a statesman, author and scientist, and an important figure in the development of modern plant morphology.

Goethe's concept of the equivalence of plant parts, whereby cotyledons, leaves and flower parts have identical relationships to the stem – '*Alles ist Blatt*'.

Goethe's idea had a profound impact, and it has influenced plant morphology to the current day. The past 200 years have seen relative disinterest in the root, and a concentration of interest in the shoots and flowers based largely on Goethe's concept of a dichotomy between leaf-like and stem-like organs. This is reflected in the most commonly applied model of plant construction, the classical or leaf–stem model.

The European fan palm (*Chamaerops humilis*). The transition of seedling leaves to mature fans in this plant was the stimulus for Goethe's ideas on the 'metamorphosis' of plants.

Applying a model of plant morphology

The classical or leaf–stem model of plant construction is the modern formal articulation of Goethe's idea (see page 35). The plant is primarily divided into roots and shoot, and the shoot is further divided into stem-like and leaf-like structures. Shoots can be either vegetative (i.e. leafy stems) or fertile (i.e. cones or flowers), but their parts bear the same relationship to one another.

shoot apices

axillary bud

axillary shoot

⋀ Magnolia (*Magnolia* spp.) flowers have elongated axes and have been used as a classical example of a fertile shoot whose floral parts are modified leaves.

⋀ In a typical plant, the stem bears leaves, which in turn subtend a bud. Buds develop into branches, which produce their own leaves. Growth continues in this iterative manner.

Phyllomes and caulomes

We refer to the two organ classes of the shoot as phyllomes (leaf-like) and caulomes (stem-like), and each has a set of characteristics that allow a structure to be attributed to it. The most important criterion is position on the plant, which reflects the organ's mode of formation. Phyllomes are formed near the tip of the growing stem, on the flank of the apical meristem. Caulomes are formed in the axils of phyllomes – the angle between the phyllome and the axis that bears it. Thus, phyllomes are lateral structures, and caulomes are axillary and subtended by a phyllome.

This positional relationship is at the heart of the repetitive architecture inherent in most plants – stems bearing lateral leaves (phyllomes), which in turn have shoots in their axils. These shoots consist of the stem (caulome), which in turn produces its own leaves, and so on. This continuous reiteration of shoots is what gives plants their modular construction and stands them apart from the growth form of animals, which typically develop from embryo directly to adult.

Other associated features

Apart from position, there are several other features associated with phyllomes and caulomes, but these can be variable and their application requires come caution.

Phyllomes are typically bilaterally symmetrical structures (i.e. flattened), and have determinate, or limited, growth. Caulomes are typically radially symmetrical (i.e. round), and have indeterminate, or unlimited, growth. Most leaves clearly fit into the category of phyllomes, and most stems are unarguably caulomes. However, when some of the features conflict, such as round leaves or flattened stems, we invoke their lateral or axillary position as the true indicator of their identity.

In the classical theory, flowers are considered to be fertile shoots; the axis is a caulome, and the sepals, petals, stamens and carpels are phyllomes. Indeed, flowers themselves are usually axillary structures, subtended by a truly lateral structure (a leaf or bract).

A SPINY PROBLEM

Spines can be either phyllomes or caulomes, depending on where they are located. Some species have spines that are in a lateral position and replace leaves; these are phyllomes, and they even have new shoot buds in their axils. Other species have spines that form in the axil of a leaf; these are modified shoots, or caulomes. In common gorse (*Ulex europaeus*) (left), both the leaves and axillary shoots are similarly spiny, but they can be distinguished by their position.

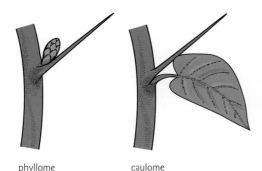

phyllome   caulome

IT'S ALL IN THE DETAIL

Plants in the genus *Ruscus*, closely related to the genus *Asparagus*, attracted the attention of Theophrastus because they appear to bear flowers on the centre of the leaves. However, if we look closely at the foliage, it is clear that each unit – which is green, flattened and of limited growth – actually sits in the axil of a small papery bract. The foliage unit is thus axillary, and from its position is a caulome; the bract is the truly lateral structure, and is interpreted as the phyllome. Note also that the flower sits in the axil of a bract that was produced by the foliage unit. We call these foliage units of *Ruscus* phylloclades, because they combine features of both leaves (flat and determinate) and branches (axillary position), but architecturally they are axillary stems that have been modified as photosynthetic organs.

⊽ Butcher's broom (*Ruscus aculeatus*). The flat photosynthetic organs arise from the axil of a bract and bear flowers – they are modified shoots, not leaves.

Alternative ways of looking at plants

(T)he leaf–stem model works well for interpreting most plants, particularly flowering plants, and even for unusual structures like the phylloclades of *Ruscus* (see box on page 37). However, it does not work for primitive land plants that have not yet evolved a level of complexity that distinguishes stems from leaves, and even within flowering plants there are examples where the leaf–stem model is unsatisfactory. To overcome these limitations, other models of plant construction have been proposed.

The phytonic and metameric models

The differences between the various models of plant morphology largely depend on how they deconstruct the plant. The classical model makes the major distinction between the leaf and the stem, but of course the leaf is continuous with the stem, and our desire to separate them is only arbitrary. Take, for example, the shoot of the Australian winged wattle (*Acacia alata*); it is not clear exactly where to delineate leaf from stem – indeed, the terms 'leaf' and 'stem' are not very useful in this instance.

Two alternative models that deconstruct the plant in different ways are the phytonic and metameric models. The phytonic model divides the shoot longitudinally into sectors; each sector includes a leaf and the adjoining stem as far as the next leaf directly below. This is a good model for describing the shoot of a winged wattle. In contrast, the metameric model divides the plant horizontally into segments; each segment includes an internode and a node, and the lateral organs arising from the node. This is a good model for plants with sympodial growth, where each segment of the axis is produced from a different apical meristem.

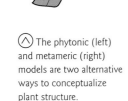

⌃ The phytonic (left) and metameric (right) models are two alternative ways to conceptualize plant structure.

⌄ Leaves of the winged wattle (*Acacia alata*) are completely continuous with the stem, making this plant amenable to interpretation using the phytonic model.

The continuum model

A third model, called the continuum model, takes a different approach. It dismisses exclusive categories such as leaf and stem in favour of a continuous morphological space. In this space, most organs tend to cluster with character combinations we associate with either phyllomes or caulomes, or other categories such a trichomes (hairs), but intermediate organs can occupy any of the space between them.

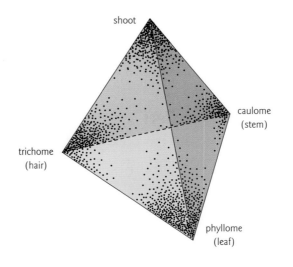

 The continuum model is a flexible approach to the conceptualization of plant structure.

Complementary views

All three models can be applied to all plants to various degrees – in other words, they are complementary rather than exclusive. None of them is right or wrong, or even more right or more wrong than another; they are just different ways of looking at plants. Certainly, for a given plant or purpose, one model may be more appropriate than another, but being aware of these different views allows one to look at plants from a variety of different perspectives.

A HOLISTIC APPROACH

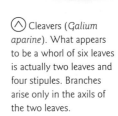

The small herbaceous plant commonly known as cleavers or goosegrass (*Galium aparine*), in the coffee family (Rubiaceae), is a familiar garden weed that scrambles through shrubs. Apart from its sticky stems and fruits, the most characteristic feature is its foliage: whorls of six identical leaves that are regularly spaced along the stem. But first appearances can be deceiving. Only two leaves in each whorl were formed as primordia at the apical meristem, and hence qualify positionally as 'true leaves' or phyllomes. The other four structures arose later, two on either side of the 'true leaf' primordia, and are considered to be stipules – appendages associated with the bases of leaves. Indeed, close inspection reveals that axillary buds only ever arise in the angle of the true leaves, not in the angles of the 'stipular' leaves.

The continuum model would emphasize the similar form of the leaves and stipules to equate these structures, while a classical approach would distinguish them based on their position. So being able to look at cleavers from both points of view allows a more enlightening and holistic appreciation of the plant.

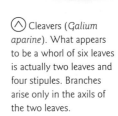 Cleavers (*Galium aparine*). What appears to be a whorl of six leaves is actually two leaves and four stipules. Branches arise only in the axils of the two leaves.

Evolution and development

The past two decades have seen enormous advances in the evolution of development, or evo-devo, and our understanding of the developmental genetics of organisms. Many of the advances have been in human and animal biology, particularly driven by medical research, but a vastly increased knowledge of the genes and genetic processes that control plant development has provided a new way of looking plant structure.

The ABC of flowers

Much of this new knowledge has resulted from experiments with thale cress (*Arabidopsis thaliana*), a small plant in the cabbage family (Brassicaceae) that is widely used in laboratory studies. It is easily manipulated and contains a relatively small number of genes, vast quantities can be cultivated in a limited space, and its speedy maturation from seedling to adult plant allows many generations to be grown in a short time.

One idea that has been turned on its head as a result of this research is the classical notion of the flower as a fertile shoot, with the leaves replaced by successive whorls of sepals, petals, stamens and carpels. Experiments with thale cress have revealed that this seemingly precise order for different floral parts is due to three classes of gene that are expressed across the four whorls of the developing flower; these genes are termed A, B and C, and hence it is referred to as the ABC model.

In a region where only gene A is expressed, sepals are produced. If only gene C is expressed, carpels are produced. Genes A and B expressed together results in petals, and a combination of B and C results in stamens. In the normal situation, the four zones express A alone (sepals), A and B (petals), B and C (stamens), and C alone (carpels). But mutant plants that lack one or other of these genes can produce flowers with missing and substituted parts. For example, plants lacking a C gene can produce only sepals (A expressed alone) and petals (A and B expressed together).

It is easy to see how an imprecise expression of this developmental pattern

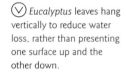

Eucalyptus leaves hang vertically to reduce water loss, rather than presenting one surface up and the other down.

In a cross section, both halves of a *Eucalyptus* leaf are anatomically identical, as they both have the same exposure to the environment.

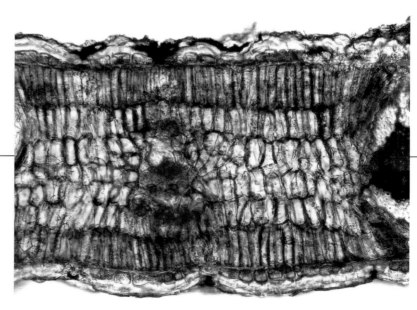

could lead to the 'mutant' flowers that are prized in horticulture, such as the multi-petalled roses (*Rosa* spp.) and carnations (*Dianthus* spp.), which are very different from their strictly five-petalled wild ancestors. Indeed, varying expressions of the A, B and C genes could explain much of the enormous variation in flower structure.

The genetics of leaf development

Another set of genes is implicated in leaf development. Typical leaves have a flat blade in which the upper and lower halves have a different cell structure; typically, the upper half has elongated, column-like cells called palisade mesophyll, which is perfect for light capture, and the lower half has stomata (breathing pores) on the surface and internal spongy cells for absorbing carbon dioxide. We now know that different sets of genes determine the identity of the tissue types in these regions. If the genes for spongy tissue are deactivated, both halves of the leaf will have elongate palisade cells, and if the genes for the palisade cells are deactivated, both halves of the leaf will be spongy. It is the interaction of the genes in the two halves of the leaf that determines its bilateral symmetry (flatness); when one gene is missing, the leaves are needle-shaped, with radial symmetry.

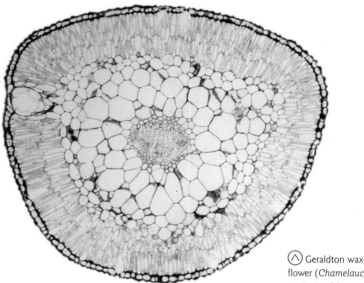

∧ Geraldton wax flower (*Chamelaucium uncinatum*) is a spectacular wildflower from arid Western Australia that is cultivated for its floral display and drought-hardiness.

▷ Geradlton wax flower has needle-like leaves with palisade mesophyll all round – the entire surface is 'upper'.

GENETIC MODIFICATIONS

The gene experiments on leaves point to possible explanations for atypical leaves found in nature. *Eucalyptus* leaves are isobilateral, and hang vertically to minimize their exposure to the hot sun. Because they do not present 'upper' and 'lower' surfaces, both halves of the leaf are the same and have elongate palisade mesophyll. Many arid plants have terete, or cylindrical leaves, which reduces the surface area-to-volume ratio and hence minimizes water loss to the environment. We do not yet know the developmental genetic basis for these leaves, but it would not be surprising to discover that it was accompanied by a modification of these or similar genes.

Plant cells, tissues and organs

Like all living organisms, plants are made up of cells. There are many different types of plant cells with different functions, and these cells are arranged into tissues that have a precise purpose. Some cell and tissue types are characteristic of particular organ structures, such as the light-harvesting cells in the leaves, but others are found throughout the plant. The cells themselves can be observed only with a microscope, but the tissues and tissue systems are obvious at a larger scale, such as wood, bark and the veins of the leaf.

Tissue systems

At the most fundamental level, plants consist of three broad tissue systems – the dermal, vascular and ground systems. The dermal system, as its name suggests, is the 'skin' of the plant. In its simplest form, it consists of a single layer of cells that covers the entire surface of the plant, and it is the main interface between the plant and the environment. It is covered with a waxy layer of cuticle to prevent the plant drying out, it contains stomata for taking up carbon dioxide from the atmosphere, and it is often expanded into hairs or prickles as a first layer of defence against herbivory.

The vascular system is the plant's conducting system, and is responsible for transporting substances around the plant. The xylem tissue transports water from the roots, where it is absorbed from the soil, to the leaves. The phloem tissue transports sugars that are made during photosynthesis from the leaves to growing regions or storage organs. The vascular system includes the wood in the stem and the veins in the leaves.

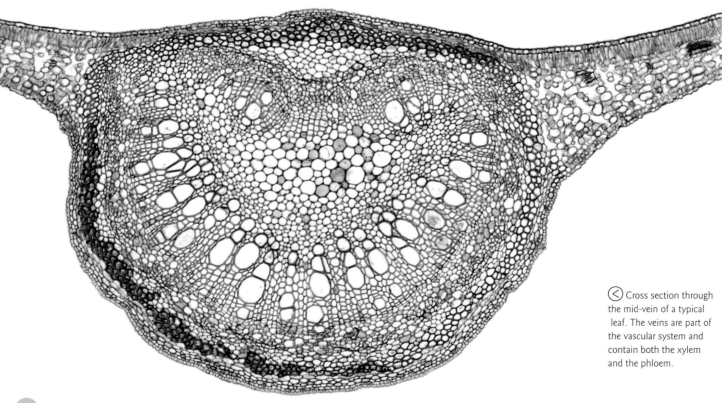

⊙ Cross section through the mid-vein of a typical leaf. The veins are part of the vascular system and contain both the xylem and the phloem.

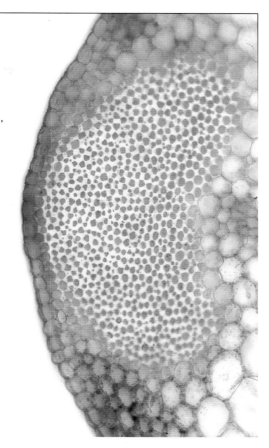

CELLULOSE VS. LIGNIN

The stringy bits on a stalk of celery are bundles of cells called collenchyma. Collenchyma cells are long and narrow, and their corners are thickened with extra layers of the cell wall material cellulose. Cellulose is a flexible material, and collenchyma is produced in places where plants need flexible structural support. When plants require rigid structural support, or a tough, hard texture, they thicken their cells walls with a layer of lignin.

⊘ Cross section of celery collenchyma bundle. The narrow cells have walls that are thickened with cellulose – they are pearly white in this fresh section.

⊘ The ridges on celery (*Apium graveolens*) stalks contain bundles of collenchyma cells that provide flexible support.

The remainder of the plant is referred to as the ground system, and consists of simple tissues that are variously adapted for storage or structural support. For example, a potato is made up almost entirely of thin-walled parenchyma cells filled with starch grains as a food storage, and the strings of a celery stalk are specially thickened collenchyma cells that provide flexible structural support for the leaf (see box).

Cell walls

The chemical composition of the plant's cell walls is one of the most important features contributing to its function and texture. All plant cells are primarily made of cellulose, which is a soft, flexible material. Some types of cells are additionally thickened with lignin, a very hard, inflexible material that makes the cell walls hard and rigid. Lignified cells primarily provide structural support, and give plant stems their woody texture.

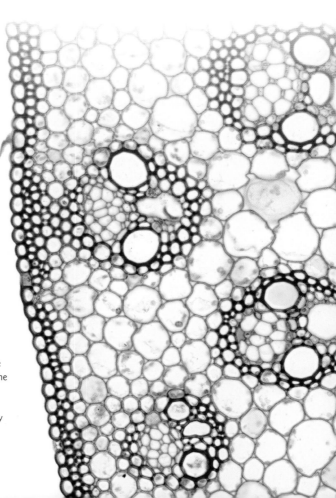

⊘ In a maize (*Zea mays*) stalk, cells surrounding the vascular bundles, and those in the outermost layers of the stem, are thickened with lignin for rigid structural support. In this section they are stained red.

Plant growth: internal anatomy

S ome of the largest trees grow from some of the smallest seeds. The tiny seedling that emerges from a *Eucalyptus* seed, with its diminutive root and stem, and only two seedling leaves, is vastly simpler than the mighty adult tree, with its massive trunk elevating the foliage up to 100 m (330 ft) above the ground. So how does this seedling become a tree, and what are the processes that produce all the cells and tissues of the plant?

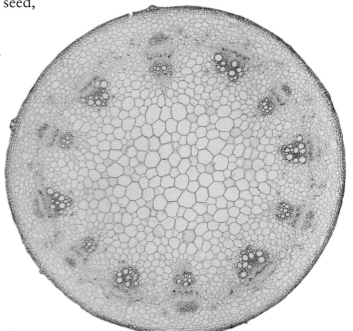

> Cross section of young sunflower (*Helianthus* sp.) stem, revealing vascular bundles in a ring. All tissues in this stem are primary tissue – in other words, they were all produced by the apical meristem.

> Longitudinal section of the shoot tip of a *Coleus* plant, with apical meristem producing leaves and primary stem tissue.

Primary growth

Plants grow from their tips. At the very ends of the seedling shoot and root are regions of cell division and proliferation known as apical meristems. The cells of these meristems are continually dividing, and they leave new daughter cells behind them that in turn mature into the tissues of the stem and the root. Thus, apical meristems add length to the stems and roots. The apical meristems are called the primary meristems, and the cells and tissues derived from them are referred to as primary growth.

Secondary growth

The primary growth of a stem is not very extensive – often less that 5 mm (³/₁₆ in) in diameter, and usually much less than the thickness of the average pencil. After all, the main function of the shoot apical meristem is to make the plant taller. But as the plant gets larger, the amount of water that must be transported from the roots to the ever more numerous leaves, and the sugars that need to be transported in the other direction to supply the ever-increasing root system, far exceed the capacity of the narrow primary stem. To overcome this constraint, the plant develops another type of meristem, the vascular cambium, which contributes more vascular tissue to the stem. This vascular cambium is in the shape of a cylinder in the stem, and it generates new xylem to the inside and new phloem to the outside. The xylem from the vascular cambium is the wood of the stem, and as it accumulates the diameter – and hence the girth – of the branch or trunk increases. Just as the apical

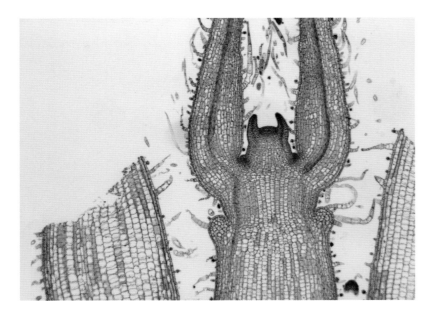

PUT A CORK IN IT

Cork has cell walls that are impregnated with suberin, a waterproof substance that gives the tissue its protective properties on the outside of the stem. This is also the feature that keeps wine in bottles. Wine corks are made from the outer bark of the cork oak (*Quercus suber*). This outer bark can be harvested sustainably; as long as the phloem and the vascular cambium is left intact, new cork meristems regenerate from the phloem.

⬃ Cork oak (*Quercus suber*) bark that has recently been peeled from the trunk of the tree.

meristem adds length to the stem, the vascular cambium adds width.

As the stem itself becomes thicker, the epidermis that forms the original skin of the plant becomes stretched and ultimately loses its integrity. It is replaced by the cork, a waterproof zone of cells that ultimately contributes to the bark. The vascular cambium and the cork meristem are referred to as secondary meristems, and the wood and bark tissue derived from them are secondary growth. A similar scenario occurs in the roots.

By the time a stem is as woody as a pencil, all of its component tissues are secondary growth; the small amount of primary tissue has either been shed with the bark or crushed into the centre.

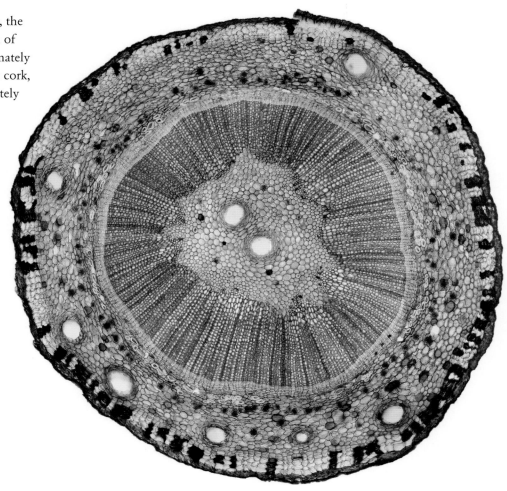

⬂ Cross section of a young ginkgo (*Ginkgo biloba*) stem in early stage of secondary growth. A zone of wood has formed internally, and the outermost cells have developed as cork.

Plant growth: external form

(W)e have already seen that the shoot apical meristem is crucially important for elongation of the stem and the development of the early cells and tissues of the plant. However, the apical meristem also plays an important role in the external form of plants.

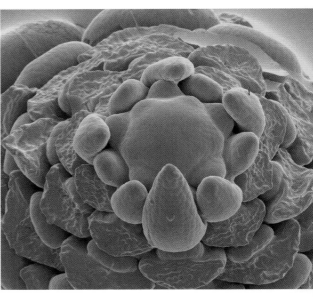

├─────────────────┤ 200 μm

(<) Campfire (*Crassula capitella*) produces leaves in pairs, each pair oriented at 90 degrees to the previous one.

(L) Sunflower (*Helianthus* sp.) florets are geometrically arranged, which reflects their precise formation at the apical meristem.

Leaf development

Leaves develop as small bumps called primordia on the side of the meristem very close to the apex. In their primordial stage they are closely, but precisely, packed around the apex, and they are spaced out into their final positions – which are not randomly arranged – as the stem elongates. This precise arrangement of leaves is termed phyllotaxy, and although there are a number of different phyllotactic patterns in plants, they are all determined by the position of the primordia at the shoot apex.

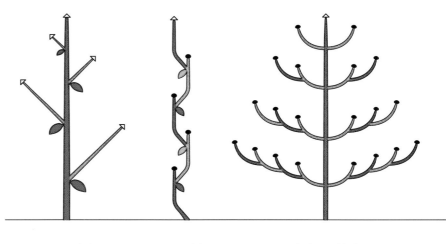

monopodial   sympodial   Fagerlind's model of tree architecture

⊘ Leaf primordia form on the periphery of the apical meristem. The position of new primordia is determined by the space available and the location of previously formed primordia.

An architectural model

New branches arise from buds that are formed in the axils of leaves (the angle between the leaf and the stem), and these branches have apical meristems that produce their own leaves. So, a combination of the arrangement of the leaves and the axillary nature of the branching both determines and constrains the architecture of the shoot.

However, not all axillary buds immediately emerge to become branches. Some abort; others remain dormant, only to emerge if the main branch is damaged; and yet others develop into flowering structures. The developmental fate of particular buds is predetermined at the level of the whole plant, and the resulting growth pattern results in a correspondingly predictable whole-plant architecture. Because there are relatively few developmental fates of a bud, the possible variation in whole plant architecture is limited. Indeed, although there are about 300,000 plant species, they conform to just 27 different architectural models.

THE GOLDEN ANGLE

Leaves are precisely arranged on stems in a regular pattern that is formed at the apical meristem. One of the most common patterns is a spiral arrangement with an angle between each successive leaf of 137.5 degrees (called the golden angle), which conforms to the mathematical Fibonacci sequence, named after a twelfth-century Italian mathematician. This sequence approximates the shape of the logarithmic spiral, which is common in nature – the shape of a ram's horn, the spiral of a snail's shell, the curve of a tiger's claw. It is the only spiral that maintains its shape as it increases in size – an important feature for growing organisms.

⊘ Phyllotactic patterns are most easily seen in plants with minimal stem elongation.

MONOPODIAL VS. SYMPODIAL GROWTH

Most shoots grow from a permanent apical meristem at the tip, and the entire length of stem is the product of this meristem; this is called monopodial growth. However, other shoots have growing tips with limited growth – they produce a segment of stem, then cease growth (often terminating in a flower), and are superseded by a branch from below that forms the next segment of stem, and so on; this is referred to as sympodial growth. These two types of growth can be combined in the same plant – for example, a monopodial main trunk with sympodial lateral branches is the basis for Fagerlind's model.

Plant morphology and classification

$\left(A\right)$ very familiar concept in biology is the fact that related organisms have a similar structure, and that the closer the relationship, the more similar the organisms are to one another. In evolutionary terms, this means that the closer the relationship, the less time the organisms have had to diverge. In fact, we use similarities between organisms to infer relationship.

⑦ Fleabane (*Erigeron* sp.) is a member of the daisy family (Asteraceae) and has both ray florets (purple) and tubular florets in the centre (yellow).

Defining characters

In the classification of plants, the major divisions such as gymnosperms or angiosperms are further divided into orders, families and genera. For example, all species of apples belong to the genus *Malus* and species of pear belong to the genus *Pyrus*. Both *Malus* and *Pyrus* belong to the rose family (Rosaceae), which in turn is classified in the order Rosales. The flowers and fruit of pears and apples are structurally quite similar, which reflects their close relationship. In the classification of plants, it is at the family level that the defining characters are most generally recognizable.

When it comes to determining relationships, reproductive structures such as cones and flowers are more reliable indicators than vegetative features such as leaves or the habit of the plant (its overall shape). This is most likely because vegetative features are permanently present, and are adapted for the particular environment favoured by the species. Flowers, on the other hand, are short-lived and temporary, and so do not become so modified through selection to such a great extent.

Peas and daisies

Two plant families that display a very uniform flower structure but widely variable vegetative morphology are the legume family (Fabaceae) and daisy family (Asteraceae). The scientific name for the legume family used to be Papilionaceae, in reference to the shape of the flower – the Latin word *papilio* means 'butterfly'.

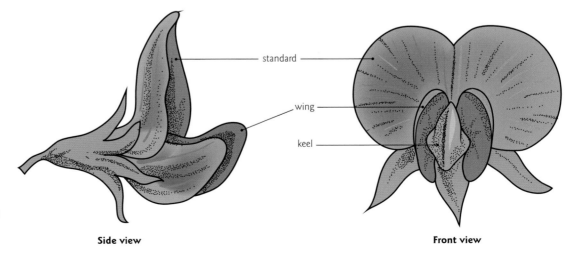

⊘ All pea flowers (family Fabaceae) are constructed on the same plan: they are bilaterally symmetrical, have a common petal structure and have a single legume in fruit.

standard

wing

keel

Side view

Front view

The flowers of species in the family are remarkably consistent, with five petals (one standard, two wings and two that fuse to form the keel) arranged so as to give the overall structure a single plane of symmetry. There are ten stamens, of which commonly nine are united to form a tube and the other is solitary. In the middle of the flower is a single carpel (female part), which develops into a fruit called a legume – pea and bean pods are typical legumes. The plants themselves differ wildly, however, and can range from tiny herbaceous clovers that stay close to the ground, to climbing lianas, arid shrubs with small, spiny leaves, and tall rainforest trees.

Flowers of members of the daisy family are small (they are termed florets) and are arranged in tight clusters surrounded by either papery, fleshy or spiny bracts. This cluster of florets, called an inflorescence, presents itself to pollinators as a coherent unit, so is functioning like the single flower of other plants. There are two different types of floret in typical daisies (although some daisies have only one type): the petal-like ray florets, often arranged around the outside of the cluster; and the more tubular disc florets, located in the centre. The habit of the plants, like the peas, ranges from small herbs to large trees, with shrubby and climbing forms in between. Again, however, the inflorescence is a common character – so much so, in fact, that the typical composite daisy inflorescence gave rise to the earlier family name of Compositae.

⋀ By aggregating their flowers into an inflorescence, members of the daisy family present a more effective floral display to pollinators.

⊓ Billy buttons (*Craspedia* spp.), in the daisy family (Asteraceae), have only tubular florets in their inflorescences.

⟨ Herbaceous white clover (*Trifolium repens*, left) and shrubby alpine podolobium (*Podolobium alpestre*, right) have very different habits, but both have typical pea flowers.

Modifications of form due to extreme function and habitat

(T)he structural features some plants have evolved in order to function in different environments are so extreme that they defy almost every attempt at interpretation. Sometimes, it is only one organ that is affected, but at other times the whole plant is modified. Some bizarre examples are seen in parasitic plants, and plants that have adapted to aquatic habitats.

Weird and wonderful

Structural modifications of plant parts as an adaptation to particular environments are common. For example, many plant families contain succulent species that are especially good at storing and conserving water. But even in the most unusual cacti, we can recognize a fleshy barrel-shaped stem with spiny leaves and axillary shoots that develop from apical meristems and associated primordia. In other words, they are weird, but we can interpret them in our normal framework. In contrast, many parasitic plants appear to have dispensed with some of their typical structures and developed, either in part or in the whole plant, an entirely new morphology to cope with the demands of their modified existence.

MINIATURE MARVELS

Duckweeds in the genera *Lemna* and *Wolffia* are among the smallest of all angiosperms. They are free-floating aquatic plants found in still freshwater environments, and their tiny size and simplicity of structure make their interpretation difficult. Each plant consists of a small elliptical module, called a frond, which buds off new modules from special growing zones. A root dangles down into the water. Most reproduction is by vegetative budding, but flowers consisting of a single stamen or carpel are produced if environmental conditions are suitable.

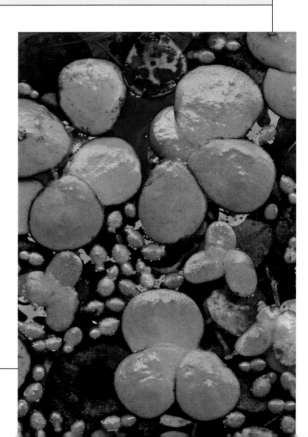

(>) *Wolffia* (small) and *Lemna* (large) plants floating on the water surface.

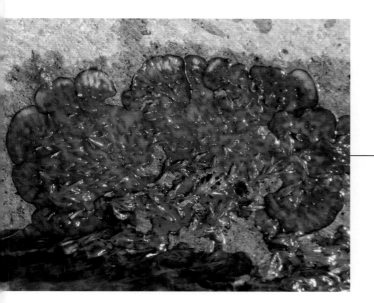

Riverweeds in the family Podostemaceae attach to rocks in fast-flowing streams and rivers. They are thalloid – in other words, they have an amorphous body that is not differentiated into stems and leaves – and are more reminiscent of liverworts than flowering plants. Even the root system is unusual, and develops to adhere the plant to the rock like the holdfast of a marine alga. Riverweeds flower only when the water level drops and their habitat is exposed, thus revealing their identity as flowering plants.

Under the mistletoe

Mistletoes, parasitic flowering plants in the families Loranthaceae and Santalaceae, have seeds that are spread to the branches of other trees by birds. The seeds germinate, but rather than producing a root system they develop structures called haustoria, or sinkers, that penetrate the host branch and connect with its vascular tissue. In this manner, the mistletoe gets its water directly from the host. Normally, host and mistletoe can coexist, but a heavy infestation of mistletoes can kill a host. Extreme drought conditions in Australia can kill mistletoes while leaving their hosts alive, because the cells of *Eucalyptus* and *Acacia* trees can cope better under extreme water stress.

The shoots of mistletoes are otherwise normal, consisting of stems and leaves, but many species exhibit a strange mimicry, whereby their leaves are similar in appearance to those of the host plant. One explanation for this phenomenon is that it makes it harder for herbivores to distinguish the palatable mistletoe leaves from those of the unpalatable host.

(⟃) Grey mistletoe (*Amyema quandang*) is a parasite of several species of wattle (*Acacia* spp.). Its leaves are similar in shape and colour to the foliage of some of its hosts.

(⟩) Common mistletoe (*Viscum album*) plant emerging from the branch of an apple tree at the point where its haustorium has invaded the host's vascular tissue.

Dealing with constraints

The function of plants is constrained by their structure, which is in turn constrained by their development. Indeed, the history of plant life on land has been one of increasing developmental and structural complexity to exploit the terrestrial environment in new and better ways. On occasion, however, plants find themselves down an evolutionary pathway that compromises some of their abilities; in these situations, they can either adapt to their new constraints, or they can develop ways to overcome them.

Arborescent monocots

In typical trees such as giant redwood conifers or mountain ash eucalypts, the small shoot apical meristem at the growing tip produces a thin, slender stem. As the plant grows larger in size and produces more leaves, the canopy needs to be supplied with increasing quantities of water from the roots, and the enlarging root system needs to be supplied with more nutrition produced by photosynthesis in the leaves. To increase the transport capacity of the stem, the plant develops the vascular cambium – the meristem that adds new transporting tissue, xylem and phloem, to the stem. This product of the vascular cambium is called

> ⊳ Palm apical meristems produce a wide zone of tissue, the primary thickening meristem, which generates the cells and tissues of the stem.

> ⌄ Palms develop an arborescent form and a thick trunk despite lacking a vascular cambium.

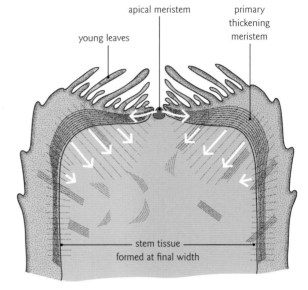

young leaves — apical meristem — primary thickening meristem

stem tissue formed at final width

secondary growth, and it leads to woody stems whose function is to maintain the connection between the canopy and the roots.

Within the angiosperms is a large group that has lost the ability to produce a vascular cambium. These plants are the monocotyledons, or monocots, and include grasses, lilies and orchids. Because they do not develop a woody stem, they are, on the whole, relatively small herbaceous plants. However, some monocots do develop into large tree-like, arborescent plants, despite their lack of a vascular cambium and their inability to increase their conducting tissue through secondary growth. Prominent among these arborescent monocots are palms and screw palms (*Pandanus* spp.), which have solved the transport problem with a combination of clever features.

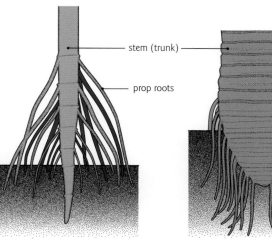

stem (trunk) — prop roots

Some palms have prop roots that support the plant and bypass the narrow base of the stem.

Date palms produce their roots from the below-ground portion of the stem.

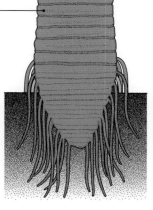

Transport solution

When palm seedlings emerge from the seed, they have a typical small shoot apical meristem, which produces a narrow stem. But as the seedling grows, the region immediately below the growing tip widens into a new zone of growth, the primary thickening meristem, which in turn produces an increasingly wider stem behind. In some palms, the shoot apex can eventually attain a diameter of 30 cm (12 in), producing the thick stem that is familiar in palm trees.

There is still an inherent constraint, however, because the first portion of the stem, produced when the seedling was small and the apical meristem was narrow, remains slender, with no mechanism for secondary thickening; there is no way for the now robust stem above to maintain an adequate connection with the roots below. Palm trees therefore dispense with the normal tap-root system, and instead produce new roots directly from the stem. As these roots appear higher up in the stem, they bypass the narrow lower region and deliver water directly from the soil to the thicker parts of the stem. In some palms, and in *Pandanus* species, the roots arise well above ground level and form spectacular prop roots that also provide structural support – sometimes even suspending the rest of the plant in mid-air. In other palms, such as date palms (*Phoenix dactylifera*) and coconut trees (*Cocos nucifera*), the seedling apex completes its widening process before it emerges from the ground, and the adventitious roots mostly emerge from the below-ground portion of the trunk.

ⓣ Palms use several strategies to compensate for their inability to widen their seedling stem. Top left: Seychelles stilt palm (*Verschaffeltia splendida*) has prop-roots that support the trunk and bypass the narrow base of the stem. Top right: Date palms (*Phoenix dactylifera*) produce their roots from the underground part of the stem.

ⓥ Roots of maize (*Zea mays*) arise from nodes of the stem and anchor the plant to the ground.

ROOTING FOR MONOCOTS

The inability to maintain a viable root–shoot continuum due to the absence of a vascular cambium is a feature of all monocots, and all solve the problem by forgoing a tap root, instead producing adventitious roots from the stem, usually at the nodes where the leaves appear. Smaller grasses have a matted, fibrous root system, but in larger species such as maize (*Zea mays*), the adventitious roots can clearly be seen arising from the basal nodes of the stem.

Roots

Roots are often an understudied component of plants. Their largely underground nature results in an 'out of sight, out of mind' attitude among people, and the inherent difficulty in accessing them makes those who study them resort to increasingly creative approaches.

However, roots and the underground processes to which they contribute should not be taken for granted. Like leaves, their above-ground counterparts, roots supply required resources to the plant, often through a number of unique adaptations to their environment. Unlike leaves, roots can, in many cases, directly modify their environment through growth, chemical interactions and associations with other organisms, making the soil environment a bustling epicentre of activity invisible to the naked eye.

Deepening our understanding of roots – including where they are located (not always where you would expect), what they are doing and how they adapt to their surroundings – will hopefully increase our appreciation of these important plant structures.

One rarely gets a glimpse of plant roots – in a way, they are like the unseen portion of an iceberg. In the case of many ficus trees (*Ficus* spp.), however, a portion of the root system can be seen growing above ground. Here, the root system spreads laterally far beyond the canopy to support this massive *Ficus* tree.

The purpose of roots

R oots serve necessary functions for the plant, including resource acquisition, storage of those resources and anchorage to the substrate. During the evolution of roots from underground rhizomes over the last hundreds of millions of years, they have adopted a multitude of adaptations allowing them to deal more efficiently with their environment. Considering the wide range of habitats in which we find current true roots of terrestrial plants, we can begin to appreciate the competitive and heterogeneous environments that affect root morphology and physiology in a way that favours one or more root functions.

Root evolution

During the Devonian period, spanning the period 416–358 million years ago, plants that occupied land underwent immense changes. In particular, the evolution of roots impacted nutrient cycling and soil formation, thus affecting plant development and spread. The earliest 'roots' within the fossil record were discovered in present-day Scotland, on such species as *Horneophyton lignieri* and *Nothia aphylla*, and are estimated to be more than 400 million years old. These early structures were most likely not true roots but rhizoids, unicellular protuberances extending from a rhizome, or underground stem. In fact, most early root structures appear to have arisen from stems that grew under or along the ground and behaved physiologically as roots. In contrast, true roots are defined by their endogenous origin (the pericycle), the occurrence of a root cap, and the presence of an endodermis and protostele. In extant bryophytes, rhizoids are still found on the gametophyte stage, and even though they lack a true vascular system, they behave in a similar way to roots (see box). Interestingly, the genes that regulate rhizoid development in mosses are similar to the gene network regulating root-hair formation in the sporophyte stage (the multicellular generation) of vascular plants.

As time progressed and new, larger, vascular plant species arose, a variety of root types evolved to serve the new functional requirements of support in addition to transport and uptake. It is quite difficult to ascertain a clear timeline for this development as roots are less frequently preserved within the fossil record. However, as roots evolved and plants became larger, particularly with the continued evolution of trees (including the prehistoric genera *Wattieza* and *Archaeopteris*), root biomass increased, contributing immensely to the formation and weathering of soil through organic acid secretion and carbon fixation.

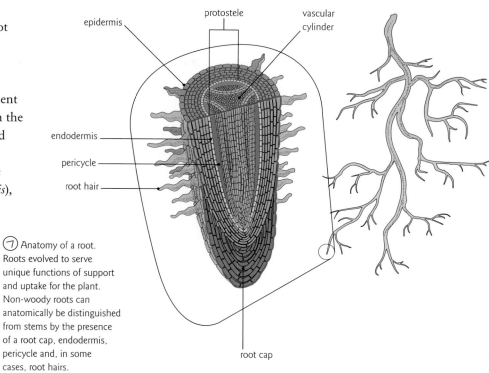

⑦ Anatomy of a root. Roots evolved to serve unique functions of support and uptake for the plant. Non-woody roots can anatomically be distinguished from stems by the presence of a root cap, endodermis, pericycle and, in some cases, root hairs.

RHIZOIDS – NOT TRUE ROOTS

Rhizoids, despite their distant origin, are still common in non-vascular plants, including the bryophytes – mosses, liverworts and hornworts. Primarily, rhizoids function to anchor the plant to the substrate, and have exhibited branching, the secretion of adhesive substances and growth in a particular direction in response to contact with a substrate. Additionally, they are involved in water absorption, most likely through capillary action, although the majority of water for bryophytes is probably taken up directly through the above-ground plant structures. Similarly, nutrients are predominately absorbed by the entire plant body, and are usually acquired though precipitation and the deposition of particulates in the air.

⋀ Evolutionary development of rhizoids. In plants that do not form true roots, filamentous plant cells at the interface of the plant and the substrate can form outgrowths of root-like structures.

The multiple functions of roots

To those who do not study roots, it may appear that a root is just a root. However, root systems are made up of diverse types of roots that serve multiple functions for the plant. Generally, root types are divided into the coarse roots, whose primary functions include stability, storage and transport; and the fine roots, the most distal (distant from the centre) portions of the root system, which are largely involved in water and nutrient uptake.

It may seem mundane to mention the importance of coarse roots for anchorage, but ecologically this function is attained through the compilation of several complex root characteristics, including shear strength, branching angles, root length and the amount of secondary thickening. In order to build a 'strong' root system that withstands uprooting, plants must invest considerable amounts of carbon into it. In other words, a strong root system comes at a price to the plant. In many cases, however, this cost is worth the sheer benefit of staying upright. For example, trees that grow on a slope must ensure proper anchorage of their root system to prevent an undesirable incline downhill. In fact, trees will change the shape of their root system in such situations to a more asymmetrical form, with enhanced root production on both the upward and downward slopes to provide extra stability in such uneven situations.

⟨∨⟩ Loading forces on the stems of trees are transferred to the roots and result in modified root growth patterns. Coordination over the entire root system is required to maintain stability in uneven environments.

⟨∨⟩ A fossilized 'root' cast from a prolific scale tree (*Lepidodendron* sp.) that grew 360–286 million years ago. The scars represent areas where small 'rootlets' were attached to the 'root', like bristles on a brush.

LYCOPODIOPHYTA – A SHOOT MODIFIED FOR ROOTING

The lycophytes are an ancient extant group of plants that resemble, and are related to, clubmosses. What is remarkable about these incredible plants is that, ancestrally, they were approximately 50 m (160 ft) in height, a vast difference from the small, 15–50 cm-high (6–20 in) species we are familiar with today. Their short lifespan (10–15 years), coupled with their large stature, contributed to the development of coal beds. Fossil evidence shows that the prehistoric lycophytes produced an unusual 'root system', termed a rhizomorph. While the poor preservation of lycophyte fossils makes it difficult to be certain, and debate still exists, it has been hypothesized (based on similar shoot and rhizomorph anatomy) that the rhizomorph was in fact a shoot that acted as a branched root system. Recently, evidence for gravitropism (directional growth in response to gravity) in the rhizomorphs of extant species and their root-bearing axis have brought this hypothesis back into the spotlight.

Coarse roots are also well equipped to transport and store water and nutrients acquired from the finer roots throughout the soil profile. Interconnected xylem and phloem cells that run the length of the plant allow the water and nutrient resources to be allocated from roots to shoots, or even from roots to other roots, as can be the case in internal water redistribution. In larger woody roots, anatomical shifts may occur, as in the production of xylem cells with smaller diameters during extended periods of drought to help protect the tree from mortality.

Of the entire root system, the finest roots are the most specialized for water and nutrient uptake. The membranes across which water moves from the soil into roots are highly permeable to allow for the largely passive process. However, not all root systems can acquire the same amounts of water. Factors such as root age, levels of suberin (a waxy substance) and connection to the soil can all drastically affect the ability of a plant to take up water. The uptake of solutes or nutrients, on the other hand, is driven

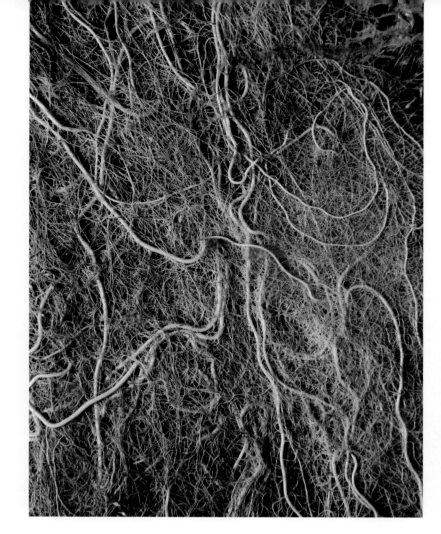

by plant metabolism and to a large extent is controlled by the plant.

All plants require the same nutrients for growth, reproduction and maintaining proper function. As nutrients are taken up by the plant, their concentration in the soil naturally decreases, resulting in a gradient that encourages the diffusion of nutrients from higher concentrations in the soil toward the lower concentrations just outside the root. The specialized proteins at the root surface that move nutrients into the root vary across plant species, and nutrient uptake itself can be quite a complex process that depends on protein regulation, nutrient concentration and the properties of the soil. Physiological changes in roots in patches of high nutrient concentration in the soil allow root production and uptake rates to be increased in order to take advantage of these resources.

⌃ Root systems are made up of diverse root types and sizes, which serve multiple functions for the plant. Small changes in root diameter can indicate large differences in function within a root system.

< Complex coordination between above- and below-ground organs regulates multiple plant functions, including nutrient and water uptake. The functional attributes of coarse and fine roots aid the plant's response to its soil environment.

Root hairs

Ⓡ oot hairs are extensions of single epidermal cells that have a very thin cell wall and so are tightly coupled to their environment, making them ideal for gas exchange and water and nutrient uptake. Additionally, the small diameters of the hairs allow them to reach soil regions that thicker roots cannot penetrate, thus making them even more important for accessing nutrients that do not readily move through the soil.

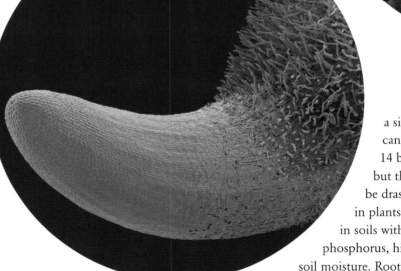

∧ The radish (*Raphanus sativus*) radicle produces dense root hairs upon germination to aid in soil contact and therefore nutrient and water uptake.

⊐ Coloured scanning electron micrograph of a germinating radish, showing the developing root (radicle) and root hairs.

The occurrence of root hairs

Despite the specialized function of root hairs, they are not found on all higher plants. For example, while most flowering plants have root hairs, gymnosperms are less likely to form root hairs and most aquatic plants have few to none. Moreover, plants that form an association with ectomycorrhizal fungi are less likely to form root hairs, probably because the fungi can encase the entire root, thus covering any existing hairs.

In the large number of species that do produce root hairs, there can be substantial variation in their size and number. For example,

a single plant can produce up to 14 billion root hairs, but that number can be drastically decreased in plants that are rooted in soils with high levels of phosphorus, high pH or low soil moisture. Root-hair formation is also influenced by the aeration of the growth medium. Submerging root hairs in water severely truncates their lifespan and usually results in disintegration. This could likely be why they are absent in many aquatic plant species.

Root hairs as a gateway

Root hairs serve as the main site of entry for microorganisms into the root system. Most likely, infection occurs here both due to the lack of a cuticle or protective outer tissue, and the higher levels of oxygen produced through respiration compared to other areas of the root system. Nitrogen-fixing microorganisms in

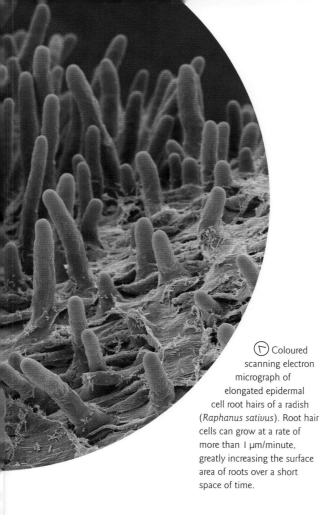

⊕ Coloured scanning electron micrograph of elongated epidermal cell root hairs of a radish (*Raphanus sativus*). Root hair cells can grow at a rate of more than 1 μm/minute, greatly increasing the surface area of roots over a short space of time.

particular infect the root through the root hairs. So-called 'shepherds' crooks', a sharp bend or curl that occurs when a root hair is infected with bacteria, are a response of the hair to hormone signalling; they vary depending on the stage of root-hair development at the time of infection.

Evolutionary remnant or specialized function?

Some researchers have reported the occurrence of stomata, the pore-like structure normally found on leaves (see pages 158–59), within the root-hair zone. Because these structures are regulated by light, humidity and carbon dioxide, the lack of light underground prevents stomata found on roots from opening and closing, and instead they remain permanently open. The function of these root stomata is puzzling. In an evolutionary context, root hairs are closely related to their rhizoid counterparts, suggesting they are a vestigial structure. Often the stomata are lost as the root ages, making it possible that they function to aid increased gas exchange within the root-hair zone during root maturation.

CLUSTER ROOTS

Proteoid roots, also referred to as 'cluster roots', are not root hairs as they might appear, but instead are very closely branched roots that are additionally covered in root hairs. This proliferation of roots within such as small area drastically increases the root surface area and therefore, like root hairs themselves, aids nutrient uptake. Frequently, proteoid roots are capable of altering their soil environment through the secretion of acids and water, which then mobilize nutrients in the soil. Iron deficiency in the plant in particular is a predominate cause for proteoid root formation.

⊝ Red pincushion protea (*Leucospermum cordifolium*), a shrub native to South Africa, develops proteoid roots to deal with the nutrient-poor sandy soils of the Western Cape region.

❯ Root system architecture can vary greatly in habitats with differing soil resource availability. No single root architecture is optimal for all resources, but in general shallow roots have better access to nutrients, while deeper roots can access more stable water resources deeper in the soil profile.

Roots are not always where you expect

Determining where roots are located can sometimes be trickier than expected. Generally, we find roots in areas where resources (water and nutrients) are ample. But because resources can vary across environments, it is important to consider root distribution as a three-dimensional structure that changes with time as resources are utilized or become available. For example, in arid environments where water is located at depth, roots must grow deep to access water. Overall, roots must explore their environment to optimize resource uptake through their root placement.

Root branching

Root architecture refers to the spatial distribution and branching patterns (topology) of a root system. There are two main forces at play in determining the architecture of any given root system: those that are a direct result of genetic makeup (endogenous), and those that are a response to the biotic and abiotic environment (exogenous). Unlike animals, plants continuously grow new organs throughout their life and adjust their growth in response to their environment. Thus, root

architecture can vary substantially both within and between species.

As a means to understanding the architecture of roots quantitatively, biologists borrow from fractal geometry to describe the arrangement of repetitive root branches. Standardizing how root branches are arranged allows studies on root responses to varying resource levels and provides a way of comparing branching structure with plant performance. Included in the parameters that are used to quantify a root system's

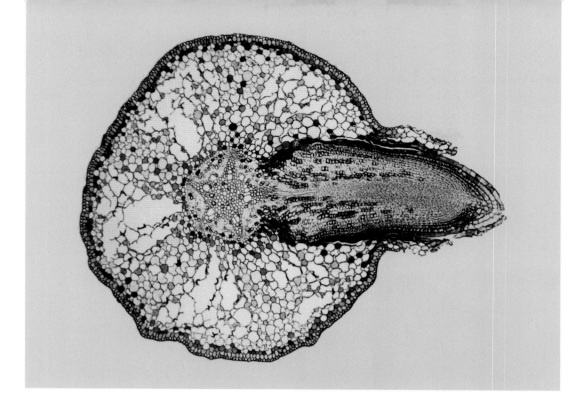

◁ Colour enhanced light micrograph of a willow (*Salix* sp.) root cross section with an emerging lateral root. The arrangement of lateral root initiation is tightly regulated by the plant's vascular tissues. Mechanical and chemical factors can alter the root's development and ultimately change its elongation rate and/or branching angle.

▽ Plants can behave quite opportunistically and establish roots in areas of high resource availability, such as shallow soil layers. Soil erosion has exposed the shallow roots of this tree, located in an urban environment.

architecture are, commonly, the growth angle of new lateral roots, the rate of root growth and the maximum depth to which the root system develops. These three traits alone can help predict root system resource acquisition strategies. By linking the root system's ability to occupy a three-dimensional space and the known resource limitations of an environment, it is possible to determine the efficiency by which roots will access water and nutrients.

Links to global food supply

Given the ever-increasing requirement for global food supplies, scientists are continually working to improve crops by selecting for traits such as root branching that will optimize a plant's ability to access resources efficiently. Natural variation within a species' root growth and branching angle can allow crops to be grown in many areas of the world where poor soils and limited fertilizer availability constrain people's access to food. However, much more work needs to be done in order to optimize the growth of plants facing multiple stresses in a field setting that requires sometimes diverse root architectures.

Root depth and spread

It is difficult to find a commonality across all species that describes the development and spread of roots throughout a plant's lifetime. Perhaps the initial germination of the seed and production of the first root is a fairly common attribute, but after that all bets are off. What we do know about root location has mostly been acquired through controlled environment studies on plants grown in pots. Large, destructive trenches have also provided clues in field-based settings, but frequently it remains difficult to locate roots precisely in their soil environment. The lack of soil uniformity is a major driver in determining where roots grow. Generally, a large fraction of a plant's root system will be concentrated in the shallower soil layers, since this is where the majority of nutrients are located in most ecosystems.

No single root architecture is optimal for all resources within the soil. Interestingly, the shallow, long root architecture found in the common bean (*Phaseolus vulgaris*) and other species, which may be desirable for reaching resources that do not move readily in the soil (such as phosphorus), is in contrast to the deeper root branching that is beneficial for accessing water. It is not uncommon to find portions of the root system of a woody species at depths greater than 5 m (16 ft). These deep roots contribute substantially to meeting the plant's daily water needs and can even add to the ecosystem's water balance. In extreme cases, roots have been found growing into deep (20 m, or 65 ft) underground caves that foster a continual water supply through a network of subterranean streams.

Horizontal root spread can also reach surprising distances. Roots of mature trees, for example, can often span two to three times the radius of their crown, allowing them to reach potential favourable environments for root growth far from their own habitat. In urban environments, this lateral root spread can become problematic for underground utilities and sewer pipes in particular. As with the natural caves mentioned above, roots can grow into breaks in water-filled pipes and soon proliferate, causing a blockage in flow and a headache for the municipality.

⌄ Major advances in the breeding of crop species such as the common bean (*Phaseolus vulgaris*) have improved root architecture to deal with low soil fertility, including in tropical areas.

The idea of the root niche

Competition between plants is a ubiquitous feature across both cultivated and natural landscapes. In the context of below-ground root interactions, the effect of competition for, or sharing of, resources among plants is poorly understood. Yet it has been argued that the placement of roots and their overlap with neighbouring plants is a dominant factor in determining how plant communities are structured. This competition for resources can play a role in where roots grow, either to avoid the competition (i.e. in a location other than where the roots of other plants are growing) or to access ample resources (i.e. proliferate despite the presence of other roots). In some plant communities, plants shift where they place their roots, moving them deeper in the soil to reduce overlap with neighbouring species. The North American burro-bush (*Ambrosia dumosa*) is known for inhibiting its root growth when an individual root contacts another root of the same species. Other species such as the creosote bush (*Larrea tridenta*), a desert shrub also native to North America, are even more extreme, completely halting root growth by inhibiting elongation when the roots of other creosote bushes are nearby.

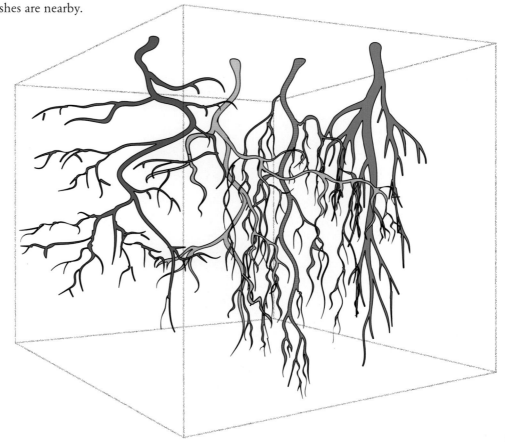

⑦ Underground caves can serve as a major contributor of water resources to woody plants. Roots of vines are seen here descending into a deep water source in Mexico.

⟩ In areas with multiple plants, as represented in this diagram by the multiple coloured root systems, differences in root system architecture inherently lead to differences in root space occupation below ground. How root systems interact depending on the proximity and identity of their neighbours is often difficult to ascertain.

Root growth

(T)he rate at which roots grow is determined by the rate at which new root cells are produced and elongate. The root cap plays a major role in determining a root's movement through the soil and can even respond to non-biological stimuli. However, in order for roots to grow into a new soil area, they must exert pressure on the soil. If the mechanical impedance of the soil is too great, then roots may increase in diameter rather than length, inhibiting their capacity for exploration.

(<) This cauliflower plant (*Brassica oleracea*) has minimal thickened roots as a result of high mechanical impedance in the soil. Club root symptoms, a result of fungal infestation, are also evident on the bulbous root in the foreground.

(^) Primary root tip and root cap of maize (*Zea mays*). The root cap protects the growing root tip and serves in gravity perception; its removal can result in random directionality in root growth.

Mechanical impedance in the soil

Roots are built to navigate their way through the resistance of the soil matrix. This soil resistance is important, however, as without it plants would not have the required stability to remain upright, particularly during windy weather. But roots can be lazy in a manner of speaking, generally taking the path of least resistance and following cracks or large pores in the soil. For roots to push through denser soil, they require a pressure that exceeds that of the soil resistance. If the soil impedes root growth, then thick, stunted and deformed roots are produced. Root functionality will also likely change if roots do not grow and develop normally, most likely resulting in stunted growth of the plant above ground as well.

THE SPIRALLING NATURE OF ROOT GROWTH

It is not difficult to appreciate the difficulty roots face in moving through dense soil, even with the production of new cells and mucilage. One way they overcome this obstacle is through 'waving' or 'coiling' as they grow. Somewhat similar to the idea of a corkscrew, roots regulate active and passive growth of their cells to produce a response that aids their ability to drive themselves into the soil and avoid any obstacles they may encounter. The roots of some plants naturally spiral as they grow, despite the density of the substrate, while others oscillate as they detect and respond to gravity. These oscillations are a result of unequal rates of cell growth on each side of the root. Contrary to popular belief, plants do not change the direction of their oscillations if they are moved from one hemisphere of the Earth to the other.

The role of the root cap in root growth

Unlike stems, roots have a group of cells produced from the actively dividing cells of the meristem at their tip. These cells, collectively called the root cap, primarily function to protect the root tip as it moves through the soil matrix. The root cap helps to protect the root by secreting a viscous substance (mucilage) as a means to aid its ability to move through the soil. But the root cap also has additional functions, including signal perception of its environment. While there is still much to be learned about how this signalling works, we know that roots are capable of growing toward water and away from toxic metals, indicating that they can change their growth patterns in response to interactions between the root cap and external stimuli.

As roots grow through the soil, border cells of the root cap are shed, leaving behind a sheath of cells and mucilage that aid the retention of root-to-soil contact and create what many refer to as the rhizosphere (see page 88). The meristematic region mentioned above continually replaces these lost cells with new ones throughout the root's growth through the soil, until the root reaches the end of its elongation phase.

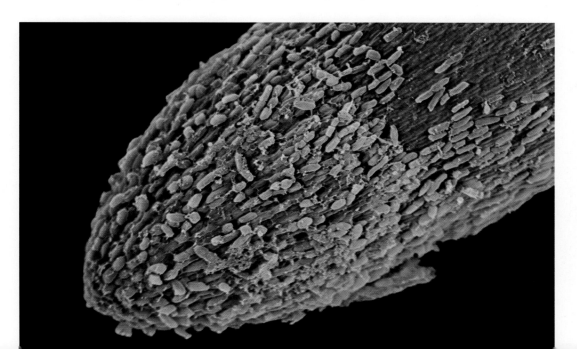

⊘ Scanning electron micrograph of the root cap of a maize plant (*Zea mays*), showing border cells being shed (magnification ×270). Border cells are metabolically active cells that aid the root in moving through the soil. This root tip has been growing for only a few days.

Cacti are well known for producing largely ephemeral root systems. During periods of rainfall, the plants produce roots to take up water, but then shed them during dry periods.

Tree root systems are quite different from those of herbaceous plants. Woody roots clearly have very long lifespans, and the plant has invested substantial amounts of energy (carbon) into their production and growth, in the form of secondary compounds such as lignin and suberin.

Root dynamics

Fine roots – those that are most distal from the plant – are the most ephemeral root type. When roots are produced, how long they live and when they die can have large implications for root system function and overall plant health. Due to the simple primary development of fine roots and the low levels of secondary compounds they contain, they tend to deteriorate with age and turn over at a higher rate than their woody counterparts, and are subsequently more prone to attack by pathogens and herbivores.

Every root has a lifespan

Production and lifespan of these finest roots are controlled by both endogenous factors (e.g. diameter and mycorrhizal associations) and environmental factors (e.g. temperature, soil moisture, nutrient availability and rooting depth), although the importance and role of these factors vary considerably over species and environments. In general, the roots of evergreen species have the longest lifespans, living for several years, whereas those of annual species and desert cacti are extremely ephemeral, perhaps living only during the season of ample precipitation to aid water uptake. In general, roots that have a mycorrhizal association (see page 82) benefit from a longer lifespan, likely due to the buffered nutrition and water status of the root.

Phenology of roots

It is easy to assess the growth and mortality of leaves, whether during major senescence events, such as the autumn months for temperate forests, or continual leaf regeneration in more tropical ecosystems. Moreover, quantifying shifts in these phenological, or life-cycle, stages in response to environmental variables can provide clues to fluctuations in leaf carbon gain – the ability of leaves to produce energy through photosynthesis – and therefore resource availability. Timing of root production and mortality, on the other hand, can be more difficult to quantify. Some level of root birth and death occurs continually, but these events can easily go unnoticed in the absence of direct root observation. While some studies on temperate tree species have found that root production and mortality are highly synchronized with foliar production, whereby root systems grow and expand prior to leaf growth in order to support necessary water and nutrient uptake, others have been unable to render clear connections in production or mortality between the above- and below-ground portions of a plant.

THE GENERAL PATTERN OF ROOT AGEING AND DEATH

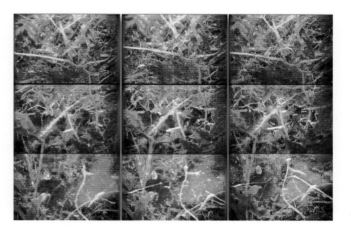

The finest lateral roots undergo a progression through their life, which, although still not well understood, can be tracked via specialized underground camera systems. Physical and biological factors play a direct role in the trajectory of a root's life and can increase or decrease its lifespan, thus influencing the root's contribution to nutrient and water uptake. While generalities that correlate plant growth rate to tissue longevity have been suggested, one should approach these with caution as such hypotheses arise from observations on above-ground organs that may not follow the same patterns as their root counterparts underground.

⌂ Specialized cameras called minirhizotrons capture repeated images of root systems through time, as shown here, to study root lifespan. Root lifespan impacts carbon, water and nutrient cycles.

Roots as underground storage organs

$\widehat{A}$ll plant organs can serve in a storage capacity to some extent, essentially reserving carbohydrates and water for growth and maintenance of the plant body. Of all plant organs, however, roots contain the highest concentrations of carbohydrates and are therefore an extremely important energy substrate. Large fluctuations in the accumulation and utilization of storage reserves occur naturally, but stressful environmental conditions can deplete them rapidly as they are used to repair or regrow tissues or organs. If stress is prolonged and reserves become critically low, the consequences will likely be detrimental to plant survival.

Water and carbohydrate reserves

Roots that store precious energy and resources in the form of both soluble and insoluble compounds and/or water can vary from morphologically indistinct (e.g. the roots of the apple tree, *Malus domestica*) to morphologically distinct (e.g. the large, swollen fleshy organs of some succulents, including the soapweed yucca, *Yucca glauca*). Starch is typically the main form of stored carbohydrate in woody perennials, but many other forms of sugar can also be utilized as a storage reserve. Regardless of the form of carbohydrate used for storage, the parenchyma cells retain it in high concentrations for the plant.

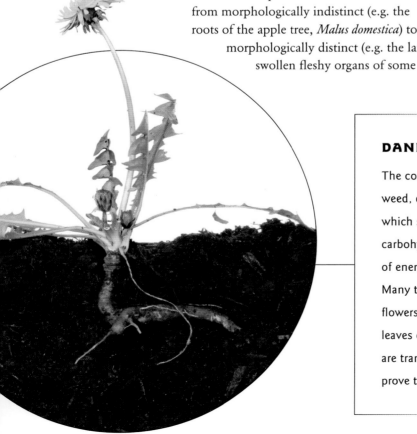

DANDELION ROOTS

The common dandelion (*Taraxacum officinale*), a familiar lawn weed, develops a deep tap root within weeks of its establishment, which serves as a storage depository for energy in the form of carbohydrates. The tap root can store a significant amount of energy as the plant ages, rendering it quite difficult to eradicate. Many tackle the tenacious weed by continually cutting off its flowers in the hope of reducing spread. However, shading the leaves or applying herbicides in the autumn, when compounds are translocated down to the root system for the winter, can prove to be much more successful management options.

Weedy species are often persistent due to the fact that they can store large amounts of carbohydrates in their root systems. Therefore, when only the above-ground portion of a weed is removed, the plant can easily regenerate from its root reserves. A successful agricultural practice to remove weeds is repeated tillage of a field, which forces the plants to resprout several times, thereby depleting their reserves and resulting in carbohydrate starvation.

In the case of water storage, roots contain increased amounts of the wax-like substance suberin, which retards the loss of water once absorbed. These specialized roots are also capable of dilating to prevent the breakage of cells or tissues during the swelling that naturally occurs during larger water uptake events.

⊘ Parenchyma cells of the cortex of a buttercup (*Ranunculus* sp.) root, containing starch grains stained purple for visualization (magnification ×100).

⊘ Many economically important food crops are enlarged root systems with ample carbohydrate storage. Here, cassava (*Manihot esculenta*) roots, which are high in starch and other carbohydrates, are harvested in Ecuador.

SUCCULENT ROOTS

The roots of plants in arid environments must endure large swings in temperatures and water availability. One way in which some arid plants deal with these limitations is by storing water within their root systems. By placing portions of their root system at extremely shallow depths, these plants also ensure that they can access any precipitation that does fall. The water is then stored in large, fleshy roots, similar in function to the fleshy stems found in many cacti. Water that is captured and stored underground in roots has the added benefit of remaining out of direct sunlight, where it stays cooler and hence is less prone to evaporation. To supply its needs, the elephant cactus or cardón (*Pachycereus pringlei*), the largest cactus in the world, can have a fleshy, water-filled tap root that is up to 18 cm (7 in) thick.

⊘ Cardón (*Pachycereus pringlei*), such as this example in Baja California, Mexico, store large quantities of water in their roots and stems. The cactus's name common name is derived from the Spanish word *cardo*, meaning 'thistle'.

Adventitious roots

Inevitably, roots can arise from two origins: the embryonic root that forms during normal plant development, or tissues such as stems or leaves that were not predetermined root tissue. It is these latter roots, which form from an organ other than the embryonic root, that are classified as adventitious. Fossil records indicate that in evolutionary terms, underground adventitious roots were likely some of the first roots that plants formed. Despite differences in origin, roots and adventitious roots are functionally similar, both structures serving to supply the plant with water and nutrients.

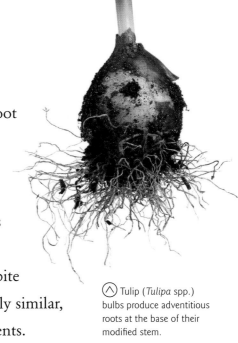

⌃ Tulip (*Tulipa* spp.) bulbs produce adventitious roots at the base of their modified stem.

⌃ A cross section of the stem of a tomato plant (*Lycopersicum esculentum*), showing the development of an adventitious root. In the tomato, adventitious roots usually develop only if the stem is buried or subjected to flooding.

As a general rule of thumb, groupings can be made between plants that can or cannot form adventitious roots, and those that can whether or not they require wounding to initiate production.

Rising from the stem

Most commonly and naturally, adventitious roots that arise from the shoot are formed at nodes and allow for increased plant spread. Common agricultural species such as the strawberry (*Fragaria × ananassa*) and the undesirable field bindweed (*Convolvulus arvensis*) are examples of plants that root abundantly from the nodes found on their stolons (horizontal above-ground stems) or rhizomes.

Horticultural advantages to adventitious rooting

Horticulturalists have taken advantage of the ability of plant cells to regenerate and form new adventitious root tissues through wounding. Plant propagation techniques can induce adventitious roots through the application of careful ratios of plant hormones (namely, auxins and cytokinins). These new roots can form on plant stems, petioles and even leaves, and play a key role in producing large numbers of new plants in a fast, efficient manner. Many gardening, forestry and food crops are produced using such rooted cuttings.

Adventitious roots of terrestrial plants

In extant species, unlike early plants, adventitious roots are frequently formed only as a response to wounding and generally above ground, therefore requiring lager anatomical changes within the plant. Above-ground root formation can clearly be advantageous to plants as these roots can exploit new environments that might not otherwise be accessible. However, the ability of plants to form adventitious roots varies quite widely from species to species, with added complexity depending on the plant age and tissue type.

ADVENTITIOUS ROOTS AND CLIMBING PLANTS

The adventitious roots of climbing plants serve two functions: to anchor the stem to the vertical substrate, and to acquire resources. In some species, anatomical differences can determine the function of the root. Roots that serve to supply the plant with nutrients and water, for example, have more developed vascular tissue than their grasping counterparts. Vines such as poison ivy (*Rhus radicans*) produce adventitious roots from their stem to secure themselves when they come in contact with a vertical surface suitable for climbing. These roots are non-damaging to the surface on which they climb, but they do produce the same toxic urushiol compound for which poison ivy is notorious.

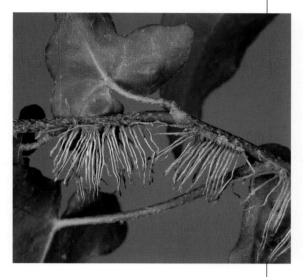

⌃ Ivy (*Hedera helix*) produces numerous adventitious roots along its stem to attach itself securely to a substrate.

By climbing up trees, or even buildings, the vine is able to reach more favourable light conditions.

ADVENTITIOUS ROOTS AND PALM TREES

Many palm trees produce a purely adventitious root system. Roots form at the base of the stem, both above and below ground, creating a large, swollen structure called a bole. As the palm ages, new rings of roots are produced from nodes. If a moist substrate is not available, the roots pause their development until growth conditions are more favourable. By the time the plant reaches maturity, thousands of roots have formed. Most likely, this adventitious rooting nature is an evolutionary one. Shifting sand and soil levels are common in the native growing regions of many palm species, and the large adventitious rooting zone – which can extend from centimetres to metres above the ground – aids their survival in these adverse conditions.

< The roots of palm trees originate from their trunk, classifying them as adventitious. If large quantities of adventitious roots are produced, the base of the trunk often flares, as seen in this image.

Root adaptations

While the soil habitat in which roots reside is in some regard buffered from environmental extremes, roots nonetheless encounter a multitude of limitations and stresses during their growth and development, including shifts in nutrient and water availability, high concentrations of heavy metals, extreme temperatures, increased salinity and mechanical resistance. Root adaptations to these stressors have resulted in a number of unique morphologies, which aid the roots in their ability either to tolerate or escape the unfavourable environment.

> Light microscopy cross section of a maize (*Zea mays*) root showing the endodermis. The casparian strip (red line) is a hydrophobic cell layer that aids in water regulation, with the vascular tissue (red circular cells) lying just inside it.

∨ Tall-stilt mangroves (*Rhizophora apiculata*) produce multiple stilt roots from their stems and branches that resemble a skirt in order to provide support for the trees in their watery environment.

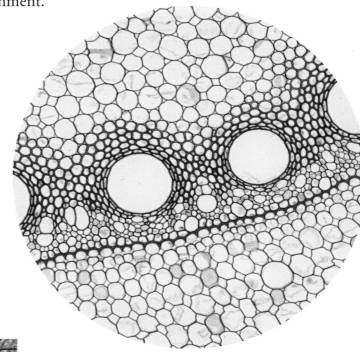

Water – too much or not enough

We can compare water availability in the soil to the story of Goldilocks and the three bears (see also page 150). For the majority of plants, either too much or too little is not good – soil water availability must be just right. However, plants that continuously experience drought or flooding have evolved mechanisms to cope. The issue of too little water commonly produces physiological adaptations rather than morphological ones. For many species, as long as the plant can access water with a portion of its root system, the passive internal movement of water driven by gradients in water potential can redistribute it to the roots that are in dry soil, thus aiding in their continued survival and functionality. For many desert plants, portions of the root system are more ephemeral in nature and regrow only during the rainy season, when water is more readily available.

Too much water, on the other hand, can cause just as serious consequences for a plant. Roots require sufficient oxygen for cellular respiration to take place, but waterlogged soils contain insufficient levels of the gas. Plants that live in areas that are prone to flooding therefore commonly produce adventitious roots from their stem as a means to extend their root system above the waterline.

WALKING ROOTS

It was first hypothesized by anthropologists John Bodley and Foley Benson in 1980 that the so-called walking palm (*Socratea exorrhiza*), found in Central and South America, could use its adventitious roots to 'walk' and essentially relocate to areas of more favourable soil or light conditions. These adventitious roots are produced to provide stability, and are referred to as prop or stilt roots. However, they have also been reported to aid the walking palm in performing what might seem like incredible feats, including the movement of the entire plant in space and the ability to realign itself should it be knocked over. Unfortunately, recent research has debunked these claims.

While the tree may produce new roots on only one side of the trunk, it remains firmly fixed in place.

⊘ Stilt roots of the walking palm (*Socratea exorrhiza*), a tree native to rainforests in Central and South America, have been reported to 'walk' up to 20 m (65 ft) a year.

Temperature

The influences of light and temperature are sometimes difficult to separate when discussing plant responses. Regardless, variation in both of these abiotic (physical) factors can affect root growth, which is important considering that soil temperature is predicted to increase in the coming decades as a result of climate change. In colder environments, this could lead to longer growing seasons and hence increased above- and below-ground growth. However, if temperatures surpass the optimum for a particular species, root damage and reduced growth could occur.

While roots typically grow toward warmer soils (positive thermotropism), under extreme conditions negative thermotropism could occur, whereby they grow away from soil that is too hot. Suboptimum temperatures are also undesirable and can result in stunted root systems with fewer lateral branches.

Metals

Heavy metals in the soil can cause myriad morphological changes to a root system. Of these, reduced root elongation and consequent stunted root growth is a primary response to

PNEUMATOPHORES

Species such as mangrove trees produce 'breathing roots' called pneumatophores, which grow vertically out of the water in waterlogged environments. Pneumatophores have small gas-exchange areas called lenticels in their outer bark, which open when water levels are low to allow for the intake of oxygen needed for root respiration, and close when they are submerged to prevent waterlogging.

high levels of such metals. Subsequently, roots often become thicker in diameter and produce greater levels of suberin, the protective waxy layer in the outer epidermal layers of the root. Lateral root branches and root hair production are also reduced. Collectively, these morphological responses reduce uptake by the roots and hence prevent accumulation of the toxic metals within the plant.

Salinity

While it might seem unusual for roots to grow in saline environments, select species have evolved ways to deal with high salt levels. In some cases, however, roots are unable to exclude salt from their tissues, and once it has entered the plant it ends up in the leaves, where it is compartmentalized in cells. While some salt is tolerable, an overabundance can prove toxic for the plant. In a response similar to that undertaken when heavy metals are present in the soil, plants that inhabit saline environments produce shorter, thicker roots.

Plants in the genus *Avicennia*, commonly known as mangrove trees, inhabit coastlines where fresh and salt water mix, and have adopted multiple adaptations to allow them to survive in such harsh environments.

⊙ Close-up of mangrove (*Rhizophora* sp.) roots showing the numerous lenticels that cover the outer bark. These lenticels act like pores and allow for gas exchange at the root level.

⊙ Mangrove (*Rhizophora* sp.) pneumatophore roots stick up from the sand like snorkels, allowing part of the root to access much-needed oxygen for at least a portion of the day.

⊲ Despite the less healthy appearance of this *Arabidopsis halleri* plant (left), the species is naturally found on polluted soils and is better able to hyperaccumulate cadmium and zinc than thale cress (*Arabidopsis thaliana*; right).

Mangrove trees can translocate salt that enters through the root system to old leaves, which are then shed, thus removing the salt from the plant. Additionally, they can hold on to fresh water by minimizing evaporation in younger leaves. The most impressive way these trees deal with their brackish environment, however, is through their modified roots. Specialized aerial roots allow the mangrove trees to 'breathe', despite the excess salt and water they encounter on a regular basis (see box).

Digging deeper

There are clear advantages for a plant to have its roots and storage organs buried under adequate soil cover. Sufficient root anchorage to ensure plant stability is one, but others include protection from temperature extremes and fluctuations in water availability. In order to dig deeper into the soil, contractile roots shrink in length by 50–70 per cent, in such a manner as to exert a pulling force that allows the above-ground plant or storage structures such as a bulb or corm to descend deeper into the soil. Because the portion of the root that does the pulling is located close to the base of the plant or bulb, the root tips and root hairs mainly stay fixed in place and only the upper plant parts are drawn down. Moreover, because the main roots have already navigated through the space into which the plant is descending, it faces less resistance and so a greater movement is attainable.

⊲ Hyacinths, including *Scilla* (shown here), along with other bulbs, produce contractile roots that aid in pulling the bulb deeper into the soil.

⌖ Aerial roots of the silk-cotton tree (*Ceiba pentandra*) can attach to numerous substrates, including man-made structures – as seen here at the twelfth-century Ta Prohm temple in Angkor, Cambodia.

Aerial roots

Ⓐerial roots do not differ functionally from their below-ground counterparts, but they have acquired a more diverse morphology to deal with their drier and more suspended environment. Unlike the roots of plants anchored in soil, those of plants that live on or climb up others must garner resources from the air or from decomposing materials that are caught in the canopies of taller plants, or, in a few cases, make the long journey down to the ground to access the resources that are available in the soil.

⌖ Strangler figs (*Ficus aurea*) start their life as an epiphyte in a tree. But, as their name implies, they grow quickly and develop roots that can take hold of other nearby plants. These roots can completely encircle their host and constrict its growth, essentially strangling it to death.

Climbing vines

Vines and lianas commonly use adventitious roots along their stem to cling to the surface of trees or other vertical substrates. These roots aid the plant in climbing toward less shady environments. Unlike roots below ground, these climbing roots can be negatively gravitropic (growing away from the earth) and in most cases garner water and nutrients from the air, leaving the plant on which they are climbing relatively unharmed. Perhaps the most extreme case of climbing vines is that of the so-called strangler fig (*Ficus aurea*), a native of southern North America, Central America and the Caribbean. As the roots of this initially epiphytic plant descend a tree toward the ground, they wrap around the host, the two fusing with each other as they grow. However, it is not until the strangler fig's roots have reached the ground and the tree is self-sustaining that its roots expand in girth, slowly outcompeting the host and causing its early demise.

Orchids, a type of epiphyte, possess aerial roots that not only capture water and ions from the air but also have a protective and insulating layer. This cell layer, called the velamen, can absorb available water within seconds and hold onto it for significantly longer periods than usual (generally several hours), allowing plants that possess it to survive in drier environments. In addition, velamentous roots contain chloroplasts, thus contributing to the plant's carbon gain through photosynthesis. When the roots are dry, they appear white due to the air-filled velamen, but upon wetting, they appear green as the velamen cells become filled with water and thus translucent, exposing their inner chlorophyll-containing layer.

⌃ This orchid cane has sprouted a new plant, known as a keiki (from the Hawaiian word meaning 'child'). Adventitious roots have developed at the keiki's base.

⟩ A cross section of an orchid root shows the outer layer of dead velamen cells, which act like a sponge when exposed to water.

The specialized roots of epiphytes

Epiphytes – plants that grow on the surface of another plant – have developed a myriad of root types that are specialized for holding onto their host as well as capturing resources in somewhat unusual substrates. Roots of these plants can capture resources from the atmosphere, including water from precipitation, dust particles suspended in vapour (termed atmospheric deposition), decomposing plant material that has gathered in the crux of the canopy or even ant nests that house ant excrement as well as the decomposing bodies of the ants themselves. The majority of epiphytes are found in tropical or humid environments, which provide the required moisture within the air to ensure the plant's survival.

Roots of some epiphytes are capable of changing their function, moving from a holdfast (a root that grows minimally in length and functions, as its name suggests, to anchor the epiphyte to its substrate) to a more absorptive capacity with the introduction of valuable resources. Adaptations such as these are what allow epiphytes to thrive in such harsh environments.

Parasitic roots

Ⓐpproximately 1 per cent of flowering plants are parasitic, meaning that they 'steal' some or all of their carbohydrates from their host. Of these, approximately two-thirds are parasitic on the roots of their host and the remaining third are parasitic on the stem. All, however, use a modified root structure called a haustorium to attach to and penetrate the host in order to fuse directly with its vascular tissue. Because parasites usurp the energy produced through photosynthesis from their host, they frequently lack chlorophyll and thus come in an array of colours.

Epicortical roots

Parasites that penetrate the stems of plants climb the host using adventitious epicortical roots, or runners. Anatomically, epicortical roots resemble stems, but they remain leafless and, more importantly, produce a root cap, a critical anatomical structure that differentiates stems from roots. Usually, the parasite plant begins its life with a primary root that acquires its resources through the more traditional route – the soil. As the parasite grows, the numerous epicortical roots it produces in turn produce numerous haustoria, which are able to penetrate the host as the parasite grows along its surface. Only a few haustoria are needed before the parasitic plant loses its primary root and thus its connection to the ground. Interestingly, the haustoria of the epicortical roots develop only on the side of the root that has contact with the host, and the frequency at which they are produced is a function of the thickness and density of the host plant's bark.

Root parasites

The seeds of parasitic plants can remain dormant in the soil for many years. Germination of root parasite seeds is initiated by a chemical signal induced when a host

▷ Colourful orange American dodder (*Cuscuta americana*), a leafless parasite, steals resources from its host without ever touching the soil by producing root-like haustoria that fuse with the host's vascular system.

▽ A photomicrograph cross section of three haustoria of American dodder (the upper orange structure), a parasitic climbing vine seen here infecting a host stem.

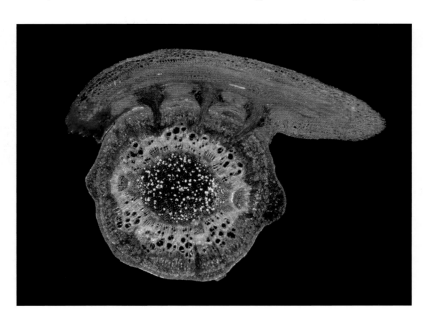

plant is near. The detected signal is a hormone called strigloactone, which plants utilize to form connections with fungi underground. Root parasites are also sensitive to strigloactones, however, and use this cue to produce an initial root that requires host contact in order to survive. If a host is found, the parasitic root will produce a substance that allows it to stick to the host's roots long enough to penetrate the endodermis. As with other plant parasites, contact with the host's vascular tissue ensures nutrient transfer. Purple witchweed (*Striga hermonthica*), a notable root parasite of crops such as sorghum (*Sorghum bicolor*), rice (*Oryza sativa*) and maize (*Zea mays*), utilizes water and photoassimilates at rates higher than its host, resulting in severe growth reduction in the host.

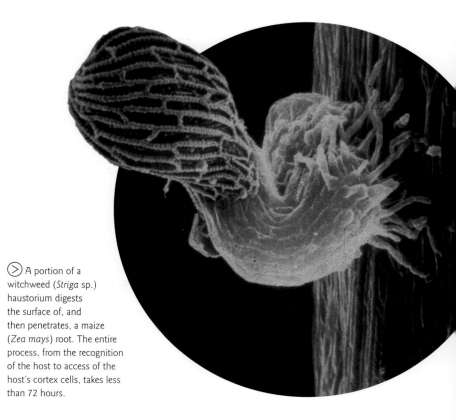

(>) A portion of a witchweed (*Striga* sp.) haustorium digests the surface of, and then penetrates, a maize (*Zea mays*) root. The entire process, from the recognition of the host to access of the host's cortex cells, takes less than 72 hours.

THE WORLD'S LARGEST FLOWER IS A ROOT PARASITE

Parasitic plants range in how much of their plant body grows either completely within or external to their host. In some cases, the entire parasite plant body is concealed within the host plant and emerges only to flower. This habit is termed endophytic and refers to organisms that spend the majority of their life cycle within another plant's body without causing it significant harm. Such plants do not photosynthesize, and instead gain all their carbon from the host.

Flowers of endophytic parasites can be quite interesting and unique, and come in many shapes and sizes, resembling everything from pine cones to rotting flesh, as is the case for the corpse flower (*Rafflesia arnoldii*).

(∧) The corpse flower (*Rafflesia arnoldii*), a root parasite, can grow up to 100 cm (39 in) in diameter and attracts fly pollinators by visual and odorous mimicry of rotting flesh.

Root associations

Roots form a number of symbiotic associations with other organisms in the soil. Mycorrhizae and rhizobia – fungi and bacteria, respectively – are the most ubiquitous and well studied of these organisms. These symbiotic associations facilitate the uptake of nutrients and water from the soil, and in the case of mycorrhizae form an underground fungal network that connects plant roots to their neighbours. The benefits of these associations outweigh the carbon costs to the plant, and allow for its establishment – and thus a competitive advantage – in habitats that might not otherwise be favourable.

Mycorrhizae

The majority of seed-bearing plants form a mycorrhiza–root association. In the simplest sense, there are two types of host inoculation with a fungus, defined by where the fungus resides within the root. In ectomycorrhizae, the long strands of the vegetative fungal body (the hyphae) are located outside the root and do not penetrate the root cells, whereas in endomycorrhizae the majority of the hyphae are housed within the root and the fungus enters individual root cells to form a close association for nutrient exchange. Depending on the plant species, a single plant can range from having one tight-knit association with a solitary fungus, to multiple mycorrhizal associations with both ectomycorrhizae and endomycorrhizae. Furthermore, those associations can shift with plant development and environmental constraints.

Ectomycorrhizae are recognizable by the visible sheath of hyphae (appropriately called the mantle) that encases the tip of the plant root. A portion of the hyphae do penetrate the epidermis and enter the root, but then simply surround the individual root cells to form a highly branched fungal network, called the Hartig net. The root's association with the fungus results in a lack of root-hair production, a visual swelling in the root tip due to the mantle that covers the outer root, and an overall stunted root length, most likely due to the fungus preferentially colonizing slow-growing roots.

There are several classifications of endomycorrhizae that, in some cases, are quite specific to their plant host. Arbuscular mycorrhizae are the most common and associate with approximately 80 per cent of plant species. More specific endomycorrhizae include ericoid mycorrhizae, specific to the heather family (Ericaceae), and orchidaceous mycorrhizae, specific to the germination of orchid seeds. A third type of mycorrhizae,

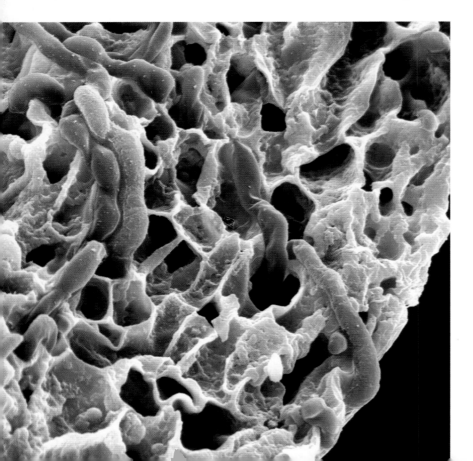

⌄ Mycorrhiza hyphae (orange) can be seen within the cortex of a plant root in this coloured scanning electron micrograph (magnification ×3,400). The fungus is able to access nutrient forms unavailable to the plant, process them and pass them on to the roots.

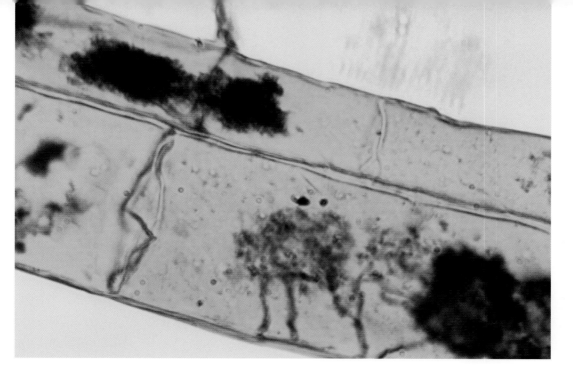

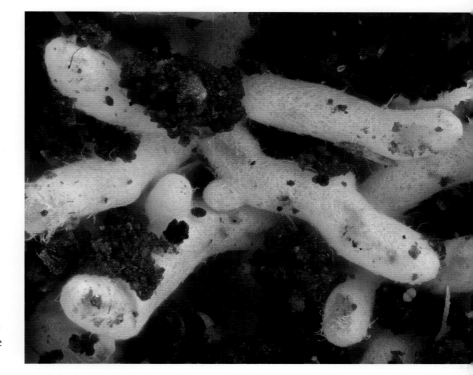

Colour-enhanced light micrograph of mycorrhizal root cells, showing arbuscules (dark lines) inside cortical root cells of a maize (*Zea mays*) plant. Arbuscular mycorrhizal fungi help plants capture nutrients such as phosphorus from the soil, exchanging them for carbon.

termed ectendomycorrhizae, are primarily found in pine, spruce and larch trees. They possess characteristics of both ecto- and endomycorrhizae, in that the fungal body is located both externally and internally within the plant tissue.

The development of the connection between the plant and fungus is an intricate process, and one that is still not fully understood. Chemical compounds produced by the roots and the fungus alike suggest that primary infection involves specific signals that either initiate the germination of the fungal spore or induce branching of the fungal hyphae. Likewise, root elongation and branching may occur during infection.

The benefits to plants that occur as a result of their association with mycorrhizae have been the focus of numerous studies. Regardless of the type of infection, the general outcome is the same, namely increased access to nutrients via the extensive below-ground hyphal network. The ability of the fungal hyphae to reach distances that the plant roots alone could not – up to several metres for ectomycorrhizae – provides the plant with access to nutrients that have slow diffusion rates in the soil, most notably phosphorus. For some plant species, the mycorrhizal association is imperative to their ability to complete their life cycle. In most cases, however, the association forms only in habitats where the plant confers a benefit,

and may not be formed at all if adequate and accessible nutrients are available to the root system. More recently, researchers are discovering that hyphal networks also connect several plants underground, thereby creating a network of carbon and nutrient transfer. This idea should make us take pause to consider the extensive and interconnected environment below ground and its role in ecosystem function.

The mycelium (white) of an ecotmycorrhizal fungus has completely enveloped the roots of this pine tree (*Pinus* sp.) (magnification ×41). Ectomycorrhizae remain on the outside of the root, and do not penetrate its cells.

Root nodules and nitrogen fixation

One of the more well-studied interactions with roots is that of nitrogen-fixing bacteria. This symbiotic relationship is of great importance to plants, in that nitrogen is the most limiting factor for the majority of species despite being required for numerous cellular processes. Rhizobia, the soil bacteria responsible for enabling plants to fix atmospheric nitrogen, are enveloped by the roots and enter via the root or root hairs after specific chemical signalling is recognized. Once inside the root, the bacteria differentiate to become millions of bacteroids and initiate cell division in the plant, causing it to produce characteristic nodules on the root that house the bacteroids.

One small problem remains with the process that converts nitrogen into a form that can be used by plants: the bacterial enzymes that fix nitrogen are extremely sensitive to oxygen. The answer is haemoglobin, the familiar protein that plays an important role in oxygen transfer between cells in animals, including humans. The root nodules, therefore, take on a red colour due to the presence of high levels of haemoglobin. In plants, haemoglobins – more precisely leghaemoglobins because they are primarily found in the legume family (Fabaceae), function to reduce the oxygen within the root nodule to protect nitrogen production and to provide oxygen to the residing respiring bacteroids.

For plants that are unable to fix their own nitrogen through this valuable symbiotic association, a small amount of nitrogen is fixed in soil through lightning, and man has developed nitrogen-containing fertilizers through the artificial Haber–Bosch process.

⌃ Bacteria nodules on this pea (*Pisum sativum*) root fix nitrogen from the atmosphere for the plant. Most members of the legume family (Fabaceae), including peas and beans, harbour symbiotic *Rhizobium* bacteria, which aid in their assimilation of nitrogen.

⌄ A coloured cross section of a broad bean (*Vicia faba*) nitrogen-fixing root nodule, showing cells filled with red-stained *Rhizobium* bacteria. The bacteria modify the plant's metabolic and molecular cellular processes to allow root cell colonization, and live in symbiosis with the plant.

Endophytes

As discussed earlier (see box on page 81), an endophyte is an organism that lives within another organism. In plants, fungal and bacterial endophytes reside within plant cells, forming relationships that range from beneficial to pathogenic. These so-called endosymbionts impart a suite of costs and benefits to the varied and globally distributed hosts they infect. For example, endophyte colonization changes host response to herbivory, competition, biomass allocation, reproductive success and ability to invade new communities. When plants are infected by endosymbionts, the resulting symbiotum has a unique phenotype that is distinct from an uninfected counterpart. For roots in particular, endophytes can impart increased nutrient uptake and resistance to soil-borne pathogens.

Root fungal endophytes are largely composed of dark septate endophytes. While this terminology is somewhat vague and simply refers to the dark pigmentation and partitions that can be found within the fungal hyphae, the characterization allows for the differentiation of fungal endophytes from other mycorrhizal associations. It is as yet unclear whether these fungi can grow through the soil to reach their host, or remain sedentary in the soil until a suitable host root is in the vicinity. What is known is that endophytes have been found in the roots of all plant species.

Endophytic bacteria, like their fungal counterparts, most likely rely on chance followed by chemical signals to gain entry into a plant root. Cracks formed at the junctions of lateral roots and/or lesions produced by insect probing are likely locations for the bacteria to gain entry into the root and begin colonization. Once inside the plant, bacterial endophytes can also confer numerous physiological and developmental benefits to the host.

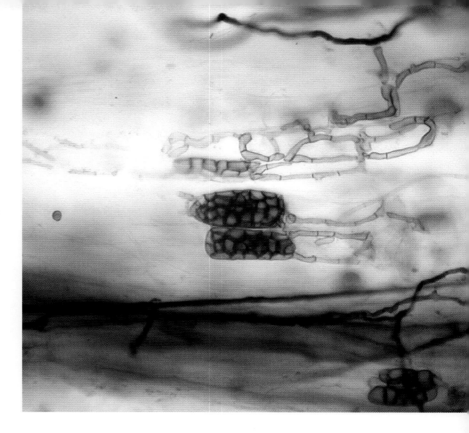

⋀ This image depicts several fungal hyphae ending in microsclerotia, which appear as grape-like clusters. These multicellular resting structures are common among plant endophytes that originate from swollen hyphae. Dark septate endophytes characteristically produce microsclerotia within the roots of their host.

⋁ Transmission electron microscope image of a cross section through a soya bean (*Glycine max*) root nodule. The bacterium *Bradyrhizobium japonicum* infects the roots and establishes a nitrogen-fixing symbiosis with the plant.

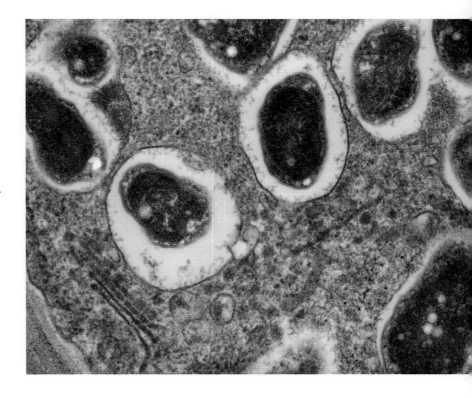

Root pests

oots encounter a plethora of biological activity below ground. Among the organisms with which they cohabit the soil, nematodes (roundworms), arthropods, insects and earthworms may affect their growth and function, either directly through feeding or indirectly through the production of abnormal cell growth. These organisms feed on both live and dead roots, affecting plant health, the cycling of nutrients in the soil, plant hormone production and plant distribution. Not surprisingly, the difficulties of observing and measuring root pests and their interactions with plants has deprived this field of adequate study.

The root aphid *Phylloxera radicicola* can cause root tip swelling and deformity, as seen in the centre of this image. When roots start growing in the spring, these asexual insects lay eggs and produce several generations through the year. Fungal infection in roots is more severe in the presence of the aphid.

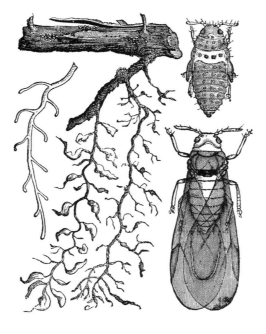

Root herbivory – modes of root damage

A number of soil animals consume roots at some point during their life cycle. While some plants can compensate for their loss of below-ground biomass with little to no effect on performance, root pests and root herbivory can cause substantial damage. Unfortunately, this loss of biomass is difficult to quantify and in many cases is not perceived until the plant's health declines – a particularly detrimental factor in crop production.

Chewing herbivores cause direct damage to roots that are consumed, or by disrupting the flow of resources through root grazing.

PHYLLOXERA

Phylloxera (*Daktulosphaira vitifoliae*) and its effect on common grape vines (*Vitis vinifera*) and the wine industry in Europe has reserved this root-sucking insect a place in the history books. When *phylloxera* was introduced to Europe from North America in the mid-nineteenth century, it became an epidemic that destroyed vast hectares of grape vines across France via the roots of the plants. *Phylloxera* infestations produce numerous root galls, which lead to root deformation and, ultimately, death. Over the span of just a few years, susceptible vines decline and wither. Thanks to the discovery that common grape vines can be grafted by combining the scions of European varieties onto the rootstocks of North American varieties that are resistant to phylloxera, catastrophe was averted and we can continue to enjoy the wines produced by vineyards throughout Europe today.

> Phylloxera aphid (*Daktulosphaira vitifoliae*) nymphs feeding on the roots of a grape vine. Necrotic spots (areas of dead tissue) develop at the feeding sites on the roots.

> A juvenile *Meloidogyne incognita* root-knot nematode penetrating a tomato (*Lycopersicum esculentum*) root. Once inside, the worm, which also attacks upland cotton (*Gossypium hirsutum*) roots, causes a gall to form and robs the plant of nutrients.

The outcome of root chewing can disrupt water uptake and carbohydrate storage reserves. Common culprits of root damage include weevils, beetles and cutworms, moth larvae that are so called for their ability to cut off roots as they feed. The relationship between the root herbivore and the specific plant species can vary, with some species being highly host specific while others are generalists, burrowing from plant to plant through the soil in search of food.

Indirect damage to roots can be just as damaging to plant functionality. Many below-ground herbivores exploit roots as a food supply by living within them or by tapping into the phloem sugar supply as a means for more sustained survival. Aphids, scales and nematodes are common root feeders that select a feeding site on the root and then penetrate it, usually inducing visible enlarged or prolific cell production and expansion called galls. Over time, as the number of galls increases, root functionality declines. Additionally, the introduction of secondary infection is increased due to the open lesions and reduced health of the root caused by the probing of the herbivore on the root itself and its ultimate entrance into the root.

The rhizosphere

The rhizosphere comprises the volume of soil that is affected by the root, which amounts to just a few millimetres beyond the root surface. Although small, it is a virtual hotspot of activity, where the interplay between the growth of roots, root chemistry, bacteria, microbial fungi and insects results in fluctuations that affect plant nutrient availability, soil water accessibility, hormone signalling and, ultimately, plant growth. In practical terms, uncovering plant–soil interactions requires the consideration of the plant and its soil environment as a single system and not separate entities.

< A scanning electron micrograph of the root tip of a sorghum (*Sorghum bicolor*) plant, with kidney bean-shaped *Azospirillum brasilense* bacteria on its surface.

∨ An artistic rendition of a root minus its soil environment, showing the diversity of organisms that live in the rhizosphere, the zone around the root.

Root exudation

Roots secrete diverse chemicals into the rhizosphere that are yet to be fully identified. Root exudates – defined as all organic and inorganic compounds released into the soil by healthy roots – are mostly composed of carbon and can influence plants, microbes and nutrient availability. Root exudates are made up of primary and secondary plant metabolites. While primary ones are essential for plant growth, secondary ones are not required to complete the life cycle, but instead are chemicals that aid in protecting plants from stress and are used in communication with other organisms.

Root exudates provide a readily available carbon resource for soil microbes. In the context of their associations with microbes, general classifications group the exudates according to their mode of utilization – for example, exudates that are easily assimilated by microbes, versus more complex compounds that must be broken down into simpler forms before they can be utilized. Depending on their chemical structure and concentration, root exudates can function as either a food source or toxin for microbes.

On a strictly chemical basis, root exudates can interact with nutrients in the soil, changing the availability of nutrients. The effects include increased nutrient availability through changes in soil pH, nutrient ion solubility or metal detoxification. The size of a root system can have an effect on the ability of roots to influence the nutrient status of the rhizosphere, and there is a direct relationship between the number of roots and the size of the zone of influence.

Root exudates have been shown to have an effect on neighbouring plant species. In extreme cases, their presence can prevent the growth of some plants. Black walnut trees (*Juglans nigra*), native to eastern North America, is the most famous species for this natural toxic herbicidal activity, termed allelopathy (the inhibition of growth by one plant on another plant). The chemical juglone, produced by black walnut roots, can persist in the soil for long periods of time, inhibiting plant growth within the tree's root zone. The Mediterranean species Johnson grass (*Sorghum halepense*) has also been known to suppress weeds, through sorgoleone root exudates produced in the root hairs. These golden-yellow allelochemical droplets are taken up by neighbouring crop species such as maize and soya bean (*Glycine max*), and are translocated into their cells, where the chemical disrupts proper cell function, leading to plant yellowing and potential death. The tree of heaven (*Ailanthus altissima*), native to China and once regarded as a choice specimen for ornamental gardens in Europe in the late eighteenth century, has fallen from favour due in part to its production of ailanthone, a toxic chemical that leaches from the root bark into the soil and inhibits the growth of other plants. The species can also sprout easily from stumps and produces large quantities of seeds, further contributing to its invasive tendencies.

Microbial community

As a root passes through the soil, chemical signalling drives a response from microbes, which may result in positive (growth-promoting) or deleterious (pathogenic) effects. While uncertainty remains regarding the mechanisms involved, the root-associated microbiome has been shown to facilitate rhizosphere responses to changing environments in ways that are potentially beneficial for the plant. For example, rhizosphere microorganisms increase plant tolerance to water stress and even improve overall performance (growth) in many systems. Some hypotheses suggest

that plants with a similar root morphology might also have similar root-associated microbial communities. For example, root systems that have little branching tend to rely more heavily on myccorhizal associations. However, this is a gross oversimplification and the field warrants research to determine other driving factors.

As mentioned above, not all microbes are beneficial to plant roots, and a plethora of pathogenic microbes also inhabit the rhizosphere. Root infection by species of pathogenic genera such as the fungus-like eukaryotes *Phytophthora* and *Pythium* is greater in root tips compared to higher-order roots. This may result from localization of infection events at the root tip, plant deployment of defences or constraints on tissue functionality. Ageing root tissues are a preferred route of entry for many generalist pathogens in the rhizosphere, so regions where this is occurring or particular patches of sloughing cells may provide potential infection sites.

⊻ *Phytophthora infestans*, the microorganism that causes potato blight, can spread through, and devastate, a crop in just a few days.

⊳ Scanning electron micrograph of *Phytophthora infestans*. The lemon-shaped structures are microscopic asexual spores called sporangia, which open to release spores in cool, wet conditions.

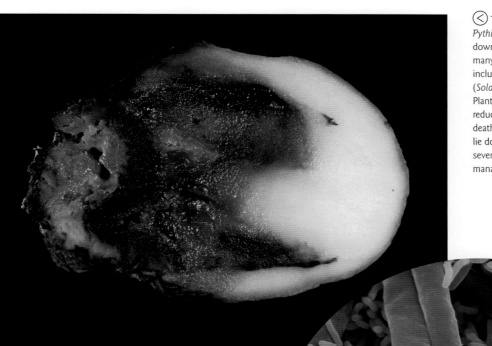

⊘ The plant pathogen *Pythium ultimum* is a downy mildew that infects many agricultural crops, including this potato (*Solanum tuberosum*) tuber. Plant infection leads to reduced plant growth and death. The pathogen can lie dormant in the soil for several years, making management difficult.

The rhizosphere and a changing climate

Climate change is altering biotic interactions in a multitude of ways. There are clear advantages to increasing our understanding of the intertwined living and non-living rhizosphere matrix and how the root–rhizosphere environment is influenced by water, temperature, toxic metals and so on. This complex environment can incur substantial changes as a result of environmental shifts induced by climate change, which can ultimately have impacts on agriculture and plant community assemblages. While this may seem hard to conceptualize as the rhizosphere of an individual root itself is very small, the rhizosphere of an entire plant or a community of plants constitutes a substantial below-ground area. As shifts in environmental variables occur with a changing climate, direct effects on the rhizosphere – including microbe interactions with other microbes and/or the plant roots – will impact on levels of carbon sequestration, decomposition and, ultimately, how plants function.

Previous studies suggest that plant controls over rhizosphere microbial communities may be more pronounced in disturbed environments and that stress may enhance plant investment in root exudation, thus potentially shifting rhizosphere functionality. Moreover, the microbial component of the rhizosphere may shift its metabolism or composition in response to altered root substrates. As if studying roots themselves were not difficult enough, research targeting the rhizosphere provides its own unique challenges. Yet, there remains ample work to be done within this focus area and in the unique feedback loops that drastically affect roots and therefore plant stability.

⊘ Soil fungi (ochre strands) and bacteria (red structures), imaged with scanning electron microscopy, are components of the microorganism community in the soil that potentially interact with roots.

Stems

No forests can exist without trees, and no trees would be complete
without stems. Woody, herbaceous, climbing or underground, stems are
integral components of the plant world, yet are rarely appreciated for all
they do. In a superficial sense, they are the scaffold that keeps a forest
ecosystem intact, but closer inspection reveals that they are more than
an assembly of inert beams. They are complex structures that conform to
essential principles of engineering and fluid flow. Water transport is essential
to all plant organs, and plant stems house the long-distance plumbing that
supplies the plant's need for water. Likewise, transport of sugars from leaves
to roots requires specific vascular tissue, an abundance of which is found
in the stem. Stems are the bridge between earth and sky.

So critical are conduction, structural support and optimal leaf exposure
to the life of a plant that natural selection has acted strongly on numerous
aspects of stem architecture. Plants cannot move about like animals, so
stems must be resistant to drought, freezing and pathogen attack. Resilient,
economical and diverse in form and physiology, plants have evolved to
function under such stress, and in the case of trees, stems faithfully record
these events as they continue to grow. Stems are a portal to the past,
and key to our planet's future.

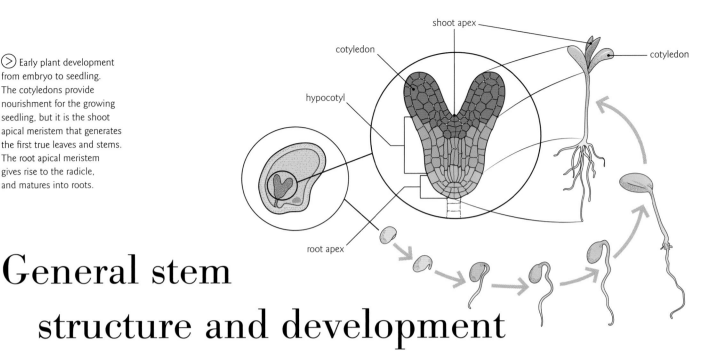

⊘ Early plant development from embryo to seedling. The cotyledons provide nourishment for the growing seedling, but it is the shoot apical meristem that generates the first true leaves and stems. The root apical meristem gives rise to the radicle, and matures into roots.

shoot apex

cotyledon

cotyledon

hypocotyl

root apex

General stem structure and development

G iven the diversity of colours, shapes and textures of the plants around us, it may come as a surprise that plant development is rather similar across species. In fact, flowering plants share a nearly identical pattern of root and shoot development once the embryo begins to emerge from the seed. Over time, a plant's appearance becomes largely a function of hormonal signals, environmental cues and genetic make-up.

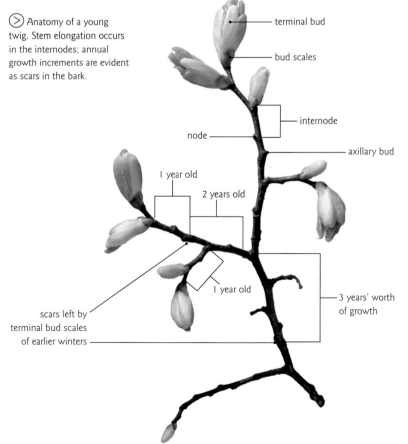

⊘ Anatomy of a young twig. Stem elongation occurs in the internodes; annual growth increments are evident as scars in the bark.

terminal bud

bud scales

internode

node

axillary bud

1 year old

2 years old

1 year old

3 years' worth of growth

scars left by terminal bud scales of earlier winters

Initial stem development

The tiny embryo within a seed bears little resemblance to a mature plant, but it contains all the information it needs to develop into one. As germination begins, the base of the embryo will develop roots, while the hypocotyl will expand into a primordial stem possessing two leaf-like appendages, the cotyledons. With time, the hypocotyl will develop into a true stem with leaves and buds, while the cotyledons, depleted of their nourishment, will drop off. At this point, the juvenile plant is the first complete module of what will become a mature tree or herb. As the stem gains in length and girth, each module will develop a node, some leaves and an internode. In the node, axillary buds are sometimes nestled in the nook between the stem and leaf petiole; these buds may give rise to lateral stems. The internode is simply a zone of elongation. Over time, the iterative growth and expansion of these modules will generate a mature plant.

The apical meristem

The terminal buds we normally see on a mature stem or twig also produce modules with nodes, internodes and leaves. More precisely, these structures arise in the apical meristems within the buds. The term meristem refers to zones of undifferentiated tissues in which growth occurs. Three basic tissue types form the developing module in the apical meristem, and these tissues ultimately comprise all of the vegetative organs in a mature plant. Epidermal tissue covers the exterior of all plant organs, while the ground tissue comprises the bulk of the stem and surrounds the centrally located vascular tissues, which transport water and sugars. The growth of vascular and ground tissues below the apical meristem is responsible for the increase in stem length and thickness.

The height of a plant, as well as the arrangement of its leaves and branches, is determined in part by the apical meristem. Apical dominance is the extent by which a leader shoot suppresses lateral branch growth.

CELLS OF THE APICAL MERISTEM

The apical meristem is the heart of the bud, and is surrounded by developing leaf tissue, whereas subtending growth may be differentiated into leaves and a node. Three cell types distinguish the apical meristem: two tunica layers, responsible for surface growth, including the epidermis and underlying ground tissues; and the corpus layer, which gives rise to core vascular and ground tissue meristems. Maturing vascular and ground tissues are evident below the first node.

⊘ The apical meristem. This structure is found within the terminal stem bud, and gives rise to new stems and leaves.

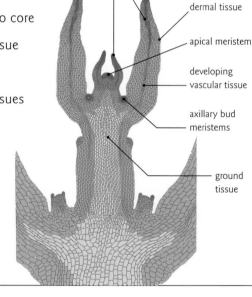

young leaves

dermal tissue

apical meristem

developing vascular tissue

axillary bud meristems

ground tissue

For example, most temperate conifers, with their central bole, exhibit a high degree of apical dominance in comparison to oaks (*Quercus* spp.), which tend to have an expansive canopy composed of numerous stems roughly equivalent in size. Controlling apical dominance is the hormone auxin, which is produced in the apical meristem; high levels of auxin suppress lateral branching, and this is indeed the case in conifers. In contrast, dicot species such as oaks are less dependent on auxin and thus benefit from a higher degree of morphological flexibility.

⊲ With its expansive canopy and multitude of branches, this valley oak (*Quercus lobata*) exhibits little apical dominance.

Evolution of stems

The first plants to colonize land had no blueprint for leaves, roots and stems, and faced a nearly constant threat of desiccation. In the face of such challenges, why would selection favour terrestrial plant life? Certainly, higher light levels, increased availability of carbon dioxide and less competition from other organisms would have offset the costs. But not for long. Crowding meant that plants needed to grow taller and put down an anchor to stake their territory. It was the evolution of stems that allowed them to grow vertically, eventually forming dense forests of ferns and other fern-like plants.

⑦ Fossil of early clubmoss genus *Baragwanathia*. Clubmosses possess a central cylinder of vascular tissue; the tiny leaves also have a vascular strand.

⬇ A rendering of *Aglaophyton major*, a Devonian-era plant that most closely resembles today's mosses. It is composed of stems, some of which terminate in spore-bearing structures.

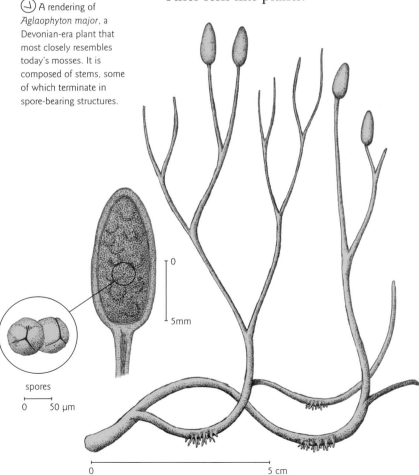

0

5mm

spores

0   50 μm

0   5 cm

Fossil clues

Early Devonian (419–393 million years ago) localities have yielded some remarkable plant fossils, giving researchers valuable insight into the evolution of plant form and structure. Northern Scotland's Rhynie Chert, once an ancient hot spring, is especially rich in fossil plants that most closely resemble modern clubmosses. The 40 cm-tall (16 in) *Aglaophyton major* lacked true roots and leaves, and was composed of creeping rhizomes and bifurcating vertical stems, which terminated in spore-bearing sporangia. Like some existing mosses, *A. major* had no true vascular tissue, but rather elongated, tubular cells that functioned in transport (see box). By comparison, larger plants such as *Asteroxylon* were vascularized and bore small leaf-like scales. Interestingly, neither deep roots nor leaves would have been essential to the survival of early land plants: hydric habitats provided essential water, while an atmosphere rich in carbon dioxide ensured relatively high levels of photosynthesis.

Support and transport

The proliferation and increasing diversity of plants during the Devonian created competition that selected for true roots, broad leaves and tall stems. Arborescent clubmoss relatives, ancestral seed plants and fern-like vegetation comprised Earth's first forests, and these were dense and complex, with an organismal array that rivalled contemporary ecosystems. But to leap from *Aglaophyton* to Carboniferous forests is to forget that in order to grow tall, Devonian flora had to contend with gravity and the need for efficient long-distance conduction. This was especially important for those plants colonizing drier habitats. Without a dependable source of water, plants needed root and vascular systems to ensure consistent hydration. Trees efficiently achieved both conduction and canopy support by way of wood, and the fossil record shows that woody stems appeared in the Middle Devonian, nearly contemporaneously with the Rhynie Chert flora. New discoveries may push this date even earlier and challenge our present interpretation of the evolution of early land plants.

The middle Palaeozoic saw fascinating variation in stem morphology, much of which has no analogues in contemporary vegetation but speaks to the diversity that is possible in plant structure and function. Countless species and scores of botanical experiments have been transformed into coal beds or have disappeared altogether, leaving us to wonder about their botanical mysteries.

⊲ *Lepidodendron*, an arborescent clubmoss. Reaching heights in excess of 30 m (100 ft), these plants were abundant in Carboniferous swamps.

PRIMITIVE STEMS

Few plants exist that can be properly called living fossils. Natural selection has rendered almost all extant taxa sufficiently different from their ancestors, but neontology (using living organisms to understand those that are extinct) can illuminate evolutionary processes. For example, today's bryophytes have no true vascular tissue, but so strong was selection for long-distance conduction that some species have modified ground tissues to serve this function. The hydroids and leptoids of a *Polytrichum* moss transport water and sugars, respectively, and the vegetative axis of this plant vaguely resembles that of *Aglaophyton* and *Asteroxylon*.

⊳ Some species of moss, such as this *Polytrichum*, have hydroids and leptoids in the centre of their vegetative or reproductive axes. These elongated cells function like the xylem and phloem of higher plants.

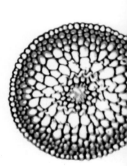

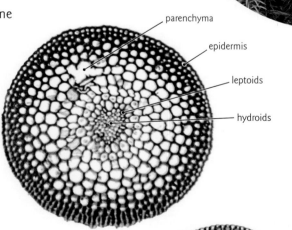

parenchyma
epidermis
leptoids
hydroids

The structure of woody stems

Wood is a marvel of economy. It is, for the most part, a dead tissue composed primarily of cellulose, pectins and phenolic constituents such as lignin, with a small fraction of living tissue. Its maintenance costs are negligible and yet it manages to achieve two tasks at once: it provides mechanical support of the canopy, and it enables the transport of water over long distances. Trees are nature's skyscrapers and their elegance lies in the adaptive structure of wood.

⋁ Water transport in conifers occurs in tracheids, single cells connected to one another with torus-margo pit membranes. Water moves from one tracheid to another through the pits.

annual growth ring

phloem

cambial region

resin canal

ray parenchyma

tracheid

border   torus

parenchyma

pit aperture

margo

ray parenchyma

vessel   fibres

vessel

axial tracheid

Wood structure and function

Conifers and angiosperms are loosely categorized as softwoods and hardwoods, although the wood of some conifers can be just as hard as that of some angiosperms, while many angiosperm species have wood that is relatively pliant. Wood is also known as secondary xylem and the seasonal production of growth rings adds to the girth of a tree. In conifers, more than 90 per cent of wood is constructed of overlapping, dead single cells known as tracheids, which conduct water to the canopy as well as support the tree (see box). The remainder consists of parenchyma tissue, which functions in the transport and storage of sugars and other constituents within the xylem. Angiosperm wood, by contrast, is more specialized. Consisting of wide vessels for water transport, narrow fibres for mechanical reinforcement (and storage), a small fraction of tracheids and more than 15 per cent parenchyma tissue, it exhibits a high degree of developmental flexibility that supports, in large part, the great diversity of canopy structures we see in flowering trees.

⟨ In angiosperms, water moves through vessels, which are composed of single hollow cells known as vessel elements. The vessels may be more than 2 m (6.5 ft) in length, and are provided with structural support by the surrounding fibre cells.

⟩ Heartwood is typically darker in colour due to higher concentrations of tannins, phenolics, oils or gums. The moisture content of heartwood is lower than in the sapwood.

Growth rings

Further attributes and distinctions can be observed in a cross section of wood. Within a growth ring of either a conifer or angiosperm, earlywood is produced at the start of the growing season, and consists of relatively large xylem conduits with thin walls. Large conduits ensure efficient water delivery to the expanding new leaves. By contrast, latewood consists of smaller cells with thicker walls and is thought to function in support. The degree to which earlywood and latewood are discernible depends in large part on the habitat of the tree. In temperate zones where the winter season enforces dormancy, these patterns are reliably observed, but not so in aseasonal or tropical habitats where trees may grow continuously. It can therefore be difficult to determine the age of tropical trees using growth rings.

Sapwood and heartwood

Tree trunks and older stems typically have regions of sapwood and heartwood. Sapwood forms the outer part of a woody stem, and as the name implies, conducts water to the canopy. Its pale colour is in contrast to the typically darker heartwood, which no longer has functional xylem or living parenchyma, and is thought to play a structural role. Permeated with phenolic constituents and resins, heartwood is prized for its colour and resilience, and lends its unique properties to musical instruments, fine furniture and artisanal wood products.

annual growth

earlywood

latewood

< Earlywood xylem conduits are larger than those produced later in the growing season. Bigger conduits, whether conifer tracheids or angiosperm vessels, move water more efficiently than small ones, so earlywood is essential for new leaf growth in the spring.

A COMPLEX TISSUE

Water travelling from roots to shoots does not move through a single conduit like water through a pipe. This would be dangerous: should the pipe break, some part of the canopy would be deprived of water. Instead, xylem is a multicellular tissue in which water moves from vessel to vessel or tracheid to tracheid in both axial and radial directions throughout the sapwood. When one conduit becomes non-conducting, many others can compensate for the loss, so the delivery of water to the leaves can continue. However, this system comes at a cost, namely an increase in friction as water must make its way from one conduit to another through structures known as bordered pits (see pages 106–9).

heartwood

sapwood

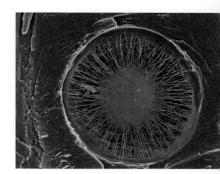

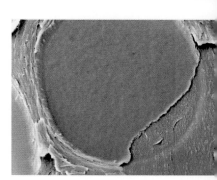

> Pits are digested regions in tracheid and vessel cell walls that allow water to move from one conduit to another. Conifers have torus-margo pitting, while angiosperms have homogenous pit membranes.

How trees grow

(F)or trees to grow tall, they must also expand in girth. Tree trunks become wider each year because annual xylem is added to supply water to the new season's leaves. Thicker trunks also improve stability, so altogether, the appropriate scaling of stem height and girth is the most basic requirement of functional tree architecture.

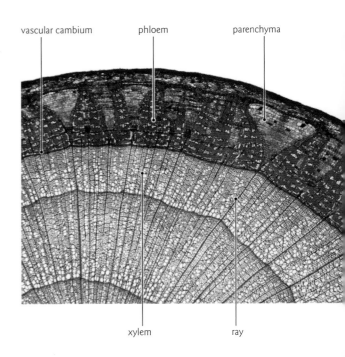

The vascular cambium generates xylem toward the centre of the stem, and phloem toward the outside.

Rays connect the phloem and parenchyma tissues with the xylem.

⋁ Herbaceous plants, such as sunflowers, can exhibit secondary growth. In young stems, the vascular cambium develops below the region of primary growth, and generates secondary growth by dividing laterally as well as radially. Lateral divisions add cells to the cambial cylinder, increasing its circumference and, ultimately, stem girth. Radial divisions of the cambium produce xylem towards the pith, while phloem develops in the direction of the bark. The cambium produces much more secondary xylem than phloem, so wood constitutes the highest volume fraction of a tree's secondary growth.

The cambial layer

Non-woody vegetation exhibits primary growth, whereas woody plants such as trees and shrubs develop both vertical extension and radial expansion, a process known as secondary growth. Tissues such as secondary xylem (wood) and secondary phloem increase a stem's girth, and it is the cambial zone, a thin meristematic layer of one or two cells, that facilitates this growth. The cambium is located on the periphery of the woody tissue not far below the bark, and is composed of fusiform and ray cells. The division of fusiform cells toward the centre of the stem produces annual growth rings of secondary xylem, while division toward the outside of the stem generates secondary phloem. Ray cells generate parenchyma tissue that penetrates both the xylem and the phloem. Because phloem is a soft tissue that compresses with the annual expansion of the xylem, it does not produce discernible growth rings.

In temperate climates, the cambium is dormant during the winter, but the expansion of stem buds in the spring initiates cambial cell division by the downward transport of the hormone auxin through active phloem tissue. Active cambial tissue then divides both radially and axially to increase the circumference and the height of the trunk. Lateral branches also originate from the cambium, and develop like the trunk.

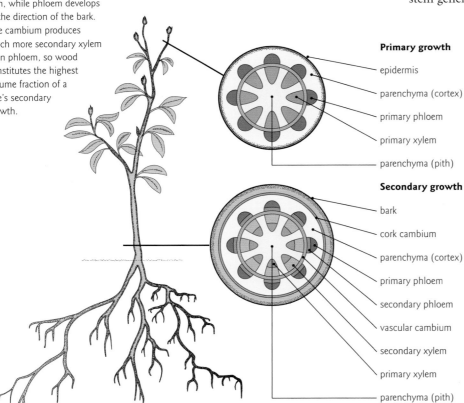

Primary growth

— epidermis

— parenchyma (cortex)

— primary phloem

— primary xylem

— parenchyma (pith)

Secondary growth

— bark

— cork cambium

— parenchyma (cortex)

— primary phloem

— secondary phloem

— vascular cambium

— secondary xylem

— primary xylem

— parenchyma (pith)

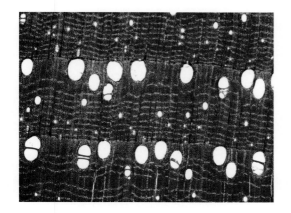

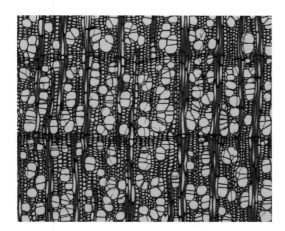

Variations in cell production

The cambium gives rise to cells of various dimensions and wall thicknesses depending on the tree species and time of year the cambium is active. In conifer and temperate angiosperm trees, the cambium produces some earlywood and latewood, although the proportion of each cell type depends on climate and water availability. For example, a relatively wide growth ring with several stacks of large earlywood cells would have been produced during moist conditions early in the growing season, whereas a narrow growth ring with only a few earlywood cells indicates a short or dry summer.

In angiosperm trees, we see further diversity of cell types. Here, the cambium produces fibres, which function in storage and canopy support, as well as vessels that conduct water. In ring-porous species such as some oaks or chestnuts (*Castanea* spp.), the earlywood vessels form a visibly distinct band in a growth ring, whereas in diffuse-porous species such as birches (*Betula* spp.) and maples (*Acer* spp.), the vessels are similar in size throughout the growth ring.

◁ Ring-porous wood (above) is characterized by the presence of large earlywood vessels, which form a discrete band in a growth ring. By contrast, vessels in diffuse-porous wood (below) are similar in size and relatively evenly distributed throughout the growth ring. Ring-porous species tend to have fewer vessels than diffuse-porous species, but these are typically wider and longer to support efficient water transport.

MULTIPLE CAMBIAL LAYERS

Some species, including members of the genus *Bougainvillea*, have numerous cambial layers, resulting in multiple bands of secondary xylem or phloem. Multiple cambial layers generate greater amounts of parenchyma tissue, which aids in the storage of water and carbohydrates. Such plants may also have more flexible stems due to greater quantities of softer, more pliant tissues placed in between the numerous layers of wood.

◔ Transverse section from a *Bougainvillea* stem showing successive cambial layers, each producing secondary xylem (large vessels) and phloem (blue stain) in a matrix of fibres.

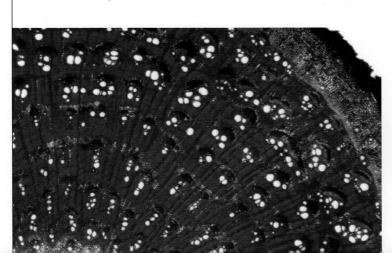

The structure of herbaceous stems

Investing in wood is an unnecessary cost for herbaceous plants with short life cycles, which instead rely on their primary tissues to provide sufficient support for floral displays and the development of seeds. It is generally accepted that herbaceous dicots evolved from woody ancestors sometime during the Cretaceous (145–66 million years ago). In fact, the capacity to produce wood has re-evolved in some herbaceous dicots found in dry, stressful habitats. Monocots, on the other hand, split from dicots earlier, possibly at the end of the Jurassic 145 million years ago.

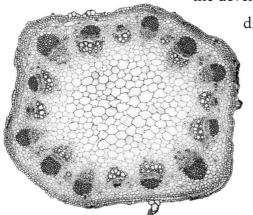

∧ Primary growth in a transverse section of a sunflower (*Helianthus*) stem. Numerous vascular bundles surround the pith; each bundle consists of xylem, phloem and a dense fibre cap.

> The sclerenchyma fibres have thick walls to protect the phloem and impart strength to the stem. They are typically much narrower and denser than adjacent cortex parenchyma cells.

Dermal tissue

The dermal tissue is typically only one cell layer thick, and covers the exterior of the stem tissue, as well as all other plant organs. In green photosynthetic stems, it may be modified to include stomatal pores to permit the entry of carbon dioxide. Extruded by the epidermal cells is a waxy cuticle, which protects against water loss and the entry of pathogens. Epidermal cells modified into glands or trichomes are not uncommon on stems, but are usually found in greater abundance on leaves. The epidermal tissue is often under tension due to the positive pressure exerted by the hydrated ground tissue.

Ground tissue

Stem ground tissue serves in storage, support, and protection from insects and pathogens. There are three types of ground tissue. Simple parenchyma comprises the bulk of a dicot stem. It is typically white and fleshy, is made up of large cells and is clearly visible in the centre of a stem cross section. Such cortex parenchyma functions in the maintenance of turgor and plant structure, as well as the storage of water, nutrients and carbohydrates. Collenchyma and sclerenchyma tissues are derived from parenchyma tissue, and function in support. Flexible support is provided by collenchyma tissue, which is composed of parenchyma cells that are much reduced in size and exhibit selective thickening of the primary cell wall. It is common in stems with ridges and corners, such as those found in mints (*Mentha* spp.). Sclerenchyma tissue is substantially more rigid than collenchyma, and is found in older stem sections that have stopped elongating. Much like wood, sclerenchyma cells have strong, lignified secondary cells walls, except that they are much shorter and narrower, forming fibres. Sclerenchyma may occur near the periphery of a stem, but more commonly it is found in bundles just outside the phloem tissue, protecting these delicate cells from injury or phloem-feeding insects.

Vascular tissue

Xylem and phloem constitute the plant vascular system, delivering water, minerals, sugars and signalling compounds throughout the plant body. Xylem functions in the transport of water, and in herbaceous dicots it is composed of vessels and fibres with lignified secondary walls. In vessels, the walls often form rings or helices, which provide the xylem with some flexibility as the stem develops. One can think of vessels as a series of short, wide, single-cell derivatives known as vessel elements, stacked to create a long, hollow tube. Fibres are narrow and short, and reinforce the vessels as well as support the stem. Xylem tissue becomes functional once cells reach maturity and die, leaving behind their cell walls. By contrast, phloem is a living tissue and serves to deliver sugars and signalling compounds throughout the plant. It is always associated with xylem.

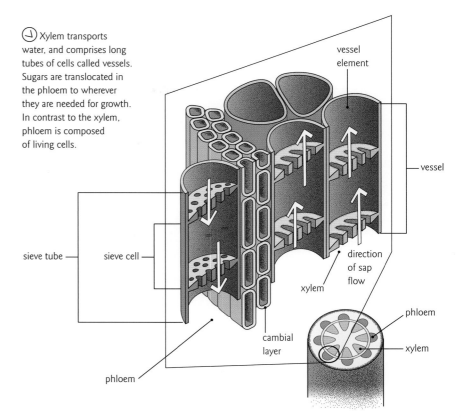

⟳ Xylem transports water, and comprises long tubes of cells called vessels. Sugars are translocated in the phloem to wherever they are needed for growth. In contrast to the xylem, phloem is composed of living cells.

vessel element

vessel

sieve tube — sieve cell

direction of sap flow

xylem

phloem

xylem

cambial layer

phloem

MONOCOT STEMS

Monocot stem anatomy differs substantially from dicots in that the xylem and phloem are packed in discrete vascular bundles that are scattered throughout the cortex parenchyma. Two to three large vessels comprise the xylem, which may also include fibres or tracheids. Thick-walled sclerenchyma forms a protective phloem cap, while additional fibres surround the bundles for protection.

parenchyma

sclerenchyma sheath

vessel element

lacuna (air space)

phloem

vascular bundle

epidermis

xylem

phloem

parenchyma cells

lacuna

sclerenchyma cells

⋀ In monocots, vascular bundles are scattered throughout the stem cross section. Xylem vessels tend to be prominent in monocot bundles.

Bamboos, palms and tree ferns

S|imilar to monocots and living ferns, bamboos, palms and tree ferns lack a true vascular cambium and are unable to produce secondary growth. This means that their trunks are generally narrower than those of woody trees and devoid of lateral branches. Despite these constraints, palms and bamboos can develop huge amounts of biomass and exceptionally strong stems very quickly, while tree ferns are recognized as resilient and are often principal entities in tropical forests.

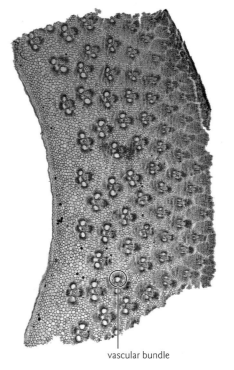

vascular bundle

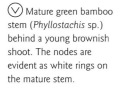

⋁ Mature green bamboo stem (*Phyllostachis* sp.) behind a young brownish shoot. The nodes are evident as white rings on the mature stem.

Bamboo stems

Bamboo stems have long been prized for their utility in building and construction, and their rapid growth rates make them a reliable and sustainable material resource. The exceptional material properties of bamboo can be attributed to several factors. Unlike wood, bamboo is a composite structure made of about 50 per cent parenchyma and 40 per cent lignified fibres, with vascular tissues comprising the balance. Fibres and vascular tissues are packed in primary vascular bundles that are scattered throughout the ground parenchyma in a standard monocot arrangement (see box on page 103), with a higher density of bundles located near the periphery of the stem, where bending stresses are highest. In addition, the cylindrical shape of the stem, in combination with solid nodes, makes bamboo especially resistant to deformation.

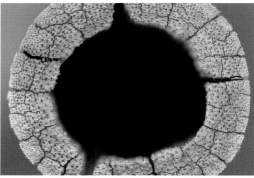

Palm trunks

The trunks of palm trees also subscribe to the standard vascular bundle arrangement of monocots, but unlike most of their relatives, they have exaggerated both the girth and the height of the trunk to achieve tree-like stature. Palm trees are able to withstand the extraordinary bending stresses imposed by hurricanes and floods by capitalizing on trunk architecture that derives its strength from a combination of cross-sectional thickness and the heterogeneous distribution of fibrous vascular bundles, similar to bamboos. Palms develop from large-diameter juvenile stems that allow the mature plant to meet expected canopy

⋀ A cross section (top) of a bamboo stem. The vascular bundles can easily be seen in products such as chopsticks or flooring. They appear as dark spots (above) due to the sclerenchyma fibres that surround the vascular tissue.

The maturing palm tree gains strength from an increase in stem density. Denser and thus stronger stems arise from a combination of greater numbers of vascular bundles in the periphery of the trunk, as well as a shift in the specific properties of the tissues in the bundles themselves. In comparison with younger vascular bundles, the older ones have fibres with cell walls that are so thick as to often occlude the fibre interior. It is common for sclerenchyma cell walls to contain one primary and one secondary layer of cell walls, but in older palm fibres these cells may generate three and even four secondary layers of cell wall.

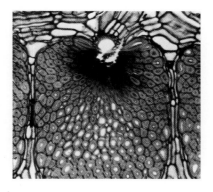

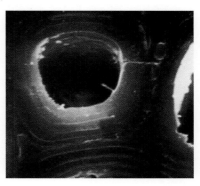

⊘ Micrographs of vascular bundles of young (top) and old (bottom) stems of the palm *Iriartea gigantea*. In older stems, the sclerenchyma cells have up to four layers of secondary walls. This anatomical reinforcement imparts the plant trunk with great strength.

support requirements; some small degree of cell division and expansion may allow for additional increases in trunk girth. However, the capacity of palms to stiffen with age is the key means by which they accommodate increased loading with height. Reduced water content and increased peripheral tissue density and stiffness together serve to strengthen the ageing trunk (see box).

Tree fern trunks

The structure of tree ferns is markedly different from bamboos and palms. The trunks of these plants derive their strength primarily from the overlapping remains of senesced petioles, which surround the ground tissues like a basket. Furthermore, the leaf bases are encased in a thick layer of sclerenchyma fibres, adding enormous rigidity and strength to the trunk. Indeed, such massive amounts of sclerenchyma have given tree ferns a reputation for ruining

saws. As in bamboos and palms, transport in tree ferns occurs via primary xylem and phloem, but unlike in monocots, these tissues form a cylinder toward the outside of the trunk.

Ⓛ A cross section of the trunk of a soft tree fern (*Dicksonia antarctica*). Overlapping frond bases comprise the exterior of the trunk. An undulating ribbon of vascular tissue is sandwiched between dark layers of sclerenchyma.

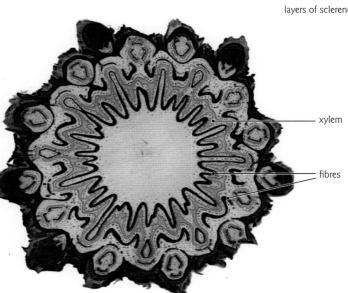

xylem

fibres

Plant water transport

Most plants consume a tremendous amount of water to sustain themselves. While the average human requires 2–3 litres (4–6 pints) of water per day for good health, a maple tree will move 200–400 litres (50–100 gallons) of water from the soil to the leaves. Why so much? It is because transpiration is an inescapable side effect of photosynthesis. Leaves are sufficiently porous to allow the entry of carbon dioxide, but at the cost of water loss through these same pores. It is the efficient delivery of water that keeps transpiring leaves from wilting.

⌄ The xylem vessels in the stem's sapwood are sufficiently conductive to deliver water from the roots to the sprawling, thirsty canopy of this leafy tree.

REINFORCED WALLS

Unlike animal cells, which are contained in a flexible, lipid-based cell membrane, plant cells are reinforced by a cellulose cell wall, like water balloons in a basket. The wall keeps the cell from tearing apart under pressure, it provides the plant with a means of structure, and it is critical to the process of water transport. Every plant cell has a primary cell wall, composed of cellulose fibres and pectin. Xylem and sclerenchyma cells also have a thick secondary wall for additional reinforcement. Secondary walls are permeated by lignins, a class of decay-resistant molecules that make wood and fibres so tough.

⌃ Here, the cell wall pectins and proteins have been cleared to reveal the cellulose structure of an onion (*Allium cepa*) primary cell wall. Layers of cellulose fibres constitute both primary and secondary cell walls.

Tug of war

The upward movement of water in a plant begins soon after the first leaves appear. Water moves passively from regions of high water availability, such as the soil, into the roots and up to the canopy, which is significantly drier. This long-distance transport takes place at practically no cost to the plant. Moisture and sunshine are all that is needed to open stomatal pores, initiating transpiration and with it the conduction of water through the xylem. To achieve this, plants take advantage of some useful properties of water. For a start, the hydrogen bonds between water molecules make water cohesive and allow it to sustain tension similar to a string pulled from both ends. So in a transpiring plant, water is effectively pulled from the ground to the leaves in a tug of war between soil particles and the relatively parched atmosphere.

Through the roots, to the stem...

Water first enters the xylem after filtration through the root's endodermis. Once there, it moves both upward and radially through the stem via vessels or tracheids. Friction from conduit walls slows the flow of sap, as can the movement of water from one conduit to another through pit membranes. This resistance adds to the tension of the water column. When water finally reaches the canopy, it enters numerous small veins, which irrigate the leaf mesophyll tissue. It is here, in the heart of the leaf, where the crux of plant water transport lies.

... and the leaves

Mesophyll tissue is moist, but as the leaf warms in the sun, water trapped in the cell walls turns to vapour, which exits through the stomata. Plant cell walls are constructed of cellulose fibres (see box), so evaporation causes the slight retraction of menisci in the cell wall pores. This is an energetically unfavourable state, so hydrogen bonding continually rectifies the curved menisci. However, levelling the air–water interface requires additional water molecules, and these are supplied by the xylem stream, which remains cohesive as long as sufficient water exists to replenish that lost from the leaf. Taken together, this cohesion–tension mechanism explains plant water transport.

⑦ Water loss is an inescapable consequence of leaf function. This is because carbon dioxide must diffuse into the leaf for photosynthesis to take place. Many more molecules of water are lost from the leaf than carbon dioxide molecules are fixed.

⟩ Plant water transport is a passive process in which water is 'pulled' up from the soil into the roots and vascular system, and into the canopy, where much of it is transpired to the atmosphere through stomata. The water thus moves from regions of high to low water availability under some degree of tension.

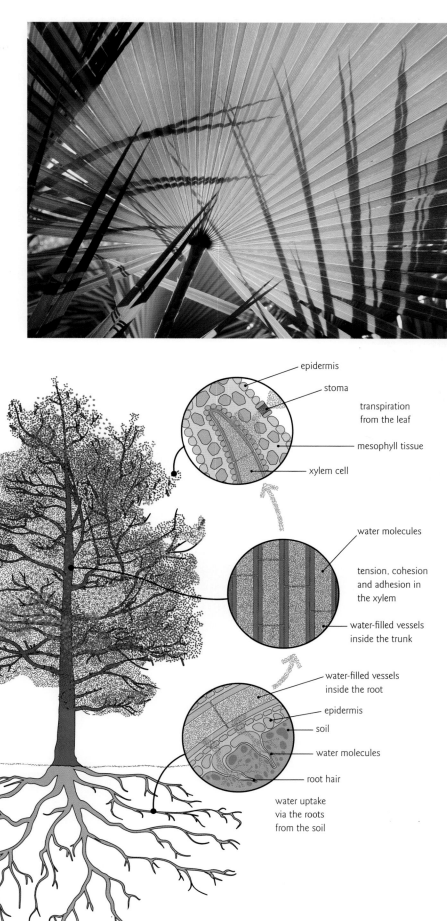

epidermis

stoma

transpiration from the leaf

mesophyll tissue

xylem cell

water molecules

tension, cohesion and adhesion in the xylem

water-filled vessels inside the trunk

water-filled vessels inside the root

epidermis

soil

water molecules

root hair

water uptake via the roots from the soil

Water transport during drought

During episodes of water deficit, it becomes more difficult to pull water molecules from soil particles to replenish what is lost from leaves during transpiration, so significant degrees of tension can arise in the xylem sap. Plants are prepared for this. First, they fortify their xylem conduits with appropriately thick walls or supportive fibres to prevent buckling of conduits under tension. Second, they have evolved ways of coping with the tension-induced transport failure that frequently occurs under drought.

$\vee$ Air-filled vessels appear as dark holes in this image of oak (*Quercus* sp.) xylem, generated using high-resolution computed tomography.

Cavitation and embolism

Maintaining functional water transport is one of the biggest challenges a plant faces during dry conditions. This is because xylem sap under tension is susceptible to cavitation, or the rapid conversion of liquid water into vapour – effectively a break in the water column. Cavitation produces embolized conduits, in which the water in the cell is replaced by a mixture of liquid and vapour, such that the cell can no longer transport water. Many embolized conduits impede water flow through the xylem; this can contribute to the dehydration and subsequent death of the plant.

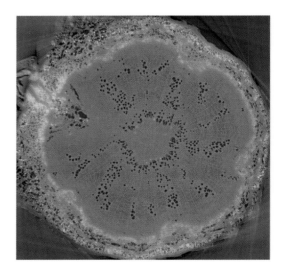

$\wedge$ Insufficient water transport through the stem may lead to wilting of leaves and flowers as seen here in a poppy.

$\vee$ Drought-induced cavitation occurs when air is pulled from a dysfunctional (embolized) vessel to one that is water-filled through the largest pore in the shared pit membrane. This air bubble can expand, creating another embolized vessel.

Cavitation is largely caused by the presence of air bubbles in the xylem stream. Bubbles expand when the tension in the xylem exceeds the inward force exerted by the surface tension of the bubble's air–water interface. The bubble then begins to expand until it fills the conduit, with the displaced water moving into a functional neighbouring cell. We know that the endodermis in the roots keeps air bubbles and impurities from entering the xylem stream, so where does the air come from? Oxygen is needed to support living stem tissues. Therefore, a plant is not hermetically sealed; air can enter the xylem either from spaces that surround the cells or from an injury to the stem. Nanobubble phenomena are also currently under investigation.

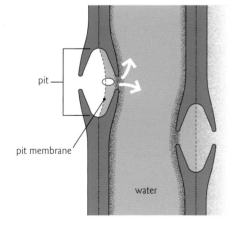

pit

pit membrane

water

drought-induced cavitation

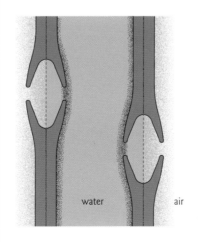

water   air

functional, water-filled conduit

TORUS-MARGO PITS

Conifer pit membranes serve the same function as the simple pit membranes of flowering plants, but their structure is completely different. Water travels from tracheid to tracheid through the pit aperture and the margo, but should one tracheid become embolized, the margo deflects in the direction of the water-filled conduit (in which the sap is under tension), such that the torus physically seals the aperture, isolating the two cells. Further embolism can occur when tensions in the water-filled tracheid are sufficiently strong to dislodge the torus from its seal, allowing air to seep through and cause cavitation.

  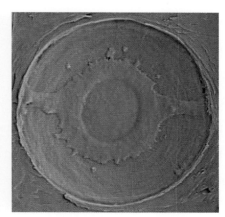

⊘ When one tracheid is embolized, the pit membrane deflects toward the water-filled conduit, effectively sealing it off from air (transmission electron microscope images at left). The torus performs the sealing function (scanning electron microscope image at right).

Adaptations to prevent cavitation

Most air bubbles enter the xylem stream through pit membranes in the conduit wall. Composed primarily of cellulose and pectins, these structures allow water to flow from one conduit to another, but because they are porous, they also permit the spread of air. In flowering plants, pit membranes have a homogenous, fabric-like structure, and they are often clustered along the vessel walls. Air bubbles spread through the largest pore in the pit membrane, a developmental lesion that is the so-called weakest link. There is evidence to suggest that species with thicker pit membranes are less likely to develop large pores, and are thus more resistant to drought-induced cavitation than those with more porous membranes. Thicker pit membranes are not always advantageous, however: reduced vulnerability to cavitation may come at the price of increased resistance to water transport.

CAVITATION CLICKS

Among many available tools to study embolism is an instrument that amplifies the acoustic clicks of cavitating xylem, allowing researchers to hear how plants respond to drought stress, much like doctors using a stethoscope. Thousands of clicks spell trouble for a drought-stressed plant.

Ecological xylem anatomy

(S)tems are core elements of plant structure and must maintain their functional integrity despite episodes of drought or freezing. The transport of water and phloem sap is key to growth and carbon acquisition, so the structure and function of plant vascular tissues, especially the xylem, has evolved to respond to climate cues. This can happen within individuals or populations, as well as across a broad range of temporal and spatial scales. Ecological xylem anatomy is as much a reflection of species' natural history, as it is an indication of physiological constraints.

(◠) Snow-covered Norway spruce (*Picea abies*) in the Austrian Alps. This species' narrow tracheids are adaptive in cold, high elevation habitats.

(◠) The chaparral ecosystem of southern California is dominated by shrubs. Challenged by drought, freezing and fire, these plants have adopted numerous life history strategies, which are reflected in their xylem anatomy.

Dealing with drought

A number of traits are typically associated with drought-resistant xylem, but species-specific variation, plasticity and varied life history can easily complicate even the simplest generalizations. For example, there is good evidence that tracheids are small and narrow in drought-adapted conifers like junipers (*Juniperus* spp.), whereas they are long and wide in water-loving species such as bald cypress (*Taxodium distichum*). Many small conduits increase redundancy, whereas fewer larger conduits favour efficient water transport. So why is it that not all conifers are as drought resistant as junipers? It turns out that xylem composed of many small tracheids is dense, making it more expensive to build. Faced with a certain carbon budget, conifers,

(◡) Examples of xylem in the cypress family of conifers (Cupressaceae). From left to right: Clanwilliam cypress (*Widdringtonia wallichii*), coast redwood (*Sequoia sempervirens*) and bald cypress (*Taxodium mucronatum*).

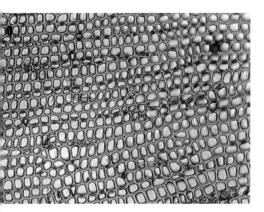

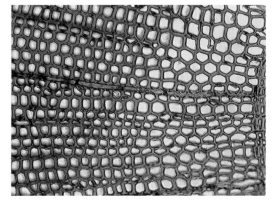

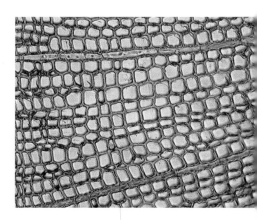

and indeed all plants, have evolved to balance the economy of stem structure with respect to climate and competition.

The angiosperm shrubs that occupy Mediterranean and high desert ecosystems employ varied strategies of xylem structure and function. On the whole, species that are drought resistant exhibit narrow vessels embedded in a matrix of thick-walled fibres, which fortify the vessels from collapse under tension; these plants exhibit high wood density. Alternatively, deep-rooted species are less resistant to embolism due to a more constant water supply, and therefore allocate less carbon to xylem reinforcement.

Coping with cold

In temperate climates, freezing temperatures selected for three general categories of wood, each associated with a unique life history strategy. This is because freeze–thaw cycles may also cause cavitation. Frozen xylem sap contains air bubbles that coalesce as the liquid sap turns into ice. As the ice thaws and tension resumes in the xylem, the largest of these bubbles may expand, cavitate the water column and embolize the conduit. Because the likelihood of cavitation increases with bubble size, which is directly proportional to conduit diameter, species with large-diameter conduits are more vulnerable to freezing-induced embolism. Ring-porous species such as some oaks and walnuts (*Juglans* spp.) have large vessels, and they abbreviate their growing season to avoid freezing. Diffuse-porous trees like aspens (*Populus* spp.) have narrower vessels, so they leaf out in early spring and lose their

HOW TREES FREEZE

Treeline conifers are adapted to extremes. Norway spruces (*Picea abies*) in the Austrian Tyrol experience more than 100 freeze–thaw cycles over the winter, but may also suffer drought stress due to water

⊼ Norway spruce (*Picea abies*) photographed using an infra-red thermal camera. The distal twigs are the first to freeze.

loss from any combination of needle heating, cuticular abrasion from ice blast and insufficient stomatal closure. By the end of the winter, the twigs may experience levels of dehydration that are comparable to those of desert plants. Furthermore, large temperature fluctuations, in combination with frozen soils and trunks, impede the even redistribution of water throughout the plant. Photos of Norway spruces taken with infra-red thermal cameras reveal that small stems are most susceptible to daily temperature swings. Even though they have very small xylem tracheids, these stems are almost completely embolized by the end of the winter.

leaves in mid-autumn. Active year-round, conifers may photosynthesize in the winter providing that temperatures are mild, and so suffer frequent exposure to freeze–thaw cycles. It is their narrow tracheids that protect them from extensive embolism (see box).

⧀ The xylem of the drought-tolerant sugar sumac (*Rhus ovata*). Fewer and narrower vessels in combination with thick fibres help make this species resistant to transport failure by cavitation.

Phloem function

The movement of sugars and other substances throughout the plant occurs in the phloem. This living tissue is always adjacent to the xylem, but unlike xylem, it can deliver contents in any direction within the plant, from a source such as a mature, photosynthesizing leaf, to where a product is needed such as new foliage or developing fruit. Because phloem is dependent on xylem for hydration, prolonged drought can significantly compromise the delivery of sugars and other substances throughout the plant.

sieve element

companion cell

parenchyma

sieve tube

sieve plate

Structure of phloem

In angiosperms, phloem is composed of two general cell types: sieve tubes and companion cells. Similar cells are found in the phloem of gymnosperms, but these are less well studied. Phloem sap moves through the sieve tubes, which are made of stacked individuals cells (sieve elements) connected to one another through porous end-wall regions known as sieve plates. Sieve elements lack nuclei as well as other essential organelles,

> The structure of phloem. Phloem sap moves through sieve tubes, for which companion cells provide metabolic support.

and are thus dependent on adjacent companion cells for metabolic support. Companion cells have the full complement of plant organelles, so in addition to assuming the sieve elements' metabolic workload, they also aid in the transport of sugars and solutes from leaf cells into the sieve elements. Copious plasmodesmatal pores connect companion cells and sieve elements, making phloem one of the most biophysically and metabolically dynamic tissues in the plant.

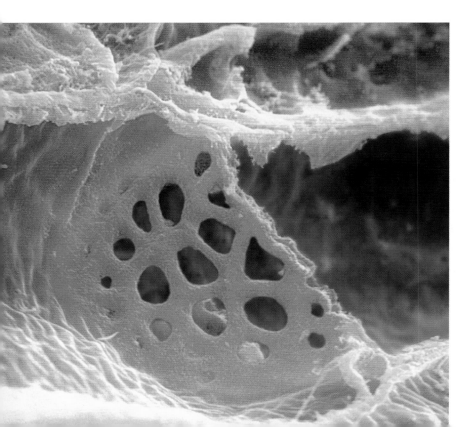

< A sieve plate connecting two sieve elements. Wounding of the phloem tissue can trigger the occlusion of the sieve plate, and thus stop the flow of sap.

Transport through phloem

There is good evidence to suggest that long-distance transport of phloem sap is driven by osmotically generated turgor pressure in what is known as the pressure-flow model. Sugars produced in photosynthesizing leaves (or other sources) are loaded into the sieve tubes via companion cells, which are in contact with leaf mesophyll tissue. Elevated sucrose concentrations in the sieve tube decrease the osmotic potential of the sap, attracting water from adjacent xylem tissue into the sieve tube. This has the effect of increasing the turgor pressure in the sieve tube. With positive pressure driving flow, the phloem sap then passively moves from source to sink tissues. Sugars are unloaded in the sinks, which causes the turgor pressure of these sieve cells to drop. As a last step, water in the phloem is recycled back into the xylem stream. This cyclical process of loading, delivery and unloading of sugars is continuous, and much like water transport, an invisible yet critical activity inherent to all vascular plants.

The loading and unloading of sugars into and out of sieve tubes can either be a passive or metabolically assisted process. In simple passive phloem loading, sugars move down a concentration gradient from the leaf mesophyll through copious plasmodesmata. In the

polymer trap model of phloem loading, sucrose in the companion cells is modified into progressively larger sugars, such that they become too large to diffuse back into the mesophyll and must move into the sieve cells. In some plants, metabolically assisted, or 'active', loading requires the coenzyme adenosine triphosphate to deliver sugars into the sieve cells against a concentration gradient. The means by which sugars are unloaded from the phloem varies depending on the sink. For example, unloading sugars into fruits or seeds requires energetic input because transport must occur against a sucrose concentration gradient.

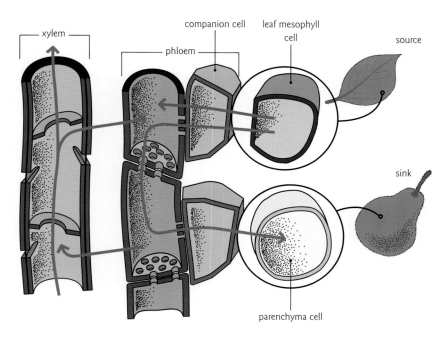

Translocation in phloem. Phloem sap (red line) moves from source to sink, relying on water (blue lines) from xylem tissue to generate positive pressure in the sieve tubes.

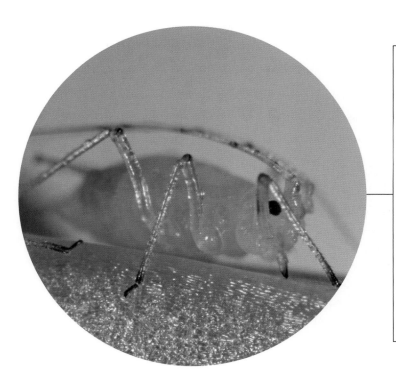

STUDYING PHLOEM

Learning how phloem works is a challenge because sieve cells respond so quickly to wounding. Sap-feeding insects such as aphids have specially adapted mouthparts that obviate this problem, often feeding continuously for hours at a time. By anesthetizing the feeding insects, then severing them from the stylets and collecting the sap exudate, scientists have learnt that while sugars comprise most of the sap contents, small proteins, amino acids, RNA and hormones can also be found in the phloem stream.

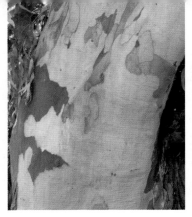

⊘ Left to right: the external bark tissue of milkwood (*Alstonia actinophylla*, Australia), *Eucalyptus* sp. (Australia), and *Ficus* sp. (lowland Costa Rica). Thick bark protects against fire, whereas peeling bark deters the establishment of epiphytes.

Structure and function of bark

Ⓐ simple epidermal layer suffices to defend the internal tissues of a herbaceous plant, but trees require bark to protect their long-term investment. While only one or two layers of cells comprise the epidermis, bark tissue is made of many different cell types in various quantities, resulting in a diversity of patterns and a range of thicknesses. In addition to protection, bark can also serve other functions, including photosynthesis, the storage of water and carbohydrates, and mechanical support.

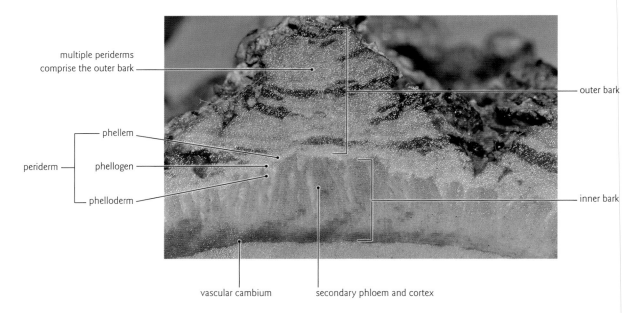

multiple periderms comprise the outer bark

outer bark

periderm — phellem

phellogen

phelloderm

inner bark

vascular cambium

secondary phloem and cortex

⑦ The internal structure of bark tissue. Multiple layers of dead tissue comprise the outer bark. The inner bark and the adjacent periderm layer are living tissues.

Development of bark

What is collectively known as bark is really an aggregation of several secondary tissues outside of the vascular cambium. The visible outer bark is composed of layers of dead tissues known as periderms that overlay the living inner bark, which includes secondary phloem and phloem parenchyma (cortex). The periderm is composed of three tissues: the cork cambium (also known as phellogen) is a meristematic layer that gives rise to cork (phellem) to the outside, and the phelloderm toward the inside. Additional periderms develop as the stem increases in girth. These periderms originate from phloem parenchyma cells, which become meristematic and form new cork cambia. Expansion of the

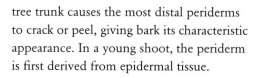

tree trunk causes the most distal periderms to crack or peel, giving bark its characteristic appearance. In a young shoot, the periderm is first derived from epidermal tissue.

Properties of bark

The form and function of bark is surprisingly varied, and researchers are beginning to uncover the protective properties of its many layers. Trees found in fire-prone systems, such as savannahs and temperate and dry tropical forests, tend to have thicker outer bark layers than those inhabiting moister habitats where fire is rare. Generally, the thicker the bark, the better its insulating properties, and the greater the likelihood that the vascular cambium will escape fire damage. Because the thickness of the bark scales closely with tree height, larger trees show higher rates of post-fire survival. Increased water content in the bark tissues is also protective against fire, as is higher bark density.

The relationships between bark attributes and biotic agents such as insects and epiphytes are more difficult to study. Some barks, such as those of eucalypts (*Eucalyptus* spp.) and plane trees (*Platanus* spp.) constantly peel or slough, making the establishment of epiphytes, lichens or parasites on the tree trunk a difficult task. Studies in the tropics demonstrate that epiphytes prefer to establish on barks that are stable, with a rougher surface and some capacity for water retention. Here, bark chemistry also plays a role. The presence of bryophytes and lichens is often related to the nutrient content of the bark, while tannins are known to inhibit the fungal and bacterial degradation of bark and wood tissues. Along similar lines, thicker barks tend to have higher quantities of liquid latex, which protects against insect attack by gumming up mandibles.

BREATHING SPACE

The insulating properties of bark can block the efficient diffusion of oxygen into the stem, yet gas exchange is critical for stem metabolism. Phloem is a metabolically demanding tissue, while the living ray parenchyma tissues in the xylem must also respire in order to function. Gas exchange through tree bark occurs via lenticels, which are porous regions in the periderm. These appear as distinct, regularly spaced markings on the surface of the periderm, and closer inspection reveals a torn structure composed of a loose assemblage of dermal and parenchymal tissues. Lenticels can arise from individual or clusters of stomata, and in green barks they may also contribute to photosynthesis by allowing carbon dioxide to diffuse into the phloem parenchyma.

▷ Lenticels. These bark perforations are evident as white dots on this poplar (*Populus* sp.) branch. Lenticels facilitate the diffusion of air into the stem.

Limits to tree height

The appearance of true secondary xylem in the evolutionary history of plants allowed trees to compete for light by growing tall while remaining well hydrated. But what sets the ceiling for tree height? Models indicate that trees are well short of what could be sustained by the material properties of wood, and the idea that photosynthetic carbon uptake cannot sufficiently compensate for rising respiratory costs with height has also been ruled out. It appears that a combination of environmental and intrinsic plant traits conspires to set the upper bounds for vertical growth.

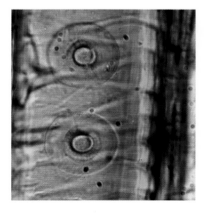

⊓ Tree ferns and woody trees compete for light in a New Zealand forest. These ferns are often taller than angiosperm trees thanks to abundant rain and high humidity.

⌄ In the Douglas fir (*Pseudotsuga menziesii*), the aperture of the torus-margo pit membrane narrows with increasing height (left, 81 m [266 ft] and right, 14 m [46 ft] above ground). Narrow apertures improve cavitation resistance but at the cost of reducing the water flow.

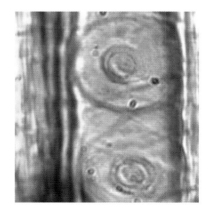

Trade-offs associated with growing tall

Numerous physiological and life history traits govern the size of trees. For example, trees growing in shallow or severely nutrient-depleted soils will not overcome stunting unless they are planted in better soil. Similarly, severe waterlogging will inhibit the diffusion of oxygen into roots and consequently limit respiration. Adequate carbon resources are needed to meet the basic demands of metabolism, pathogen defence, wound repair and reproduction, but because surplus carbon is required for growth, photosynthetic activity must be sufficient to meet the costs of stem construction.

Consequently, tree height is greatest in bright, fertile habitats in which a mild climate and abundant precipitation can support healthy leaf function.

Currently, the most convincing explanations for limits on tree height centre around the connection between plant water transport and photosynthesis. The biggest obstacle for tall trees is the decline in leaf hydration with increasing height. This is because water transport is challenged by the effect of gravity in combination with greater friction along a longer transport pathway. Stomata in leaves at the top of the canopy may respond to drought stress by closing and thus reducing photosynthetic rates, while turgor pressure may be insufficient to support cell expansion for stem and leaf development. Simply put, the branches at the top of a tree typically experience drought stress to a greater degree than those at mid-canopy. Drought stress also raises the likelihood of xylem embolism, which could further impede the delivery of water.

Researchers surveying some the world's tallest redwood (*Sequoia sempervirens*) trees have discovered that the physiology of a tree at the

IN A FOG

Foggy climates create ideal growing environments for tall trees because high humidity staves off water stress. Coast redwood trees derive further benefit from fog by directly absorbing fog water into their leaves. Foliar absorption is thought to occur through the cuticle when the surface of the leaf is wet, and may be sufficient to drive reverse water flow back into stems during heavy fog events. This water can also be stored within the leaves to buffer water loss.

top of its canopy is markedly different from its branches near the base. On a sunny day, leaves at the top of a 110 m-tall (360 ft) redwood exhibit a nearly twofold drop in leaf hydration relative to leaves at 30 m (100 ft), a response coupled with substantially reduced rates of photosynthesis. Others have shown that as leaf water stress increases with height, so do the costs associated with preventing embolism. In the

Douglas fir (*Pseudotsuga menziesii*), the aperture of the torus-margo pit becomes narrower with height, presumably to generate a stronger torus seal. This eventually creates a problem, because such narrow apertures generate additional friction for water flow. The propensity for greater water stress at the top of the canopy may help explain why the tallest trees in the world inhabit wet, often foggy climates.

⊙ Despite being more than 80 years old, pygmy trees in coastal California barely reach 2 m (6.5 ft) in height due to the extremely nutrient-deficient soil on which they grow.

Climbing plants

The movements and habits of climbing plants have fascinated many naturalists, including Charles Darwin, who first published an extensive essay on the subject in 1865. Having devised a way to reach sunlight at a fraction of the cost faced by trees, climbers seem ingenious. They are also structurally efficient: the leaf area of climbing plants occupies roughly one-third of the total leaf area of a forest, yet represents less than one-tenth of the total forest biomass. This economy of form explains both the success and the limits of these structural parasites.

Holding on by loop, hook and cranny

Vines, lianas and climbing monocots have evolved clever mechanisms that allow them to find vertical support, then quickly ascend. Many such plants bear tendrils, which upon reaching a certain length begin to oscillate in all directions in what is known as circumnutation. Common to all plant stems but amplified in tendrils, circumnutation is the result of coordinated directional changes in cell volume. After establishing a contact coil, the tendril squeezes the host structure to avoid slipping off. A free coil formed earlier may then help draw the stem closer. Leaf climbers such as *Clematis* species use modified leaf petioles to wrap around support structures, while others employ irritable organs – modified branches and peduncles that swell in response to touch. The herbaceous stems of hops (*Humulus lupulus*) and twining palms use sharp hooks for support, while clinging climbers, such as vines in the genus *Parthenocissus*, secrete a polysaccharide-based cement for support, or force their holdfasts into crevices, where they thicken and remain tenaciously wedged.

Internal structure

The internal stem structure of a climbing plant assumes very different shapes and mechanical properties from self-supporting upright plants. Woody climbers such as lianas are slender, and often have anomalous cambial architecture, giving rise to lobed secondary xylem and flattened, ribbon-like main stems. Lianas may fall off their host, so these complex structures are thought to prevent compressive damage to

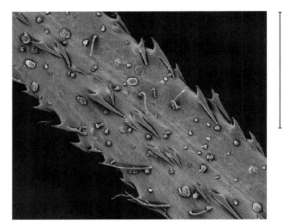

1mm

⊙ A scanning electron micrograph of hooks on a herbaceous stem. These hooks catch on the surface of the host, and also increase friction to prevent slipping.

⊘ An aggressive climber, the trumpet creeper (*Distictis buccinatoria*) produces trifurcating tendrils that coil in response to contact with a host. Once established, the young and flexible apical stem will become thick and woody with time.

vascular tissues. In the absence of a cambial layer, monocot climbers have fewer options with which to modify their mechanical attributes, but it has been suggested that the scattered vascular bundles found in the pith may act as cables within a composite structure. Leaf sheaths and aerial roots may confer additional mechanical benefits to the climbing monocot. In both monocots and dicots, the younger, distal regions of a climbing stem are more pliant and less reinforced with fibres or secondary growth than older portions of the stem. This strategic allocation of resources ensures that younger stems respond adaptively to their support structure, while the basal sections establish long-lasting structural and resource connections to the host and the soil.

⊳ A woody climber winds its way up to the canopy. The smooth vines are actually roots of canopy-dwelling monocots. Woody vines such as this are common in the tropics, where abundant precipitation supports high rates of water transport. Vines may have xylem vessels over 2 m (6.5ft) long.

⊴ The adventitious roots of ivy (*Hedera helix*). Root hairs secrete polysaccharide and protein glue that hardens, becoming a strong and permanent adhesive.

Transition trade-offs

The transition from a self-supporting to a climbing growth form (and back) has occurred numerous times across several plant lineages. Invariably, these transitions invoke structural and functional trade-offs. For example, erect forms of Pacific poison oak (*Toxicodendron diversilobum*) build stems that are, on average, three times as wide as those of climbers; wider stems are more structurally stable. Other plants invest more in fibres and secondary tissues prior to their transition to climbing, with higher fractions of parenchyma and wider vessels comprising the 'cheaper' climbing stems they subsequently develop.

COILING CUCUMBER TENDRILS

Recent work on cucumber (*Cucumis sativus*) tendril anatomy has revealed specialized lignified cells that form a ribbon within the tendril. Variations in cell wall structure, in combination with water content, contract and alter the shape of the fibre ribbon, which then drives coiling. Furthermore, the presence of the fibre ribbon encourages 'overwinding', whereby helices can extend and even add additional coils as the tendril is stretched. This strengthens the tendril under tension.

The importance of stem shape

Because reproductive structures are ultimately more valuable to plant success than the shoots and leaves per se, selection builds plants that are robust yet economical. Despite comprising more than 95 per cent of a plant's biomass, stems and leaves are simply a means to an end, so from the plant's point of view there is every reason to be spendthrift. Hollow stems in horsetails (*Equisetum* spp.), bamboo and even old trees are fine examples of structural economy, because the plant loses little mechanical resilience by divesting carbon from the centre of the stem.

Stem shape and resistance to bending

The vast majority of herbaceous stems are round, but hollow, square and triangular stems are also common. To appreciate the significance of these shapes, it is important to understand how geometry alters bending resistance away from the geometric centre, a property known as the second moment of area. Consider a dandelion (*Taraxacum* sp.) stem with a flowerhead perched on top. Built hollow like a straw, the stem is resistant to deformation, but if one were to arrange the stem tissue into a solid cylinder of equal length, then the stem would become narrower, and thus much more bendy – and certainly more expensive to build. Perhaps a very short, solid stem could do the job, but this might compromise seed dispersal. The tall, hollow stem benefits from having a higher second moment of area than its hypothetical solid equivalent, and is a much more

economical structure. Given these trade-offs, one might rightly wonder why not all stems are hollow. One idea is that hollow stems are prone to buckling (crimping) under lateral and top loads, such as those imposed by canopies. This, along with developmental constraints, may explain why large branches are notably absent in bamboos, horsetails and grasses.

Stem symmetry and resistance to bending

Symmetry also has a profound effect on bending. Round stems, whether solid or hollow, can flex in any direction, but the bilaterally symmetrical square stems of mints (*Mentha* spp.) and triangular stems of sedges (*Carex* spp.) have fewer bending planes. In the case of mints, square stems may not present a biomechanical advantage so much as improved light acquisition by virtue of the alternating leaf arrangement, which may reduce self-shading. In contrast, triangular sedge stems may be useful for stabilizing the sizeable leaf and floral displays at their tip. At the far end of the spectrum are climbing plants, whose specialized stems exhibit extreme variation in their second moment of area. For example, some members of the genus *Clematis* have oval stems that twine around a trellis much like a ribbon, because they are more pliant in one bending plane than the other. The cost is that in the absence of external support, *Clematis* might never get off the ground.

The hollow stem-like peduncles of dandelions (*Taraxacum* spp.). Cylindrical stems are economical to build and are resistant to bending stress, but preclude storage in the absence of a pith; they may also crimp irreparably.

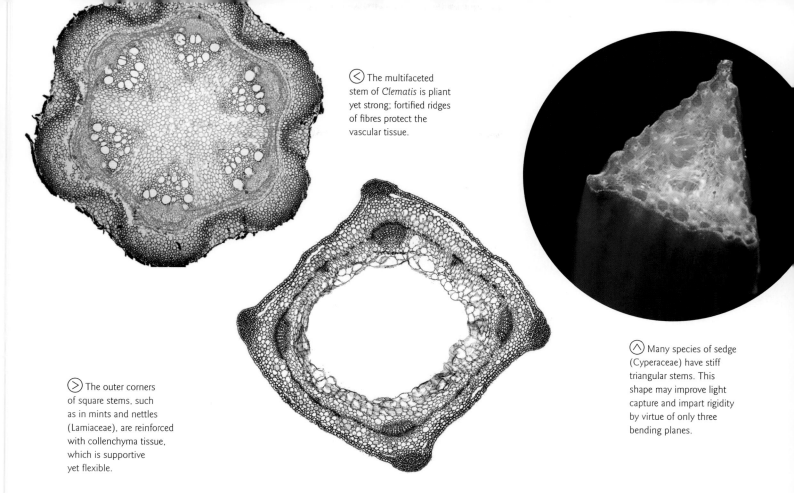

< The multifaceted stem of *Clematis* is pliant yet strong; fortified ridges of fibres protect the vascular tissue.

> The outer corners of square stems, such as in mints and nettles (Lamiaceae), are reinforced with collenchyma tissue, which is supportive yet flexible.

∧ Many species of sedge (Cyperaceae) have stiff triangular stems. This shape may improve light capture and impart rigidity by virtue of only three bending planes.

HIDDEN SHAPES

Bananas (*Musa* spp.) are herbaceous plants in which the main axis is a pseudostem composed of overlapping leaf bases. Both strong and flexible, banana plants withstand gale-force winds yet support a large leaf area in the absence of wood or even a true stem. The remarkable properties of the banana plant are due in part to the presence of stellate parenchyma in leaf petioles.

Present in numerous monocots, this tissue provides internal support while allowing the leaf axis to twist and reconfigure in the wind without buckling.

> Stellate parenchyma in a banana (*Musa* spp.) leaf petiole. Individual cells arranged in a star pattern impart mechanical strength economically. Experiments have shown that removal of this tissue significantly weakens the petiole.

Underwater stems

(A) quatic plants have evolved physiological traits that allow them to occupy surprisingly stressful habitats. In contrast to terrestrial species, they may deal with low light levels, turbulence, drag, low carbon dioxide and oxygen diffusion, and the completely anaerobic conditions in which their roots and rhizomes are submerged. Taken together, water lilies, ferns and rushes have arrived at several convergent solutions to these problems. In contrast, species such as water hyacinth (*Eichhornia crassipes*) and duckweed (such as *Lemna* spp.) have avoided them altogether by becoming buoyant rafts.

(∧) Water lily (*Nuphar* sp.). The submerged rhizomes of aquatic plants can grow in habitats depleted of oxygen. Their leaves and stems have aerenchyma tissue to support gas exchange and buoyancy.

Gas exchange in water lilies

Water lilies (*Nymphaea* spp.) are perennial, and rely on extensive underwater rhizomes for anchorage and carbohydrate storage. In temperate climates where lakes freeze, the plants shed their leaves, leaving the rhizomes to overwinter at the bottom until the spring, when they produce new growth. Critical for the lilies' survival, these modified stems may be many metres in length and up to 10 cm (4 in) in diameter. In some cases, the rhizomes are buried in soil 2 m (6.5 ft) below the surface of the water, but even in shallower waters they almost always occupy a stagnant, anaerobic environment in which the absence of oxygen in combination with microbial toxins creates a noxious habitat. The survival of the rhizome thus depends on an efficient means of acquiring oxygen and eliminating respired carbon dioxide.

It is thanks to the evolution of a passive, thermally driven mechanism of ventilation delivering atmospheric oxygen to the rhizome that water lilies can thrive in anoxic substrates. Gases move within the plant through lacunae, which are large, empty spaces in the plant organs. The lacunae are connected, and comprise more than 60 per cent of the volume of petioles and 40 per cent of rhizomes (see page 323). Gas flow originates in the lacunae of the young, recently emerged lily pads, which are pressurized in response to heating by the sun. The pressure-driven downward movement of gases through the petiole creates a vacuum within the leaf, which again draws air into its air spaces, and thus keeps the flow of gases

fresh air

oxygen-depleted air

young leaf

old leaf

mud

rhizome

(>) The flow of air through a water lily (*Nuphar* sp.). Fresh air (blue arrows) enters the young leaves, moving into the rhizome via pressure-driven flow. Oxygen-depleted air (red arrows) exits through older leaves.

continuous. It has been suggested that more than 22 litres (46 pints) of air may perfuse the rhizome over the course of a warm day. The efflux flow of gases occurs through older leaves, which over time become quite porous and unable to sustain pressure. Old lily pads thus act as a vent, and are the point source of much of the air bubbling up in lily ponds.

This type of convective throughflow ventilation has been found in other aquatic rhizomatous plants, including lotus (*Nelumbo* spp.), the perennial *Phragmites* grasses, bulrushes or cattails (*Typha* spp.) and spikerushes (*Eleocharis* spp.). Research has shown that sun-induced stomatal opening facilitates the uptake of air, and that the low internal resistance to gas flow gives these plants a competitive advantage over those relying on diffusive gas flux. Convective flow is also ecologically important. The rates of methane and carbon dioxide flux are up to 15 times greater in wetlands that harbour species with convective ventilation than in systems where the plants rely on gas diffusion. Global warming may be consequential for wetlands.

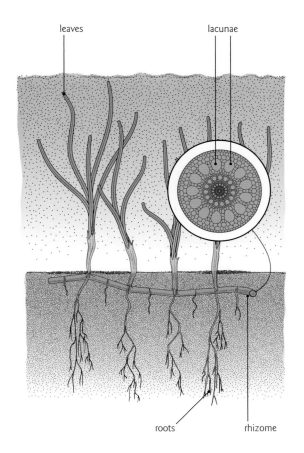

leaves

lacunae

roots

rhizome

AERENCHYMA TISSUE

Aquatic plants achieve both buoyancy as well as gas diffusion through specialized tissue known as aerenchyma. Aerenchyma is found throughout the stems, roots and leaves of aquatic plants, as well as the cortex tissues in the bark of frequently flooded trees. It a specialized kind of parenchyma that enhances the diffusion of air throughout the plant, thereby raising levels of oxygen to support respiration. Aerenchyma can be formed schizogenously, that is by the separation of cells to form voids, or lysigenously, by the dissolution of cells. Ethylene gas has been implicated in the lysigenous formation of root aerenchyma. Anoxia promotes the release of ethylene, which then triggers programmed and controlled cell death. A similar process may occur in the formation of rhizome aerenchyma.

⌃ Cross section of a leaf petiole of common water-hyacinth (*Eichhornia crassipes*) showing aerenchyma. These lacunae facilitate gas exchange in stems and submerged roots.

⌃ Seagrasses are monocots that are fully submerged in salt water. Forming extensive beds in bright, shallow habitats, seagrasses are anchored in the sand by a creeping rhizome, often creating large colonies. Water transport is not critical for these plants, so the xylem tissue in the rhizome is highly reduced in comparison with the phloem. Most of the rhizome is occupied by either parenchyma or aerenchyma tissues.

Stem modifications

Plants have managed to modify nearly any organ to serve a need beyond its original purpose. Some stem modifications occur above ground, some are subterranean, and still others are found in epiphytic or aquatic plants. In many cases, it can be difficult to differentiate between a modified stem and a storage root without knowledge of a species' life history, development and even anatomy. Some generalities can help, however. Modified stems give rise to leaves, flowers, roots or other stems, they may have lateral nodes, and importantly, they are always found between roots and leaves.

Form and function

If the overall gestalt of a plant lies in stem architecture, then modified stems greatly extend the possibilities of diversity in plant form and function. In some climbers such as the common grape vine (*Vitis vinifera*), distal stems are modified into clasping tendrils that can produce leaves and flowers upon maturity. Cladophylls, which are stems that evolved to resemble foliage, form the edible paddle in *Opuntia* cacti; the small, insidious spines are actually modified leaves. In addition to spines, physical defences include thorns and prickles, but of these, only thorns are proper stems because they arise in the axils of leaves. Prickles, such as those found on roses (*Rosa* spp.), are outgrowths of the epidermis and cortex.

THE INFLATED STEM

The desert trumpet (*Eriogonum inflatum*) is a perennial of the American Southwest, characterized by inflated stems emerging from a leafy basal rosette. Most desert ephemerals are active in the spring, setting seed well before the onset of the summer heat, but the desert trumpet extends its growing season by virtue of its green stems, which remain photosynthetically active after the basal leaves have died. More tolerant of high temperature and dry air than the leaves, the stems lose little water during photosynthesis, probably because their stomata can remain closed while they fix the carbon dioxide that is stored within the hollow stem. Incredibly, carbon dioxide concentrations inside inflated chambers are at least 20 times greater than atmospheric levels on account of the respired root and soil carbon dioxide that enters through the elongated stem cavity.

Stems modified for the storage of water and food may be familiar to most. Columnar cacti such as the saguaro (*Carnegiea gigantea*) have swollen stems, in which parenchyma cells with large vacuoles store copious amounts of water. Shallow-rooted and subject to infrequent precipitation, saguaro relies on its stem for hydration, support and photosynthesis. Water-filled stems can also be seen in the spurge family (Euphorbiaceae), and in the trunks of baobab trees (*Adansonia* spp.).

Below-ground modifications

Stems vary considerably below ground as well, forming tubers, bulbs, corms and rhizomes, all functioning in the storage of starch. They are highly reduced in the case of bulbs (for example, garlic), which are aggregates of succulent leaves adhered to a flat stem disc. Globally, the most widely consumed tubers are potatoes. These develop from the tips of slender stems known as stolons, or runners, which may grow in both the air and the soil. The 'eyes' of a potato (or a yam) are the nodes that produce shoots, and these must be included in potato cuttings for successful propagation. Rhizomes may or may not be subterranean. In *Iris* species, the rhizomes extend underground, whereas in ferns, they can occupy terrestrial or epiphytic habitats.

The rewards of runners

The survival value of most stem modifications is self-evident, but the functional role of stolons remains open to scrutiny. Studies have shown that they are advantageous in beachy habitats where the instability of shifting sand dunes threatens to dislodge or bury slow-growing plants. Stolons quickly extend a plant's reach, allowing it to establish and tolerate frequent disturbance. Furthermore, stolons facilitate resource sharing in clonal species whereby nutrients, water and sugars are translocated from thriving individuals to other parts of the clone. The adaptive flexibility of stems has allowed plants to thrive in the most challenging conditions.

⌃ The trunks of the baobab (*Adansonia* sp.), a Madagascan endemic tree, store copious amounts of water, which supports the flush of new leaves at the end of the dry season.

⌄ Fishhook cactus (*Sclerocactus parviflorus*). Composed primarily of parenchyma tissue, the stem of this cactus is a reservoir of water that allows it to survive prolonged episodes of water deficit and heat.

< Pygmy iris (*Iris pumila*) showing the root-like rhizome. Rhizomes are elongate, perennial non-woody stems modified for storage or support. The meristems in the nodes give rise to shoots.

Response to mechanical stress

(P)lants respond to an onslaught of mechanical stress in addition to the challenges presented by herbivores, too little or too much water, and changing light levels. Common examples of physical disturbance include heavy winds, rain, snow, deformation from falling canopy debris, and animal mischief such as climbing and swinging from branches. The physiological and structural responses of plants to such interference is called thigmomorphogenesis, and differs between herbaceous and woody plants. However, all plants have one thing in common: exposure to some stress makes them more resilient.

⌃ Exposed trees, such as this pine (*Pinus* sp.), are subject to persistent mechanical loads from wind, rain or snow. Lower leaf area reduces drag on the canopy, while reaction wood strengthens the stem.

What a (wind) drag

The general response of plants coping with wind and mechanical agitation is to shorten their overall stature and increase the flexibility of their canopy. At the cellular level, herbaceous plants develop higher collenchyma content in their stems, as well as greater amounts of lignified sclerenchyma fibres. Smaller, hardier leaves are produced from shorter, thicker stems. Whether the stems become more pliant or stiffer appears to vary depending on the context and species: plants exposed to multi-directional wind may develop more flexible stems, while point loading may result in greater stem stiffness. Building resilience comes at a cost, however – usually a decrease in flowers, fruits and seeds.

Responses in woody plants

Thigmomorphogenesis in woody plants depends on the response of the cambial layer. Xylem is a dead tissue and thus unreactive to mechanical or chemical stimuli, so only the most recently generated wood can exhibit a response to stress. Taken together, both conifers and angiosperms display shorter and smaller xylem conduits, and an increase in wood density in habitats where winds are unidirectional and cause the stem to bend. Wider, asymmetrical stems and reconfigured canopies that reduce wind drag are a common sight in windswept environments. Some species also shed or curl their leaves to increase air flow through the canopy and thus reduce stress on the trunk. In very stressful habitats such as high-elevation treelines, krummholz trees adopt permanent deformation, stunting and canopy asymmetry. The twisting of xylem in many krummholz trees is believed to increase axial flexibility under heavy loads.

Mechanosensing

The exact means by which plants sense and react to pressure is not yet fully understood. Generally, the response to stimuli is dose dependent, systemic and saturable. Studies with thale cress (*Arabidopsis thaliana*) have shown that more than 3 per cent of the plant's genome – corresponding to several hundreds of genes – becomes active within 30 minutes of mechanical agitation. Mechanosensing may have its origins in cell plasma membranes, in which specialized proteins such as stretch-activated channels respond to stimulation or changes in turgor pressure, and therefore induce a cascade of signal-transduction pathways involving calcium and a suite of plant hormones.

REACTION WOOD

Static loads such as heavy rain, snow and even the weight of the leafy canopy impose a stress on branches that, over time, causes them to resume their original angle and position on the tree. Behind this thigmomorphogenic response is specialized xylem referred to as reaction wood. In conifers, reaction wood is known as compression wood, whereas in angiosperms it is called tension wood. Compression wood forms on the underside of a top-loaded conifer branch, where it is darker and more lignified than regular xylem, and is composed of short, round, thick-walled tracheids. The formation of compression wood leads to asymmetrical stem development, which *pushes* the loaded stem back up. In contrast, tension wood in angiosperms develops on the upper side of the branch. Composed of shrunken, gelatinous fibres, tension wood *pulls* the stem back up to its proper position. Reaction wood compensates for the tilting of entire trees due to landslides or windstorms, reorienting them to grow straight again.

> Normal (opposite) wood and reaction wood in a pine and a poplar. Note that cells become rounder and cell walls become thicker in both compression and tension wood, relative to the opposite wood. Compression wood is denser and stiffer than opposite wood (lower left panel). By contrast, variation in the thickness and chemistry of the gelatinous layer, in combination with variable fibre fractions, makes tension wood a more mechanically complex tissue (lower right panel).

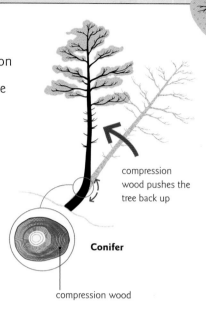

tension wood pulls the tree back to its original position

Angiosperm

tension wood

compression wood pushes the tree back up

Conifer

compression wood

< ∧ Reaction wood in conifers and angiosperms. Conifer compression wood forms on the underside of a conifer stem, whereas tension wood in angiosperms forms on the top.

∨ Opposite wood

∨ Opposite wood

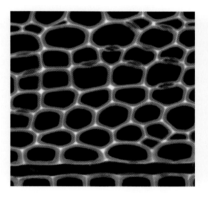

Pine (*Pinus radiata*)

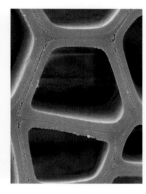

Poplar (*Populus* sp.)

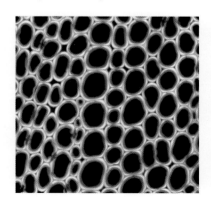

∧ Compression wood

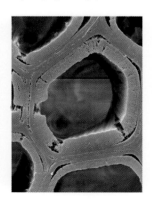

∧ Tension wood

Dendrochronology

(D)endrochronology is the study of tree rings over time, and is an expanding, multifaceted discipline that seeks to date and provide chronologies for particular events, as well as understand and contextualize natural and human phenomena. It is centred around the fundamental premise that tree growth and xylem development are a reflection of aggregate signals, including climate, ageing, stand dynamics and disturbance, as well as some degree of unexplained variation.

(↗) The growth rings of a tree. Annual variation in growth ring thickness and cell attributes can be linked to climatic signals. Records from several individuals are needed for a statistically robust interpretation.

(↙) Long-reference chronologies can be generated by cross-dating live trees with ones that have died at different points in time. In this example, repeated growth ring patterns among disparate trees align several samples; this produces a continuous chronological record for the area.

Dating events

As trees mature and develop, their xylem records numerous events and conditions, some of which will be of interest, some not. The underlying aim of dendrochronological methods is to eliminate spurious or irrelevant information through statistical analysis in order to amplify a desired signal, such as climate. To this end, researchers take great care to select appropriate sites, to replicate their sampling sufficiently within the sites, to core the trees properly, and to cross-date the rings in order to assign the correct chronology. Proper interpretation of tree rings then hinges on cross-dating. Here, the position and thickness of growth rings of an undated tree core are visually reconciled against a vetted specimen, first to determine if the sample contains all of the expected tree rings (some rings may be missing, other rings may be false), and second, to ascertain if the rings share some relevant attributes. Repeated iterations of cross-dating on progressively older trees thus yields a chronologically accurate record, the interpretation of which depends on the known relationships between the growth rings, and variations in climatic and biological factors that could affect tree-ring development.

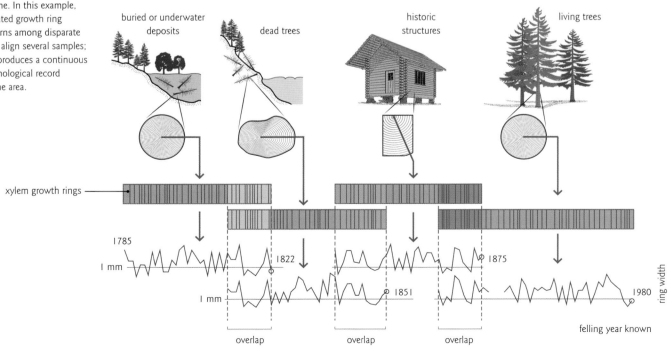

buried or underwater deposits

dead trees

historic structures

living trees

xylem growth rings

1785

1 mm

1822

1875

1 mm

1851

1980

ring width

overlap

overlap

overlap

felling year known

DENDROCHRONOLOGY AND VOLCANIC EVENTS

Expert knowledge is required to interpret the signals conveyed by tree rings. Shown here are examples from Great Basin bristlecone pines, among the longest-lived trees, which were sampled from individuals in the mountains of eastern Nevada. The black dots indicate cross-dated years, and the white circles correspond to tree rings that were formed in 1836 and 1838. These years, along with 1840, are frequently reported as ring-width minima, believed to be driven by climatic cooling and reduced sunlight following several major volcanic eruptions, including the 1835 Cosiguina eruption in Nicaragua. Considerable differences can be seen in the absolute width of each individual's growth rings, but consistent variation within each reveals a pattern when the samples are chronologically aligned.

⌄ Bristlecone pine (*Pinus longaeva*) rings from the 1830s to the 1840s. Ring width minima (white circles) observed across several individuals provide strong evidence of a climatic event that reduced tree growth.

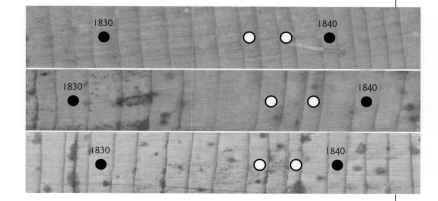

Understanding events

The powerful approach of cross-dating and tree-ring analysis has yielded many important insights about human activity and asteroid impacts, among other phenomena. One of the most notable contributions to archaeology was made by American astronomer A. E. Douglass, who in the early twentieth century assembled a 700-year tree ring chronology to determine the occupation dates of numerous Native American settlements in the arid American Southwest. He achieved this by sampling *in situ* construction beams as well as wood artefacts from museum collections, cross-dating the rings, and subsequently anchoring this record with samples from living trees of determined age. A great deal was known about the culture of these ancient peoples, but assigning time frames to their site residency allowed archaeologists to unlock core facets of their history.

On the opposite side of the globe, Russian and American scientists have used tree rings to learn about the Tunguska event, which occurred in Siberia on 30 June 1908. Evidence suggests that a celestial body either impacted or exploded mid-air, causing an extraordinary release of energy, so much so that trees were felled over 2,000 sq km (770 sq miles) of forest. Close examination of the annual rings from trees that survived revealed that the impact most likely defoliated these individuals, as judged by their narrow and poorly developed annual rings from 1909 onwards. The researchers were then able to speculate about the magnitude of the heat impulse created by the explosion.

Stem exudates

Plant secondary metabolic processes give rise to stem exudates such as resins, latex and gums. These viscous substances are important for plant defence, but they are not critical for survival like sugars and carbohydrates, which support respiration and growth. In nature, the complex chemistry of stem exudates, in combination with their sticky, viscous quality, provides formidable protection against herbivores and fungal pathogens. Humans, on the other hand, are attracted by their useful properties and appealing aromas, and often go to great lengths to find resin-producing trees and shrubs.

⊙ The sticky, milky latex that is characteristic of the Euphorbiaceae family is seen here exuding from the cut stem of petty spurge (*Euphorbia peplus*). The latex serves primarily to defend the plant from herbivores, kill pathogens and seal damaged tissues.

⊙ Milky latex extracted from a rubber tree (*Hevea brasiliensis*) as a source of natural rubber. The latex is collected by first slashing the tree's bark in a cross-hatch pattern, and letting the liquid pour down the side of the tree into a bucket over a period of several hours.

The biology of stem exudates

The chemical composition of exudates varies greatly among species, as does the structure of the tissues that produce them. Resins are common in conifers and several angiosperm groups, and are broadly defined as variably viscous, lipid-soluble mixtures of terpenoid and phenolic secondary compounds that have differing degrees of volatility. For example, the terpenoid alpha-pinene is volatile, and responsible for the characteristic scent of pine pitch, whereas abietic acid has low volatility, serving as the resin base. The degree of fluidity is determined by the ratio of volatiles to non-volatiles; volatile constituents tend to thin the resin. In stems, resins are produced in canals, ducts or pockets, which can be found in both the xylem and bark. These structures have an epithelial lining, which secretes the resin into the lumen.

In angiosperms, the most common stem exudate is latex, with some groups such as members of the spurge family generating amounts in such abundance as to be economically important. The colour of latex can range from milky white to red, with terpenoids, proteins, carbohydrates, tannins and other constituents comprising the bulk of the fluid. Latex occurs in specialized cells known as laticifers.

Lastly, gums are primarily found in angiosperms. Gum exudates are produced by parenchyma cells that form a cavity, which contains the polysaccharide-rich fluid. Members of the Rosaceae), such as plum (*Prunus* spp.) and apple (*Malus* spp.) trees, often secrete gums.

⊙ Pine resin. Resins serve as insect deterrents, but also help seal wounds that might otherwise allow bacteria and fungi to enter the tree. Economically important, pine resins are used to treat and finish wood products.

⊙ Resin canals in the bark (large voids), and in the xylem of a pine (*Pinus* sp.) stem (lacunae lined with blue-stained epithelial cells). The canals are under positive pressure, so damage by feeding insects will release the resin and trap the insect, or irritate its mandibles.

The role of stem exudates

Stem secretions play a critical role in protection against pathogens and herbivores. Damaged stem tissue does not heal, so the sealing of cuts by resins and exudates presents a physical block against fungi and bacteria. External tissues are often a major source of resins, so it is critical to leave a collar of bark when pruning branches; this allows the tree to seal the cut.

In conifers, the release of 'pitch' is often related to attack from insects such as bark beetles, whose larvae bore into the cambial region to feed on phloem. In fact, so great is the selective pressure exerted by pine beetles (*Dendroctonus* spp.) that many members of the pine family (Pinaceae) produce resins constitutively in extensive canal ducts, and upregulate resin flow in response to an attack. Contact with the resins is usually sufficient to immobilize the insect and interfere with its mandibles. Numerous studies with ponderosa pines (*Pinus ponderosa*) reveal that the chemistry of the resins becomes more complex during the growing season, and yields increase in time with insect life cycles.

MYRRH AND FRANKINCENSE

The gum resins of frankincense and myrrh have enriched religious narratives and rituals around the world. Frankincense is derived from the exudates of *Boswellia sacra*, a resilient shrub endemic to the southern Arabian Peninsula and northeastern Africa. The most prized harvests are from the third bark scrapings, when the resin – a mixture of polysaccharides and terpenoids – is most sweetly fragrant. Myrrh is another gum resin, collected from the bark of trees in the diverse genus *Commiphora*, including *C. myrrha*, whose distribution overlaps with that of frankincense. Historically, the astringent qualities of myrrh made it desirable for funereal purposes, and it continues to have value in modern perfumery and medicine.

⑦ ⊗ In a desert environment with relatively little vegetation, the frankincense tree (*Boswellia* sp.) is greatly dependent on its resin to defend itself from herbivores.

Stress resilience

(H)ealthy forests are critical to the ecological and climatic balance of our planet. Forests comprise about 30 per cent of the land surface, sequester at least as much carbon dioxide as mankind's emissions, and influence climate via energy exchange, water transport, nutrient cycling and the release of volatiles. Up to 50 per cent of the rain in the tropics is the result of transpiration, so even modest forest removal can create shifts in humidity and air temperature. Following decades of deforestation, questions are now being raised about how we tend to our planet's greenery.

(↻) Healthy aspen forest. Quaking aspen (*Populus tremuloides*) is a key deciduous species in high-latitude/high-elevation environments. Its characteristic white bark and trembling leaves are emblems of western North American forests.

(∨) Quaking aspen stands affected by drought in southern Colorado. Because aspen is adapted to temperate climates and higher-elevation habitats, it is vulnerable to drought stress during prolonged episodes of heat and water deficit.

Warming and drought stress

Increases in climate-driven forest mortality have been observed around the globe, with warming and drought being the biggest culprits. Precisely how trees die varies from region to region, but there is growing consensus that two non-mutually exclusive processes may be at play: carbon stress and xylem embolism.

Carbon stress can present itself when warming and/or prolonged water deficit causes stomata to close, thereby limiting photosynthesis. Continued respiration in the absence of carbon dioxide uptake reduces starch and sugar stores, leaving the plant with few resources to fight off pathogens and recover from the drought. Evidence for 'carbon starvation' is limited; it is unlikely to kill trees outright.

On the other hand, the evidence for embolism-induced hydraulic limitation is gaining strength. Here, the combination of warming and soil-water deficit amplifies plant drought stress, thereby increasing sap tensions in the xylem, rendering the network extremely vulnerable to embolism. Inability to transport water to the canopy spells catastrophe for the tree. Research on aspens shows that trees now frequently exceed their hydraulic stress limits in regions struck by drought and warming, making this the most compelling explanation for the large swathes of aspen forest mortality observed in the American West. The story is far from resolved, however, especially since drought-stricken trees are more susceptible to pests.

Insect infestations and fire

Recent bark beetle outbreaks have devastated millions of conifers in forests of western Canada and the United States, causing huge economic losses and increasing atmospheric carbon dioxide levels in the process. Alongside the loss of forests often comes the loss of wildlife, and once again, the combination of warming and drought contributes to this calamity. Summer drought and higher temperatures weaken trees, all the while speeding up the insects' life cycles. Furthermore, winter temperatures may not be sufficiently lethal to kill the larvae that persist near the cambial layer.

The degree to which pine beetle outbreaks interact with forest fires adds to the complexity. Infested, dying trees may contribute to fuel loads and alter the characteristics of fires, or alternatively, bark damage from earlier burns may make trees more susceptible to infestation. Corresponding with increased fire activity is the greater secretion and terpenoid content of resin, which finds its way to the forest floor and increases the flammability of the understorey litter. The interconnectedness of these events is perhaps inevitable, but forests are resilient and do recover, if given enough time.

◁ Forests stressed by drought and beetle infestations are more vulnerable to intense fires. Increased fuel loads drive greater fire frequency and severity.

RESISTING FIRE

Species inhabiting fire-prone regions have evolved strategies that allow them to survive a burn. Several such shrubs – including sumacs (*Rhus* spp.), banksias (*Banksia* spp.), many eucalypts (*Eucalyptus* spp.) and even some conifers – rely on lignotubers to generate new crowns. Lignotubers are swellings at the base of a stem that contain buds and abundant starch reserves. But even in the absence of prominent lignotubers, many fire-adapted tree species are exceptionally resilient thanks to their thick bark and persistent cambial activity. As long as a large fraction of a tree's cambial layer remains intact, there is good potential for its recovery after a fire.

◸ The Greek strawberry tree (*Arbutus andrachne*) inhabits fire-prone Mediterranean regions. Its prominent lignotuber develops new sprouts after a fire and thus supports rapid recovery of the canopy.

▷ A fern-leaved banksia (*Banksia oblongifolia*) lignotuber. Endemic to eastern Australia, this species has adapted to bush fires by producing sprouts from its lignotuber and by releasing seeds from its cones.

Leaves

Leaves are so ubiquitous that it is almost possible to overlook them. Green leaves are the canvas of life on Earth, and inside these organs the silent hum of photosynthesis powers most of that life. Photosynthesis is a remarkable piece of chemical wizardry that uses the energy from sunlight, along with nutrients, carbon dioxide and water, to make useful food, fibres and other molecules like drugs. These materials are of primary importance to the plants themselves, but they have a huge secondary importance to the animals that consume the plants, including humans.

However, while most leaves are flat, green, photosynthetic organs on an aerial stem of a plant, a leaf is not *defined* as a flat, green, photosynthetic organ on an aerial stem of a plant. In biology, the parts of a plant are not defined by their function but by their position on the plant. Leaves are organs that grow on stems, and where the leaf meets the stem, there is a bud. This axillary bud can grow into a side shoot or maybe into flowers. Leaves can therefore be more than just photosynthetic organs, and some of them can even turn the tables on members of the animal kingdom and digest them.

Bryophyte leaves

The mosses, liverworts and hornworts live, often quite literally, in the shade of the vascular plants. These plants – known collectively as the bryophytes – are the closest things alive today to the first land plants that emerged from the water 470 million years ago. Because they do not produce woody tissue and lack 'plumbing', they are unable to grow more than several inches tall, and yet in some cases the structures that house their photosynthetic apparatus look like the leaves of species of flowering plants.

Not true leaves

The leaves of the common haircap moss (*Polytrichum commune*) and those of the tree heather *Erica scoparia* are, on first glance, superficially similar. However, the leaves of the moss do not have a bud in their axil where they meet the stem, so strictly speaking they are not true leaves and instead are what some people call phyllodes. Internally, the leaves of these two plants are very different. The leaves of mosses are generally just one to a few cells thick and have a slightly swollen central rib, known as the costa. Inside this rib there may be cells known as hydroids and leptoids that are specialized in conducting either water or the products of photosynthesis.

⌄ Appearances can be deceptive, and what appear to be leaves in the common haircap moss (*Polytrichum commune*) are in fact phyllodes, since they do not have an axillary bud.

∧ Mosses such as peat moss (*Sphagnum*) can hold large quantities of water in their cells to avoid dehydration due to the lack of air spaces in their cells.

< Although superficially similar in appearance to *Polytrichum* phyllodes, the leaves of *Erica scoparia* are in fact true leaves since they have a bud in their axil.

Water loss

Unlike the leaves of vascular plants such as ferns and angiosperms, those of mosses do not contain air spaces. This is because they are so thin that all the cells are in contact with the air and gases can therefore move in and out of the leaves easily. The leaves of all other land plants are much thicker, and so contain air spaces to allow gases to enter and leave all cells. Mosses absorb a great deal of their water, nutrients and gases through their leaves, which means that they cannot have a thick waterproof cuticle. They are therefore very prone to water loss

ALL LEAF

The mosses are just one of three groups that make up the bryophytes, and compared to the other two they are very leafy. The hornworts are rare and today number only about 200 species worldwide. In comparison, there are approximately 12,000 species of moss, many of which are very familiar (see box below). Almost as speciose, at approximately 9,000 species, but much less well known, are the liverworts. Meaning literally 'liver-like plant', the liverworts do resemble very sick-looking pieces of liver. They are well known to gardeners as a weed of pots sown with seeds, because the conditions that are ideal for seed germination are also ideal for the growth of species such as the common liverwort (*Marchantia polymorpha*). These leafy liverworts are basically all leaf except for small rhizoids and their reproductive structures, which look like umbrellas.

∧ The leaves of the common liverwort (*Marchantia polymorpha*) are a familiar sight in pots of seedlings.

during dry spells. Despite this apparent weakness, mosses are often found growing in places that are subject to severe drought, including the tops of walls. This is because they can rehydrate their leaves better than most land plants. This was a very important early adaptation in the colonization of land by plants, because there was no soil and, like the mosses of today, the early land plants had no roots to speak of.

∧ Although prone to water loss, mosses are extremely resilient and can survive periods of drought, managing to survive even on the top of stone walls.

BEAUTIFUL AND USEFUL

Sphagnum mosses grow all over the world in acidic bogs. Despite this rather mundane habitat, these beautiful plants have saved countless lives. Because their leaves can absorb many times their own weight in fluids, they make a very good wound dressing. In fact, they are even better than a sterile bandage because they possess an antibiotic that prevents infection.

⟩ Prior to the discovery of penicillin, in the First World War peat moss (*Sphagnum*) was used in wound dressings for its absorbent and antiseptic properties.

Fern fronds

The majority of plants on Earth have an internal system of lignified pipes known as vascular tissue, and are therefore known as vascular plants. Most extant vascular plants produce seeds, but the pteridophytes – comprising the lycophytes and the ferns (sometimes referred to as the monilophytes to indicate that they include the horsetails and whisk ferns) – do not. The lycophytes were once far more numerous and ecologically important, but their numbers are now on the decline. Ferns, however, are still very much with us.

Similar but different – convergent evolution

The leaves of all vascular plants have four things in common: they have veins containing pipework, they do not keep growing forever but have a final size (known as determinate growth), their upper and lower surfaces differ, and they are arranged in one of only a very few different ways. Despite the fact that vascular plant leaves have these four features in common, it now seems that they have evolved more than twice, and perhaps many times. If the same word is used to refer to a structure that has evolved many times, the fundamental differences between the various groups of plants are ignored. Their internal structures all look very similar, implying that there is really only one way to be a flat photosynthetic organ.

Ferns have fronds

The foliage of ferns is often referred to as fronds, which implies quite correctly that they are different from the leaves of seed plants. For one thing, fronds do not have a bud in the axil where they join the stem of the fern plant. Furthermore, on their undersides they have structures known as sori, which produce the spores that are a stage in the reproduction and dispersal of ferns (see Chapter 5). A third difference between true leaves and fronds is that is some species of fern (e.g. the New Zealand hen and chicken fern, *Asplenium bulbiferum*) produce plantlets on their fronds. Finally, the fronds of ferns unfurl, and so in bud they look like a bishop's crozier.

⌄ Clubmosses, including this marsh clubmoss (*Lycopodiella inundata*), are an ancient group of plants that are declining in number.

⌄ As young fronds of the green ostrich fern (*Matteucia struthiopteris*) unfurl, they resemble a bishop's crozier, distinguishing them from the leaves of flowering plants.

⟨ Unlike true leaves, the fronds of some ferns such as the hen and chicken fern (*Asplenium bulbiferum*) demonstrate vivipary, an adaptation to increase plant numbers rapidly. Horticulturalists take advantage of this property to propagate new plants.

WHAT ARE SPORES?

The spores produced from the underside of fern fronds are partly analogous to seeds, because this is how ferns are dispersed to new habitats. However, they are more than just a survival capsule for dispersal and have an important role in sexual reproduction (see Chapter 5). There are many beautiful and different arrangements of the spore-producing structures, and they are very useful when you are trying to identify fern species – as these pictures show.

⟨∨⟩ Ferns show a remarkable variety of spore-producing structures on the underside of their fronds, which can be used to identify the different species. Left, Hawaii fern (*Microsorum* sp.); middle, giant fern (*Angiopteris evecta*); upper right, spotted fern; lower left, *Blechnum* sp.; lower right, interrupted fern (*Osmunda claytoniana*).

Gymnosperm leaves

For a relatively small group of species, the gymnosperms have a remarkable diversity of leaves, and the familiar pine needle is anything but typical. Among the most instantly recognizable leaves of any plant are the fan-like leaves of ginkgo (*Ginkgo biloba*), which seem to be identical to those of members of the genus fossilized in rocks dating back 185 million years. At the other end of the scale are *Ephedra* species, which seem to have no leaves at all and be made up entirely of stems.

∧ Cycad leaves resemble those of the coconut palm, and are some of the hardest and sharpest in the plant kingdom.

∨ Conifers are the most familiar and well-known gymnosperms, but as a group they demonstrate great variation in leaf shape.

The types of gymnosperms

There are four groups of gymnosperm species living today. First, there is ginkgo, which is perhaps one of the most isolated species on the tree of life because it is the only species in its class. To give you an indication of just how strange this is, the class Aves contains all 10,000 bird species. Next, there are the cycads, which have leaves that look a bit like those of the coconut palm (*Cocos nucifera*). Third, there are the most familiar and biggest group – the conifers – but even here there is variation, from the scale-like leaves of the giant redwood (*Sequoiadendron giganteum*) to the needles of the pines (*Pinus* spp.). Finally, there is a group of oddballs, the Gnetales (pronounced 'knee-tale-ease').

◁ Ginkgo (*Ginkgo biloba*) is a classic living fossil. Although this fossilized leaf is 49 million years old (left), the actual age of this species has been calculated to be 185 million years, pre-dating the dinosaurs. It has been able to survive an extraordinary range of climate change during that period, and the leaves on living trees (right) appear indistinguishable from those in the fossil record.

The most un-conifer-like gymnosperms

The order Gnetales consists of three groups with very different leaves. *Gnetum* is a large genus whose leaves could easily be mistaken for those of a flowering plant. They are similar to the leaves of a cherry laurel and have a thick cuticle to reduce unnecessary water loss. *Ephedra*, the second genus in this group, contains plants that live in seasonally arid

The needles of pine trees come in bundles of one to eight at the end of a short shoot, with the needle number per bundle normally very constant within a species. Internally, the arrangement of the cells is almost intermediate between a stem and a leaf. This is in part due to the fact that the needles are an adaptation for living where water is scarce for some of the year, either because the soil is bone dry, or because the water is frozen in the soil. To a plant, water frozen as ice is inaccessible and so equivalent to having no water at all.

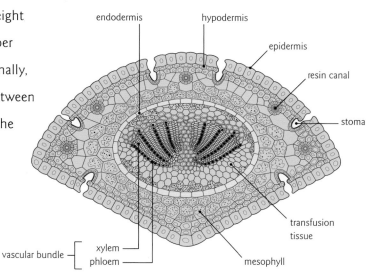

endodermis   hypodermis   epidermis   resin canal   stoma   transfusion tissue   mesophyll

vascular bundle { xylem   phloem

regions and so have dispensed with leaves of any size in order to conserve their water supplies. The final member of this extraordinary trio is the genus *Welwitschia*, comprising the single species welwitschia (*W. mirabilis*) from the Namib Desert. Plants of this species produce only two leaves, which grow continuously at their base for decades. The result is a twisted tangle of two strap-like leaves several metres in length, making the whole plant look like the head of an old goat on a very bad horns day. The leaves of welwitschia are built not only to photosynthesize and store and retain water, but also to collect water. The majority of the plant's water comes in the form of dew that condenses on the leaves overnight and then runs down them to the soil below, where it is gratefully absorbed by the surface roots of the plant.

(L) Although *Gnetum gnemon* might look like a flowering plant with broad, flat leaves, it is in fact a conifer.

(V) Welwitschia (*Welwitschia mirabilis*) survives in the intense heat of the Namib Desert due to its strap-like leaves, which optimize water retention and storage.

The parts of a leaf

The majority of leaves have two main parts. There is the obvious flat green section, known as the leaf lamina or blade, and the stalk, known as the petiole. The petiole is the start of the main vein that runs up the centre of most leaves. However, plants are almost infinitely variable, and thus it is unsurprising that the leaves of many do not conform to this 'default' setting.

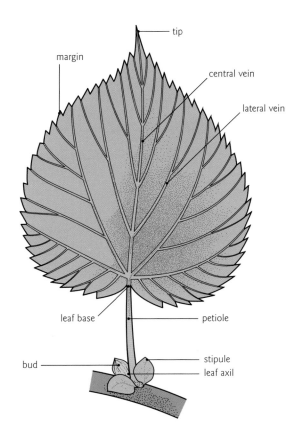

Structure of a leaf. Although leaves are generally made up of a leaf lamina and a stalk, the variations upon this theme are almost infinite.

HAIR EVERYWHERE

Many parts of many plants are covered in many hairs. There are at least 16 different types or shapes of plant hairs, from long and thin single-celled hairs, to hairs that look like combs and are actually hairs with hairs. Some hairs, as we shall see later, have very obvious functions. However, much of the time one is left wondering why on earth a plant invests energy in a structure that is completely useless, much like the spandrels in the roof of a cathedral. Among the most beautiful of plant hairs are the stellate hairs on the leaves of the Turkish sage (*Phlomis russeliana*), a native of southwest Asia.

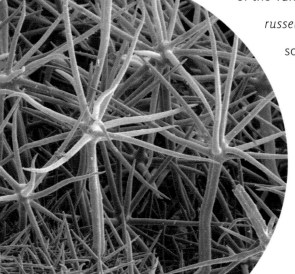

Many leaves bear hairs to help reduce water loss, the beauty of which can be appreciated only through a microscope. These stellate hairs are on the leaves of a *Phlomis italica* plant.

The leaf lamina

The lamina is the business part of a leaf because this is where the majority of photosynthesis takes place. For this to happen, the cells in the leaf need to be supplied with water and minerals. These are taken up by the roots of the plant and then moved up its stem through the vascular tissue. This network of pipes continues into and throughout the leaf in the form of veins, the patterns of which vary and can be very useful in plant identification. The two major options are either for the veins to form a net, known as reticulate venation and seen, for example, in maple (*Acer* spp.) leaves, or for the veins to be parallel, such as those seen in grasses. However, the diversity does not end there, because what the veins do when they reach the edge of the leaf also varies. Some veins keep going right to the edge of the leaf, while others turn back and form a loop, as can be seen very clearly in leaves of magnolia trees (*Magnolia* spp.). The bulk of the cells in leaves form the mesophyll; these cells house the chloroplasts, the chlorophyll-containing organelles responsible for photosynthesis. In the surfaces of the leaves are stomata, which open and close to allow gas exchange, and thus also regulate water loss.

The leaf petiole

The petiole is a rather mundane part of a plant. In many species there is no petiole, the leaf instead growing directly out of the shoot. The advantages of possessing a flexible petiole are varied, but it is easy to see that leaves with a petiole are better able to move with the wind and not snap. However, as often happens in plant biology, an otherwise unremarkable part of a plant can become adapted for other functions. For example, the petioles of many climbing species of *Clematis* can respond to touch and thus twist around the branches of the plants over which they are scrambling.

⟨ In many leaves, the veins extend right to the margin and very often fluid leaks out of the ends, a process called guttation. The veins in *Magnolia* leaves avoid guttation, because their veins never quite reach the edge of the leaf.

A LEAF STALK THAT CHANGED THE TREE OF LIFE

A humble leaf stalk was the first clue that led to the discovery of one of the most remarkable evolutionary relationships on the tree of life. In the nineteenth century, the French botanist Henri Ernest Baillon (1827–1895) noticed that the leaf petioles of the oriental plane tree (*Platanus orientalis*) and the sacred lotus (*Nelumbo nucifera*) look very similar. However, he decided not to publish this observation for fear of having his sanity questioned, because the oriental plane is a spreading tree that reaches 60 m (200 ft) in height and the sacred lotus is an aquatic herb. It was not until 150 years later that the evidence of DNA sequencing showed that these two species are each other's closest living relatives. It should be noted, however, that their common ancestor lived 70 million years ago.

⟨ The sacred lotus (*Nelumbo nucifera*) is an aquatic marginal herb with peltate leaves. The wax surface of these leaves has inspired engineers to develop dust-repellant surfaces for flat-screen televisions.

⟩ Although it has palmatifid leaves, the Oriental plane tree (*Platanus orientalis*) is, surprisingly, the closest living relative to the sacred lotus.

The molecule visible from space

(W)ith the exception of water, the molecule that is visible from further away than any other is the chlorophyll that makes plants green. Green is the canvas on which life is painted. Human eyes are more sensitive to the variations in green than any other colour, and red, the traditional colour for danger, stands out even more against a contrasting background of green.

(V) Cyanobacteria, or 'blue-green algae', which develop at the surface of slow-flowing freshwater rivers or lakes in the summer, can be harmful to people and animals.

(⌐) Single-celled cyano-bacteria are photosynthetic powerhouses filled with chlorophyll.

The 2.4 billion-year-old molecule

Photosynthesis as we know it seems to have appeared about 2.4 billion years ago in small single-celled organisms called cyanobacteria. For hundreds of millions of years they absorbed the sun's energy using chlorophyll. The energy they absorbed was then used to drive photosynthesis. This gave these cyanobacteria a great advantage over other organisms in the sea at the time, because they had an almost limitless supply of energy, as long as they were near the surface of the water. However, other organisms discovered that these cyanobacteria made very nutritious meals.

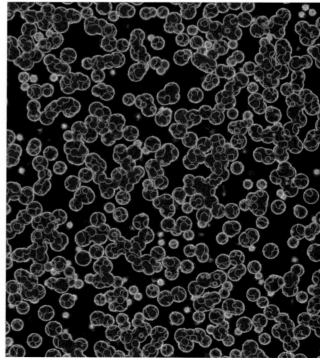

Press-ganged into working for others

The chloroplasts that are so ubiquitous in plants today started life as a free-living cyanobacterium going about its daily routine. A much larger single-celled organism engulfed this tiny organism, but the cyanobacterium fought back and for some unknown reason it was not digested. Instead, a partnership was forged, whereby the cyanobacterium handed over the products of its photosynthesis in return for board, lodging and protection. This ingenious theory was championed by the American biologist Lynn Margulis in the 1960s and is now widely accepted as the best explanation for the origin of chloroplasts. The evidence includes the fact that a chloroplast has two membranes, one from the cyanobacterium and one from its original attacker. The chloroplast also contains some of its own DNA, although some of this has also been handed over to the nucleus of the plant cell.

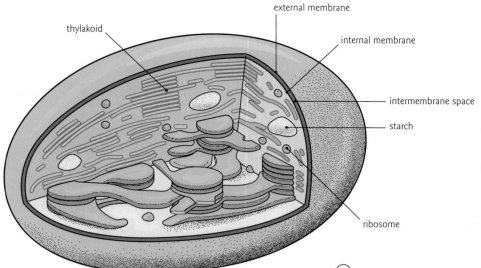

thylakoid
external membrane
internal membrane
intermembrane space
starch
ribosome

⌐ Formed through an endosymbiotic relationship between cyanobacteria and an early eukaryote, the chloroplast is one of the most important structures in biology. It is packed with the apparatus to power the planet and has proved impossible to replicate no matter how hard bioengineers have tried.

Outwardly calm, inwardly busy

Plants may be defined as immobile green organisms to distinguish them from the animals, fungi and other major groups. However, while plants may be outwardly immobile and calm, internally they are very dynamic, their chloroplasts constantly jostling for position in what looks like a carefully choreographed ballet. This enables the plant not only to optimize light capture when the conditions are dull, but also to avoid burning out when the light is too bright.

THE SILENTLY HUMMING ORGANELLE THAT POWERS THE WORLD

There is no requirement for biology to be beautiful – it just has to work. With the invention of the electron microscope in the second half of the twentieth century, a beautiful world became even more beautiful. It turned out that the patterns and colours that we see at the macroscopic level are replicated at the microscopic scale. And as well as looking beautiful, this tiny organelle still does what humans cannot: it photosynthesizes.

⊘ Colour-enhanced transmission electron micrograph showing a chloroplast, the organelle in plants responsible for photosynthesis (magnification approximately ×14,500).

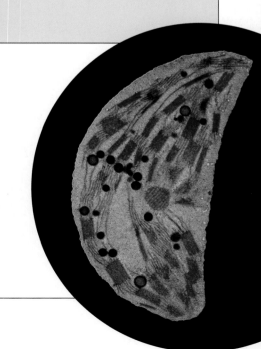

Leaf shapes and sizes

$\left(T\right)$ here are more than 400,000 species of plants, so it is therefore not surprising that there are many different leaf shapes and that these leaves vary in size. However, the range of shapes and sizes is surprising. The smallest leaves are about 1 mm ($^1/_{32}$ in) long and belong to floating aquatic plants in the genus *Wolffia*, while the largest are probably those of the African oil palm (*Elaeis guineensis*), at more than 7 m (23 ft) long. The shapes of leaves are often split into two groups – simple and compound.

Infinite variety

Leaves are described as either simple or compound. Broadly speaking, a leaf is simple if its blade or lamina is continuous up both sides of the central vein. This does not mean that the width of the leaf blade has to be evenly distributed – it can be highly variable, leading to many different-sized lobes. In the main, however, leaves are symmetrical on either side of the central vein, and when they are not, their shape is a very useful diagnostic character (e.g. in elm trees, *Ulmus* spp.).

Compound leaves

If a simple leaf is one that has a continuous lamina, then a compound leaf is one where the central vein and often some lateral veins do not have laminae on each side. This often

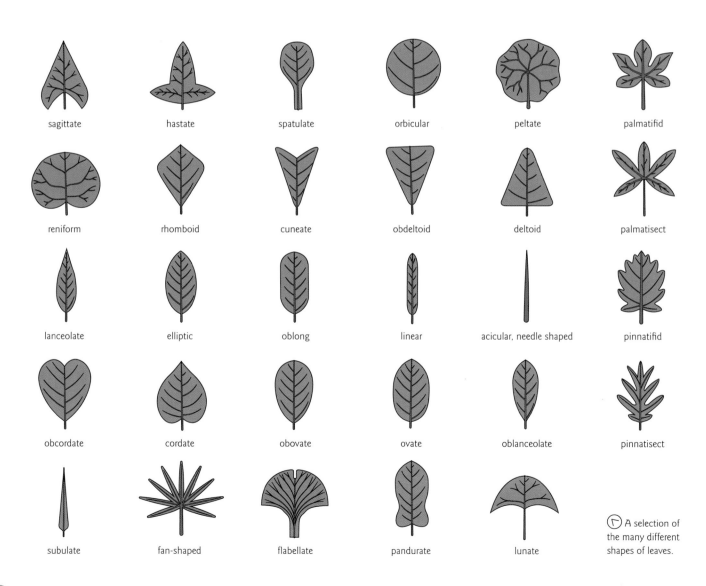

sagittate	hastate	spatulate	orbicular	peltate	palmatifid
reniform	rhomboid	cuneate	obdeltoid	deltoid	palmatisect
lanceolate	elliptic	oblong	linear	acicular, needle shaped	pinnatifid
obcordate	cordate	obovate	ovate	oblanceolate	pinnatisect
subulate	fan-shaped	flabellate	pandurate	lunate	

$\left(T\right)$ A selection of the many different shapes of leaves.

MUST EVERY FEATURE HAVE A FUNCTION AND CONVEY AN ADVANTAGE?

Ever since Charles Darwin's theory of evolution was published and accepted in the mid-nineteenth century, people have tried to find a reason for every feature on every organism. This is based on the assumption that life is so competitive that there is no room for waste, extravagance or even mediocrity. However, this is clearly not true. Most organisms simply muddle through, being just good enough. Nowhere is this more clearly seen than in leaf shape. There are no good explanations for the diversity of leaf shape, except when there is a function or stress to be overcome. While we do not know why this or that plant has this or that shaped leaf, we do know that leaf shape is controlled in most plants by the same genes, which implies that the leaf evolved just once in flowering plants.

results in an arrangement that looks like a stem with pairs of opposite leaves and a single terminal leaf. However, these are actually the leaflets of a compound leaf, because they do not have buds at their bases. The only bud is at the base of the exposed central vein (the rachis) where it meets the stem. If you cannot see a bud, then the leaf may be compound.

Tips, bases and margins

While the overall shape of a plant's leaves can be, and is, described, of equal importance when identifying species are the tip of the leaf and its base and edges (technically known as margins). The different shapes of these structures have evolved in response to various factors. For example, the extended tip of some tropical tree leaves is postulated to help with draining rain off the leaf. The genes that control the development of leaves are being identified and patterns are being correlated with habitat, but we do not yet fully understand what drives and selects leaf shape.

(∧) At first sight it would appear that each lamina in the rowan or mountain ash (*Sorbus aucuparia*) is a single leaf, but in fact they are leaflets. The whole structure is the true leaf.

(<) Leaves of cannabis (*Cannabis sativa*) are another example of individual leaflets making up the true leaf. Despite its controversial use by humans as a recreational drug, it is now the subject of intense medical-related work.

Leaf arrangements

There are two main options when it comes to the arrangement of leaves on a stem, with just a few exceptions to these. In most plant species, the leaves are either arranged in opposite pairs or 'alternately'. In the latter, each leaf is positioned about 120 degrees further around the stem than those above and below it. If the leaves are close together, then this arrangement can look like a spiral.

⊘ Although they superficially resemble small leaves, the structures at the base of these *Lathyrus* leaves are stipules. This does not mean that they are unimportant for photosynthesis since they are still green and contain chloroplasts.

⊗ Goosegrass (*Galium aparine*) has whorls of what appear to be leaves, but of the six 'leaves', four are actually stipules.

⊙ Leaves can be arranged in pairs, opposite each other on the main stem. Each pair may be directly in line with the pair above and below or at ninety degrees to them.

⊗ Birch (*Betula* sp.) leaves demonstrate a different arrangement, with each leaf positioned alternately on the main stem stem at approximately 120 degrees to the one above and below.

Stipulating the exceptions

The exceptions to the opposite or alternate dichotomy are sometimes interesting for the fact that they are not exceptions. The most commonly quoted 'third way' is for the leaves to be arranged in rings or whorls around the stem. Occasionally, these whorls comprise three or more leaves coming off the same point on the stem, known as the node. However, what sometimes looks like a whorl of six leaves is in fact an opposite pair of leaves and four stipules – this is often the case in plants in the coffee family (Rubiaceae). Stipules are bracts that grow from the base of the leaf stalk, and in members of the coffee family they can look just like the leaves except they do not have buds at their bases. This is very easy to see in species of *Galium*.

BASAL LEAVES

Many plant species do not grow stems that are covered in leaves in the manner of trees and shrubs, and instead produce a basal rosette of leaves and a tall shoot, on which the flowers are borne. The leaves in this basal rosette are generally arranged alternately and very closely on a short stem. Basal rosettes are commonly found in species that have an underground perennating organ such as a bulb or corm (e.g. the onion, *Allium cepa*), and are also common in ephemeral species that have a very short life cycle, including the Eurasian thale cress (*Arabidopsis thaliana*), which can pass through one generation in six weeks. Conversely, they are additionally found in plants with a very long life cycle, such as *Agave salmiana* var. *ferox*, which in northern European gardens can take 100 years before it flowers and then dies.

⊼ Appearances can be deceptive. Basal rosettes are actually alternately arranged leaves crammed together on an extremely short stem.

▽ The century plant (*Agave americana*) is a giant rosette of alternately arranged leaves that can persist for many years before the plant finally flowers and fruits once and then dies.

Leaves and bracts

The definition of a leaf is simple and unambiguous, and yet there are many times when the word bract is used instead of leaf. This generally happens when either the structure in question looks like a leaf but is, in fact, a bract (e.g. as in *Galium*), or it is a leaf but does not look like one. The latter is more easily understood with examples. If you look at the flowering spike of a foxglove (*Digitalis* sp.), you will see a green leaf-like structure at the base of every flower in the inflorescence. The flower developed from a bud at the base of this structure, but the structure itself is an inflorescence bract and not a true leaf.

◁ The flowering spike of the common foxglove (*Digitalis purpurea*) may look like it has leaves on the reverse, but these are bracts because they subtend a flower.

Deciduous or evergreen?

Leaves are expensive structures, both to build and to maintain. The resources available to a plant vary with the time of year, its location and the stage it is at in its life cycle, and most of the time one or more resources are finite and thus limiting. To deal with the differing circumstances they face, plants have made strategy choices, including whether to build robust leaves and hang on to them year-round, or to build disposable leaves that can be discarded when conditions get tough.

The water problem

Life first emerged and evolved in water, and living organisms have never been able to shake off their reliance on this essential resource. Most plants get most of their water from the soil via their roots. However, some of that water is needed in the leaves, where it is used in photosynthesis. There are two possible mechanisms plants use to move water upwards. One solution is an energy-expensive pumping system that pushes the water up the stem. The other is a low-tech, low-energy solution that exploits the laws of physics and sucks water passively up a capillary tube, with the 'suck' being provided by the evaporation of water from the leaves (see Chapter 3). Low tech it may be, but this system can successfully move water 100 m (300 ft) up a giant redwood tree.

> The evergreen needles of conifers are covered in a thick waxy cuticle to help resist water loss and extreme cold temperatures in winter.

> Even when the stem is as tall as that of a giant sequoia (*Sequoiadendron giganteum*), simple capillary action is sufficient to draw water from the base to the very top.

The Goldilocks syndrome

Water is abundant on our planet, but there is a proviso here: for plants to access it, the water must be at the right temperature – neither too hot nor too cold (see also page 74). If it is too warm, it will evaporate too quickly and the supply system will not be able to keep up. And if it is too cold, it freezes and cannot be moved up or down a plant's phloem and xylem tissue. If water is constantly available to plants at the right temperature, then the ideal strategy is to build permanent leaves and be evergreen. However, such places are very rare. In most parts of the world, plants face a period of water stress at some point in the year.

Coping with cold

At high altitudes and high latitudes, there is normally a period in the year when the water in the soil is frozen. One way plants cope with this problem is by building inexpensive leaves that can be 'thrown away'. If these soft leaves were kept on the plant, they would eventually desiccate and die, even at cold temperatures. Many woody plants and also many bulbous plants adopt this strategy. Another way plants cope with long periods of frozen soil is to build tough leaves that do not desiccate easily, a strategy adopted by pine trees and many other species.

THE CHALLENGES OF THE MEDITERRANEAN CLIMATE

The Mediterranean basin (and four other regions with a similar climate) is one of the richest areas in the world in terms of plant species. This is partly because its challenging climate makes it impossible for any one species to be a winner all the time, and none is successful enough to dominate. One of the problems for plants here is the extreme seasonality, with winters that are mild and damp, and therefore good for growth, but summers that are hot and dry. To get through the summer, the plants are either deciduous and drop their leaves during this season, or they are evergreen but build extraordinarily tough leaves using a very resilient tissue called sclerenchyma. This is the hidden secret behind the success of many evergreen sclerophyllous plants found in Mediterranean-type regions. It should be noted that it takes a lot of energy to make such a resilient material and so this strategy works only when plenty of energy is available – as is the case in the Mediterranean and other regions with a similar climate.

⊘ The rigid, sclerophyllous leaves of the European olive (*Olea europaea*) reduce water loss during the hot, dry summers of the Mediterranean climate.

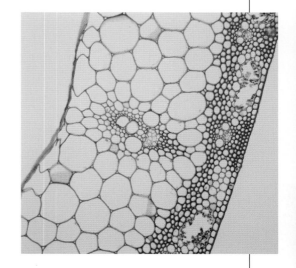

⊘ Minimising water loss in a hot climate is so important that evergreen Mediterranean plants invest a lot of energy in the production of woody sclerenchyma tissue shown here.

⊘ Having tough, persistent leaves that are undamaged even by extreme cold allows pine trees to survive long periods of low temperatures when water cannot be taken up by their roots.

Inside a typical leaf

T here may be an infinite variety of leaf shapes and sizes, but when it comes to the internal structure of photosynthetic leaves there is far less variation because form is dictated by function. The function of a leaf is to produce sugars through the process of photosynthesis. In order to do this, it must be exposed to the sun and it must have a supply of water, nutrients and carbon dioxide.

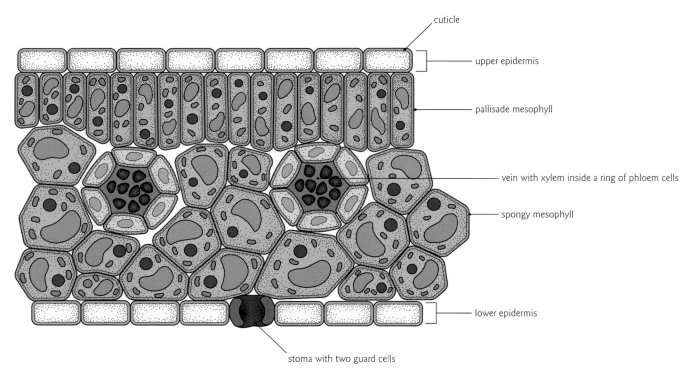

cuticle

upper epidermis

pallisade mesophyll

vein with xylem inside a ring of phloem cells

spongy mesophyll

lower epidermis

stoma with two guard cells

⋀ Cross section through a leaf of a C3 grass. Although leaves can be found in a vast array of shapes, a cross section through many leaves will reveal remarkable conformity, since the need to optimise photosynthesis dictates the structure.

⊳ The leaves of the Mediterranean shrub gum cistus (*Cistus ladanifer*) rely on their rigid sclerenchymatous scaffolding to optimise the photosynthetic surface area presented to the sunlight.

Maintaining rigidity
Leaves need to retain rigidity so that maximum sunlight can be absorbed by a sufficient number of cells to drive photosynthesis. This rigidity can be achieved in two ways. In evergreen species, the leaves are harder, with cell walls that are rigid, particularly the walls of the cells in the veins. The sclerophyllous leaves of Mediterranean plants (see box on page 151) are an extreme example of this, remaining rigid even when the plant is severely water stressed. In deciduous plants, the rigidity of the leaves is maintained by the turgor of the cells. When a plant is well watered, everything is fine, but when water loss exceeds water uptake, the leaf becomes flaccid and wilts, and the damage

may become irreversible. When the leaf is rigid, its uppermost layer of cells is like a well-laid line of bricks standing on their ends. This layer is known as the palisade mesophyll.

Supplying water and minerals

We have already seen how leaves are the passive 'suck' that draws water and minerals up the vascular tissue. Of particular importance among these minerals are manganese, which is an essential component of the chlorophyll molecule, and iron, which is also essential for the activity of chlorophyll. The veins in a leaf have to be extensive enough to supply the leaf cells efficiently with both water and minerals.

Supplying gases

Carbon dioxide in the air provides the carbon that is incorporated into the more complex carbon-based molecules that ultimately supply us and every other animal with our basic needs. In order to absorb sufficient carbon dioxide quickly enough for the photosynthesis process, a large leaf surface area is required. To achieve this, the spongy mesophyll in the leaf comprises a loose collection of cells with lots of air space between them, balancing the need for structural strength with the need for cell surfaces to be exposed to carbon dioxide.

⋀ ⊓ Well-watered pumpkin (*Cucurbita pepo*) leaves present the maximum leaf surface area to sunlight (left), but the leaves of water-deprived plants wilt (right), reducing their photosynthetic surface area and often resulting in irrevocable damage to the leaves.

▷ Leaf veins act as a transport system, distributing essential water and nutrients. Eudicots commonly have reticulated, or net, venation.

▷ In contrast, monocot plants have parallel leaf venation, although the function of transportion remains identical. It is tempting to try to explain the difference between monocots and eudicots on the grounds of function, such as efficiency, but it is just as likely to be an accident of evolution.

EVOLUTION IN ACTION?

As recently as the 1980s, carbon dioxide was a limiting factor in the growth of plants, and so in the United Kingdom extra carbon dioxide was pumped into commercial greenhouses. As the level of carbon dioxide in the air has risen over the past four decades, however, this is no longer required and horticulturalists have been spared the expense. The leaves of greenhouse plants have become more productive because carbon dioxide is no longer a limiting factor. It remains to be been seen how plants will change and evolve as the level of carbon dioxide in the air increases further.

Inside an atypical leaf

While there is a very definite 'typical' leaf, this is not necessarily the best-adapted form for all conditions. In the next few pages we look at the many adaptations to leaves that enable them to function in different habitats. Most of these adaptations are visible to the naked eye or with a hand lens, but one group of modifications is not visible because they take place inside the leaf and much of the change is at a molecular level.

The problem with RuBisCO

There is a big design fault in the middle of the photosynthesis process. To understand this, we first need to remind ourselves that photosynthesis as we know it first emerged at least 2.4 billion years ago, when the Earth's atmosphere contained very little oxygen but plenty of carbon dioxide. One of the enzymes involved in photosynthesis, and perhaps the most important, is ribulose bisphosphate carboxylase (RuBisCO to the word-count conscious, or *rbc* to the lovers of acronyms). This very effective enzyme has changed the Earth, but it evolved in different times, when carbon dioxide was at much higher concentrations than today. These high levels of carbon dioxide and lack of oxygen masked the problem that at temperatures below 26 °C (79 °F) this enzyme fixes carbon dioxide, but as temperatures rise above this threshold the enzyme reacts increasingly with oxygen instead and thus releases carbon dioxide. This is known as photorespiration and is an issue for many plants in hot environments, especially where water is in short supply.

The enzyme ribulose biphosphate carboxylase, or RuBisCO, is essential to photosynthesis, fixing carbon dioxide from the atmosphere, but it is also a limiting factor for plants growing in hot climates.

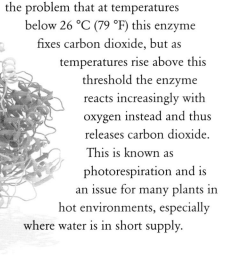

Evolution to the rescue

One solution to this flaw in the functioning of RuBisCO is to create a situation where the enzyme is subjected to elevated concentrations of carbon dioxide, thereby preventing oxygen from competing successfully for its attentions. This involves a mix of anatomical changes and neat biochemistry in leaves. The adapted anatomy is known as Kranz anatomy and is reminiscent of strands of electric cable running through the leaf (see box opposite). The sheathing layer of cells in these strands incorporates clever biochemistry that allows the cells to grab carbon dioxide even at high temperatures. The molecules that include the fixed carbon dioxide are then transported to the cells in the core of the strands, comprising normal mesophyll cells with lots of chloroplasts. Here, the carbon dioxide is re-released and the RuBisCO grabs it, because the elevated concentrations of the gas overcome the temperature-related problems. This is known as C4 photosynthesis, as opposed to the more common C3 photosynthesis. Nearly 80 per cent of C4 plants are grasses or sedges, but they are also found in more than a dozen families and include maize (*Zea mays*), sugar cane (*Saccharum* spp.), sorghum (*Sorghum bicolor*) and a number of species of *Euphorbia*.

Sugar cane (*Saccharum officinarum*) is grown commercially in tropical regions of the world and is able to optimise its productivity through C4 photosynthesis.

Sorghum (*Sorghum bicolor*) is the fifth-most important cereal crop in the world and is another plant that has evolved C4 photosynthesis.

Kranz anatomy in leaves – internal bundles of concentric layers of tissues – has evolved many times since plants first colonized the land. This is because RuBisCO, the enzyme at the centre of photosynthesis, only works well up to 26°C in our current levels of atmospheric carbon dioxide. The Kranz modification enables the plant to create a much higher concentration of carbon dioxide within the leaf and thus work around the design fault.

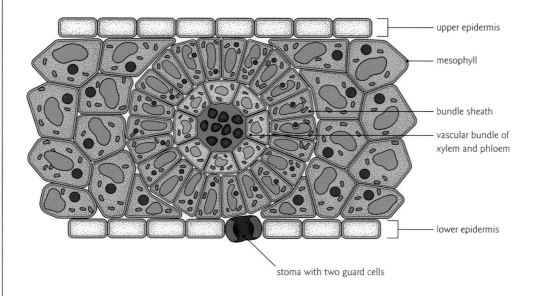

upper epidermis

mesophyll

bundle sheath

vascular bundle of xylem and phloem

lower epidermis

stoma with two guard cells

◁ A cross section through the leaf of a C4 grass showing the specialized Krantz anatomy, which is super-efficient at carbon dioxide capture at high temperatures. This anatomy is unique to all C4 plants and is an extraordinary example of parallel evolution.

High benefits but high costs

The only downside of the increased efficiency of C4 photosynthesis is that the plant has to invest more energy to survive, so it is only a benefit at high temperatures. However, it has been recognized that rice (*Oryza sativa*) could be a more productive crop if it were a C4 plant like sugar cane and maize. Researchers are working hard to achieve this conversion and, if they are successful, C4 rice could help to feed an ever-growing human population. C4 plants have evolved independently many times, so the question is: if natural selection has created C4 plants many times, then how difficult can it be to create them artificially?

Leaf colour

As we have already seen, the majority of leaves are green thanks to the presence of the green pigment chlorophyll. However, not all leaves are green, and those that are green are not always so, having a different colour in spring or changing hue in autumn.

⌃ The bright red young foliage of flame of the forest (*Pieris japonica*) may act as a deterrent to browsing herbivores.

⌃ The soft young leaves of *Photinia* would be a tempting feast for herbivores if it were not for the warning red colour.

Spring colour

Autumn colour is a well-recognized yet poorly understood phenomenon, exemplified by the legendary fall in New England and Hokkaido in northern Japan. And yet there are also many plants that have different-coloured leaves in the spring. For example, members of the genera *Pieris*, *Photinia* and *Derris* have young leaves that start off as various shades of red and then turn green later in the year. Is it a coincidence that these are all evergreen species? Why would a plant go to the expense of synthesizing a pigment that it later replaces unless the first pigment has a function and leads to an increase in fitness? How can these plants survive alongside others that do not produce this interim colour? No one knows the answers to these questions for sure, but it may be that the young leaves in these plants do not turn green until they are protected against herbivory by toxins or the development of a thick, unpalatable surface. As soon as they turn green, this is a signal to herbivores that they contain sufficient nutrients to make them worth eating. It has therefore been suggested that the red pigments protect the plants from herbivores, which have learnt that non-photosynthetic leaves contain fewer nutrients.

Autumn colour: a continuing enigma

While spring colour is a trait found only in evergreen plants, autumn colour is found only in deciduous species. However, it is variable, with the colours fluctuating from year to year depending partly on the weather during the preceding summer. The colour varies from species to species and sometimes between varieties of the same species, and it also varies with soil pH. There are two possible explanations for autumn colour, both of which may be true in different situations: either it is a meaningless artefact of necrosis, or it is a process that is advantageous to the plants. One common explanation is that as the green chlorophyll is broken down with the slowing of photosynthesis as day length reduces, the red pigments are revealed. In addition, some pigments are synthesized in the autumn, although the reasons for this have not yet been identified.

IN THE EYE OF THE BEHOLDER

Primates' eyes are more sensitive to the variations in green than any other colour in the visible spectrum, and female primates are particularly skilled at discriminating between the different tints, hues and shades of green. It is believed that this is because of the role females play in caring and foraging for their young. The ability to harvest only ripe fruit saves energy, and the colour of grass can also indicate the presence of water.

The autumnal rainbow of colours in deciduous plants is the result of the nutrient recycling and reclamation that takes place before their leaves are discarded.

One of the most spectacular autumnal leaf displays occurs in Hokkaido in northern Japan, where the conifers provide green to contrast with the stunning reds and oranges of the deciduous trees.

A reasonable explanation

One logical explanation for autumn colour is connected with the recycling of nutrients and good housekeeping of limited resources. As already mentioned, plants go to a lot of trouble and expense to take up and transport nutrients, in particular iron and manganese for photosynthesis. The ability to reclaim these nutrients before the leaves fall therefore seems sensible. It has been suggested that the pigments that characterize autumn colour protect the tissue in the dying leaf while the iron and manganese, and perhaps also nitrogen, are being reclaimed, and that they may be intermediate compounds in the recycling process. All we can say for sure is that we do not yet know why Nature puts on such a spectacular show in some years in some places.

ABSORBING OR REFLECTING?

Leaves are green because they reflect the wavelengths in light that we perceive as green, so in fact they are rejecting the green, throwing it back at the sun. The wavelengths that the plant uses are the reds and the blues, as shown in the graph below.

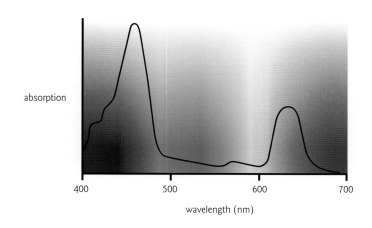

Water loss and stomata

$\left(\text{T}\right)$ranspiration, or the loss of water from the leaves of plants, is a necessary evil because the movement of water and nutrients from the soil and the roots to the leaves, where they are needed, is driven by the suction created by this water loss. Water movement in plants is therefore free of cost when water is abundant, but as soon as it is in short supply plants needs to conserve it by limiting transpiration, which is where stomata come in.

The fail-safe position

Stomata – the pores on leaves and other plant organs that allow gas exchange – generally open at dawn and close at dusk. However, as any engineer will tell you, when designing a system you try to ensure that if it stops working it defaults to the position that does the least damage. We see this principle in the design natural selection has favoured for regulating water loss from leaves. The stomata comprise an aperture bordered by two cells shaped like short sausages, called guard cells. Their mechanism for opening is very simple and can be understood by the analogy of two balloons. Take two uninflated cylindrical balloons and put a strip of sticky tape down one side of each. Place the balloons next to each other with the strips of tape touching. As you gently inflate the two balloons simultaneously, an oval-shaped aperture will appear between them.

In the case of the two guard cells of the stomata, they are inflated not with air but with water. Thus, the opening of the stomata is inextricably linked to the turgor of the leaf: if the leaf – and the guard cells with it – is flaccid, the stomata will automatically close, reducing further loss of water and the risk of irreparable damage to the plant.

This scanning electron micrograph shows a stoma on the surface of a tobacco (*Nicotiana tabacum*) leaf. It is open because the guard cells are inflated and turgid (magnification ×640).

When the guard cells are flaccid, the stoma is closed. Thus, when the plant is water-stressed, the stomata default to shut (magnification ×640).

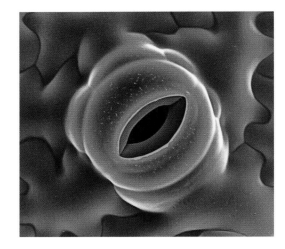

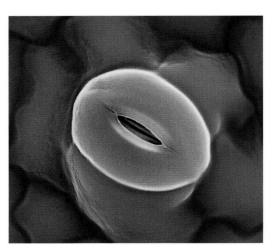

A HIDDEN BENEFIT

While transpiration is responsible for drawing water and nutrients up the stem, it also fulfils a more subtle function. When water evaporates, heat is lost; this is the principle behind sweating in humans. This latent heat loss can cool the leaves of some desert plants up to 15 °C (27 °F) below the ambient temperature, and without it, the leaves could be up to 20 °C (36 °F) above the ambient temperature. Of course, this mechanism can be exploited only if there is sufficient water available to the plant.

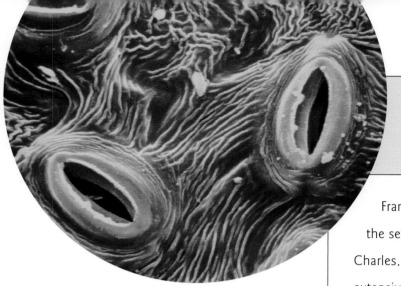

STUNNING STOMATA

Why not open stomata only at night?

If water loss is greatest during the day, when temperatures are highest, why don't plants close their stomata then and instead open them at night? The answer is that plants need high levels of carbon dioxide – which enters through the stomata – for the photosynthesis enzyme RuBisCO to be most effective. As we saw earlier (see page 154), C4 photosynthesis gets around the problem of high temperatures compromising the efficiency of the enzyme RuBisCO. Elevated temperatures are also often accompanied by water stress, and evolution has come up with another modification to the photosynthesis process to combat this, known as crassulacean acid metabolism (CAM). Here, carbon dioxide fixation and the work of RuBisCO are not separated physically by Kranz anatomy, but instead are separated temporally. CAM plants keep their stomata closed during the day, when water loss would be at its highest. Instead, the stomata open at night and carbon dioxide is grabbed using the same enzyme we saw in C4 plants. The molecule containing the fixed carbon is stored in cell vacuoles until the sun comes up in the morning. At this point, the stomata close to reduce water loss and the carbon dioxide is released at a high concentration into the air spaces in the leaf, enabling RuBisCO to work efficiently at high temperatures and photosynthesis to be most effective.

Francis Darwin (1848–1925), the seventh child and third son of Charles, was also a naturalist. He wrote extensively about stomata and in 1898 delivered his seminal paper on the subject to the Royal Society in London. Stomata evolved soon after plants colonized the land 470 million years ago. They are essential because they enable the plant to control water loss, which was the first major problem that had to be overcome by the early land plants.

△ The density of the stomata over a leaf surface varies between plants in different habitats. The drier the conditions, the fewer the stomata (magnification ×1200).

▽ A cross section of leaf tissue showing an open stoma with an air space below it, from which gases can be absorbed by the surrounding cells.

◁ Francis Darwin, the third son of Charles Darwin, devoted much of his time to the study of stomata.

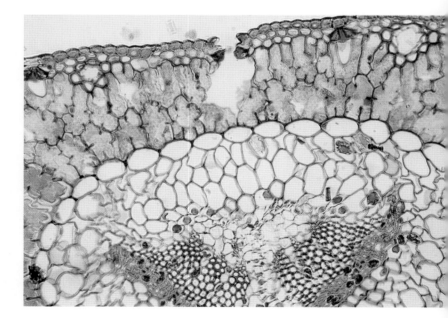

Water conservation and storage

ater is vital for life on Earth and should not be taken for granted. The leaves of plants show many modifications and features that ensure a prudent balance between photosynthesis and water loss. These modifications come into play on different timescales depending on the permanence of the water shortage threat.

Stomatal options

The opening and closing of stomata can change the amount of water a plant loses through transpiration almost instantly. However, the amount of water lost can be reduced if the stomata are situated at the base of a sunken pit in the leaf, which reduces the drying effect of the air passing over the leaf surface. Stomata tend to be smaller on plants that are permanently subject to water shortages. There is also a very clear relationship between the density of stomata and latitude: the further a plant grows from the Equator, the fewer the number of stomata on its leaves. There are probably several reasons for this, of which water conservation is just one.

Superficial solutions

A very simple way to reduce the volume of water lost from leaves is to have fewer, smaller leaves. The cross-sectional shape of a leaf can also be modified to reduce the surface-to-volume ratio. Another common solution to water stress is to coat the leaf with a thicker waterproof cuticle than is found on the surfaces of leaves where water is abundant. The leaves of evergreen species tend to have a thicker cuticle than the leaves of deciduous species, which instead drop their leaves to reduce water loss during a predictable, seasonal period of water stress. Reflecting the sunlight is a strategy used to reduce the absorption of heat by the leaf, and hence water loss – shiny, silvery-grey leaves are more common in areas of water stress, especially high altitudes. Hairs on the surface of a leaf can achieve two aims: silvery hairs reflect light away from the leaf, and the same hairs can reduce water loss through evaporation by slowing the speed of the air moving across the leaf surface. This is a particularly common adaptation seen on the undersurface of leaves.

(∧) Edelweiss chamomile (*Anthemis* sp.) employs a common strategy used by plants where high levels of ultraviolet light could damage cells and there is a danger of excessive water loss: its leaves are covered with silver-grey hairs that reflect light away from the surface.

(>) Silver-grey hairs on the leaf surface of plants such as silver sage (*Salvia argentea*) reduce the air movement across the leaf surface resulting in a saturated micro-environment that reduces evaporation from the leaf surface.

Living off your savings

A common problem facing plants is unreliable and irregular rainfall. Storing water is therefore one way to make it available for photosynthesis even when it has not rained for days or even weeks. Water can be stored in a number of organs in the plant, including the leaves. These succulent leaves are found in a wide range of plant groups, including *Crassula*, *Agave* and *Aloe* species. There is, however, a problem associated with storing water, which is that it is an attractive temptation for animals. It is therefore very common for plants that store water to have armour and spines to deter herbivores from stealing their supplies.

The succulent leaves of *Agave* species are made less palatable to herbivores with the addition of sharp spines.

The jade plant (*Crassula ovata*) stores water in its succulent leaves in order to survive periods of drought.

PLUMPING FOR SUCCULENTS

If rain is infrequent, and falls in large volumes when it does eventually arrive, as a plant you need to have storage tanks with very waterproof walls. Not surprisingly, this is what we see inside many of the world's succulent plants, which have an impervious outer layer.

Aloe vera (*Aloe vera*) leaves are essentially water-storage vessels and have thick skins to prevent the water from escaping.

One of the world's worst aquatic weeds, water hyacinth (*Eichhornia crassipes*) employs buoyancy tanks in its petioles to keep it afloat on rivers.

Floating leaves

Plants with leaves that can float on water are living in a potentially comfortable environment, because two of the basic requirements for life as a plant – water and light – are abundant. However, they have to stay afloat to maximize the amount of light they receive, and they have to prevent water from entering and diluting the cellular contents. The surrounding water may lack nutrients, so plants might need to fix nitrogen, or it may be so full of neighbouring plants that these need to be bulldozed aside (see box on page 169).

Airbags

A common method for keeping leaves afloat is for some part of the plant to become an air sac. This can be achieved by evacuating all of the cell contents, leaving a cellulose box either containing a waterproof sac or several cellulose boxes within a waterproof sac. The leaf petioles of the Amazonian water hyacinth (*Eichhornia crassipes*) and Asian watergrass (*Hygroryza aristata*) are modified in this way. Another method of 'capturing' air is between hairs on the leaf surface. This tactic is employed by the water lettuce (*Pistia stratiotes*) and *Salvinia natans*, a floating fern with the enchanting common name of water butterfly wings. It should be noted that many of these plants are not enchanting at all, and have invaded many countries following the movements of humans, overwhelming habitats and outcompeting native species.

Asian watergrass (*Hygroryza aristata*) also relies on trapped air to keep it afloat, but in this case the air is trapped in its leaf bases.

Water lettuce (*Pistia stratiotes*) keeps its head above water by trapping air between the hairs on its leaves.

AN EXTRAORDINARY ASSOCIATION

Most of the plants listed in the main text have been involved in biological invasions and one of them carries a secret device. Living inside the fronds of the mosquito fern (*Azolla pinnata*) are colonies of the cyanobacterium *Anabaena azollae*, which is capable of fixing nitrogen and hands some of it over to the plant in return for shelter. What is extraordinary about this is that, 1.2 billion years ago, the chloroplasts in the *Azolla* fronds were also cyanobacteria. Are we witnessing evolution repeating itself? Will the *Anabaena* undergo an endosymbiotic event and become a nitrogen-fixing plastid in the cells of the *Azolla*? And further more, could this be promoted in the leaves of any of our crop plants?

◁ The mosquito fern (*Azolla pinnata*) is light enough to allow it to float without any additional devices.

Small plants

Air sacs can also be located within the leaf itself, but there is a conflict here because the leaf needs to take in carbon dioxide, but if it opens its stomata water will also enter. Duckweed (*Lemna gibba*) has stomata on the upper surface of its leaves, but there is some evidence that these are non-functional, remaining either continuously open or closed.

The same is true for the floating mosquito ferns (*Azolla* spp.), whose tiny fronds are divided into a lower floating lobe on the water surface and an upper aerial photosynthetic frond. Because these fronds and leaves are small, like those of mosses, gases may enter directly by diffusion across the leaf surface rather than through stomata into an internal air space.

▽ A convenient hiding place for frogs and other aquatic animals, duckweed (*Lemna gibba*) quickly covers the water surface and stays afloat due to air sacs in the leaves.

Leaves that move

Ⓗaving said on more than one occasion that plants can be described as green, immobile organisms in order to distinguish them from the other major groups of living organisms, it might seem contrary to claim now that leaves are not, in fact, motionless. One of the most famous plants in the world – the Venus flytrap (*Dionaea muscipula*) – has reached its status by virtue of the fact that its leaves actually do move, and faster even than the flies they trap (see pages 166–67).

⌃ Cornish mallow (*Lavatera cretica*) tracks the sun's path across the sky so as to optimise photosynthesis.

Follow that sun!

The movement of leaves was yet another topic studied in depth by Charles Darwin. It was he who coined the term diaheliotropism to refer to the movement of leaves during the day as they track the sun. The understanding of this sun tracking has since been refined, and we now know it as cupping or paraheliotropism. This is where leaves maintain a constant angle to the sun's rays, thereby optimizing the absorption of light and reducing the proportion that bounces off them. These movements have been observed in a wide range of plants, including Cornish mallow (*Lavatera cretica*), upland cotton (*Gossypium hirsutum*) and desert lupine (*Lupinus arizonicus*), and it is easy to see why they would be a worthwhile investment of the plant's energy if the benefit derived from the increase in photosynthetic productivity is greater than the cost of moving the leaves during the day and back again at night.

⌄ Desert lupine (*Lupinus arizonica*) is another species that can move its leaves to optimise photosynthesis. Throughout the day the leaves are kept at a constant 90 degrees to the sun's rays.

⌄ Upland cotton (*Gossypium hirsutum*), an exceptionally valuable crop plant, can increase productivity by tracking the sun to optimise photosynthesis.

POTASSIUM IS THE KEY

It is currently thought that the rapid seismonastic movements (non-directional responses to touch) of the sensitive plant (*Mimosa pudica*) and the slower sun-tracking movements and night movements of other species are due to the same molecular mechanism. This sees a sudden loss of turgor in the cells of the leaf petiole, resulting in their asymmetrical collapse and hence a bending of the petiole. It is believed that the active and rapid movement of potassium ions is involved in the process, and that the same mechanism may be responsible for the opening and closing of stomata.

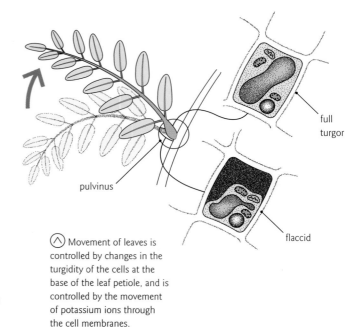

full turgor

pulvinus

flaccid

⋀ Movement of leaves is controlled by changes in the turgidity of the cells at the base of the leaf petiole, and is controlled by the movement of potassium ions through the cell membranes.

Frightening the herbivores?

The fact that the leaves of lupines move as they track the sun should not come as a complete surprise, because some members of the legume family (Fabaceae), to which these plants belong, are notorious for leaf movements. The telegraph plant (*Codariocalyx motorius*, syn. *Desmodium gyrans*) has two small leaflets and one much larger leaflet in its trifoliate compound leaves. It is thought by some that the incessant flickering movements of the smaller leaflets help the plant to 'decide' if it is energetically worthwhile moving the much larger leaflet to a more beneficial angle to the sun. Others think that the movements are aimed at deterring herbivores. This latter explanation is also a commonly quoted reason for the rapid movement of the leaves of the sensitive plant (*Mimosa pudica*). However, why herbivores do not learn that the plant is just trying to fool them has yet to be explained.

Thank you and good night

The stomata of most plants are located on the underside of the leaves (or leaflets in the case of compound leaves). It is not uncommon for plants in the legume family to fold their leaves in half at 6 pm and then open them up again at 6 am the following day. In this way, the undersides of opposite leaflets are brought together, further reducing the possibility of water loss through poorly shut stomata. This is such an advantageous strategy that even plants facing no water stress will fold their up leaves in the evening – *Calliandra* species do so even if the plants are being grown in perfect conditions in a greenhouse.

◑ Any herbivore intending to take a mouthful of the sensitive plant (*Mimosa pudica*) is likely to be scared off when its intended meal suddenly moves. The left image shows the natural position of the leaves. The right picture shows what happens when a leaf is touched.

Carnivorous leaves

The leaves of carnivorous plants, which generally grow on nutrient-poor substrates, are some of the most highly modified in the plant kingdom. These modifications, including the enzyme soup that digests the prey, all come at a high cost, but there is a higher benefit: the ability to harvest nitrogen and potassium from the animals that are trapped. It is important to note that carnivorous plants do not digest animal prey to derive energy, but continue to rely on photosynthesis for this.

The Venus flytrap (*Dionaea muscipula*) is an 'intelligent' plant, being able to count, remember and discriminate between objects.

Unsuspecting prey investigating the purple pitcher plant (*Sarracenia purpurea*) fall victim to its slippery surfaces and end up as a nutrient supplement to the plant.

The Venus flytrap

The two leaves that form the trap of the Venus flytrap snap shut around prey probably as a result of a collapse of turgor in the cells in the hinge. This may well involve potassium, as in other types of leaf movement, but in addition there is clearly an electrical component. Importantly, memory is also involved, because it prevents the trap shutting when no prey is present.

The surface of the leaves are covered in special hairs called trichomes. When one of these is touched, a specific amount of a chemical is released into the leaf. Theoretically, this chemical can trigger an electrical pulse that will shut the trap, but not enough is released by just one touch of the trichome to trigger this pulse. Following the release of the chemical, its concentration declines slowly. However, if the trichome is stimulated again soon enough after the first stimulus, the second release of the chemical is enough to push its concentration over the threshold necessary to trigger the electrical pulse and thus shut the trap. The leaf of the Venus flytrap therefore not only has a memory, but it can also count, and it can discriminate between an inanimate object (such as a drop of water or a falling twig) and a living animal that might become its lunch.

Evolution repeating itself – again

There are several basic ways in which leaves are modified to attract, catch, retain and digest animals. Snap traps are the most spectacular, but there are also pitfall- or pitcher-type traps

and flypaper-type mechanisms. Pitfall traps have evolved independently at least three times (in *Nepenthes*, *Cephalotus* and *Sarracenia*) and flypaper mechanisms at least four times (in the sundews, *Roridula*, *Pinguicula* and *Byblis*).

Pitfalls

The first stage in the construction of a pitfall trap is to create the 'bucket' by folding the leaf towards its upper surface. This is not a trick confined solely to carnivorous plants, and has been recorded happening spontaneously in the leaves of the common box (*Buxus sempervirens*) and garden croton (*Codiaeum variegatum* var.

pictum 'Nepenthifolium'). The next stage is to secrete a digestive fluid into the trap. Further refinements include making the sides of the trap slippery with wax (although a thick waterproof cuticle could already be sufficiently shiny) and secreting nectar from extra-floral nectaries near the rim of the trap to attract insect prey.

Flypapers

The construction of the flypaper-type trap of the sundews and other species is more straightforward than pitfall traps because hairs are a very common feature of leaf surfaces and the hairs of carnivorous plants are simply glandular hairs that secrete one of two substances. The first of these is a sticky substance that holds onto the insect. The more it struggles, the more the animal is coated with the gloop, until it is so covered that it dies of suffocation or exhaustion. The second substance comprises enzymes that break down the animal prey to release the nitrogen and phosphorous it contains. It appears that the glandular hairs on the different groups of flypaper plants are very different in structure, so this is a good example of convergent evolution in which different structures produce the same result for the plants. The leaves of some but not all species of flypaper plants curl around the dead animals to improve contact between the leaf surface and the prey.

⊲ The elaborate pitcher traps of *Nepenthes* plants commonly trap insects, such as ants and flies, but some species are large enough to trap small mammals, such as rats.

BEAUTIFUL BUT DEADLY

The sun catching the drops of sticky fluid secreted by the hairs of sundews (*Drosera* spp.) earned the plants their common name. But the many *Drosera* species that possess these photogenic hairs are actually attracting dinner. The sticky gloop is there to detain and often suffocate the insect as it struggles to escape.

⊳ Once trapped on the sticky leaves of the common sundew (*Drosera rotundifolia*), insects have no chance of escape and suffocate under the 'glue'.

Leaves for defence

(L)eaves are of almost inestimable importance in supporting and generating the Earth's biological diversity. In fact, they are so important as food for animals that it is easy to forget that plants did not evolve leaves to feed animals but to feed themselves. Once plants started producing nutritious leaves and animals started eating them, thus began an arms race in which leaves evolve defences and animals evolve counteractions to these. In addition to defending themselves, some leaves and their associated stipules have evolved into structures to defend the whole plant.

Leaves defending themselves

Perhaps the first conscious interaction that children have with plants is when they fall foul of a stinging or common nettle (*Urtica dioica*), poison oak or poison ivy (*Toxicodendron* spp.), or any of the other plants that secrete or contain chemicals that caused a mild to very serious reaction on our skin. In the case of the nettles, it is the plants' modified and adapted hairs that provide its defence. One of the most painful activities in an English garden is clearing up underneath holly trees (*Ilex aquifolium*) or mahonias (*Mahonia* spp.). Both of these have leaf margins furnished with spines that can penetrate every gardening glove on the market. Many leaves defend themselves more passively by producing toxins or having a very thick cuticle that makes them hard to digest.

(<) Under a scanning electron microscope, the modified hairs that provide the vicious armoury of the stinging nettle (*Urtica diocia*) can clearly be seen.

(∧) The seemingly innocuous stinging nettle (*Urtica diocia*) delivers painful and persistent stings to anyone who inadvertently brushes against the plant.

CACTI VS. SPURGES

One of the most quoted examples of convergent evolution is the striking similarity between the cacti found in South America (e.g. *Cereus* spp.), and the spurges (*Euphorbia* spp.) found in north, eastern and southern Africa. Both groups have evolved the same strategies for living in areas of very low, irregular rainfall: lose the leaves, become succulent and photosynthesize in the stem. The leaves of the cacti have not actually been lost, but instead have been modified into spines to deter herbivores. The spurges have taken a different route. They have jettisoned their leaves unless there is a lot of rainfall, and their stipules have developed into hard spines instead.

Leaves defending the plant

Leaves that defend themselves with spines and irritant hairs are also protecting the whole plant. However, in some plants the leaf has become modified into a structure whose primary purpose is to inflict injury of some form on grazing animals. The legume family has already been mentioned several times in this chapter, perhaps because this large group of 17,500-plus species is more 'imaginative' in its use of leaves than any other. If you want to see an example of this, inspect a European gorse bush (*Ulex europaeus*). It is very difficult to work out whether the short, pointed stems that make up much of the plant are leafless stems, pointed stipules or pointed leaf rachides. It is quite possible that all three are modified into the spines.

ⓔ European gorse (*Ulex europaeus*) forms an impenetrable bush, with incredibly hard, sharp spines. Almost every part of this plant becomes modified into a spine at some point.

ⓣ Cacti such as these *Cereus* species are armed against attack with spines derived from leaves and shoots.

ⓥ Euphorbias, such as *Euphorbia canariensis*, are similarly armed, but their defensive spines are modified stipules.

VEGETABLE BULLDOZER

The upturned thorny edge of leaves of water lilies in the South American genus *Victoria* has confounded botanists for many years. It has been suggested that it helps to keep water and/or animals off the leaf surface. However, when you see these plants in a crowded situation it is clear that the upturned edge works like a very efficient bulldozer, pushing other leaves out of the way or simply mounting the competition.

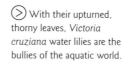

ⓢ With their upturned, thorny leaves, *Victoria cruziana* water lilies are the bullies of the aquatic world.

Leaves for climbing

(L)eaves have yet another role: helping plants to climb. Plants grow upwards towards the light they need for photosynthesis in an effort to outcompete their neighbours. Some plants become woody and stand up on their own unaided, while others exploit these woody plants by climbing up them. There are two basic strategies for climbing: twining and twisting the stem around the 'host' plant, or climbing through the host and every now and again grabbing hold using tendrils or hooks. This latter strategy involves leaves.

⌃ Brambles (*Rubus fruticosa*) use their vicious thorns to hook themselves in place as they scramble through and over other plants.

⌃ The backwards-facing thorns on the underside of the leaves of the cockspur coral tree (*Erythrina crista-galli*) serve not only as a defence mechanism, but also as an aid for support.

Hooking up

Hooks are very commonly found on the stems of such plants as roses (*Rosa* spp.) and brambles (*Rubus* spp.). The mechanics of this strategy are very simple. The downwards- or backwards-facing hooks pose no obstacle when a shoot is growing through other plants. However, when the weight of the climbing plant starts to pull the whole plant down, the hook latches onto anything in the vegetation, thus preventing a fall. Many plants have such hooks on their leaves, on both the petioles and leaf veins. One of the most vicious examples is the South American cockspur coral tree (*Erythrina crista-galli*), which has unbelievably sharp little thorns on the central vein on the underside of its leaves. These are particularly effective at preventing humans from stealing the seedpods. You reach into the plant to pick a pod blissfully unaware of the thorns, but as you retrieve your hand and the seeds, your hand is sliced and diced by the thorns. This is yet another example of one feature having two functions – in this case protection and climbing.

Twisting petioles

The humble leaf petiole is usually just seen as a support for the leaf, if it is considered at all. But another function some petioles cover is holding up the whole plant. One of the three functions of the leaves of carnivorous pitcher plants in the genus *Nepenthes* is to support the plant (see image on page 167).

IN A DISSIMILAR VEIN

Species in the legume family perform more tricks with their leaves than any other group of plants, and one of these is climbing. Here, it is not the leaf petioles that do the twining and hanging on, but modified leaflets that twist and curl, attaching themselves very tenaciously to anything they come into contact with – including each other. As this picture shows, the leaves in fields of peas (*Pisum sativum*) form a tangled alliance as they hold each other up.

ⓥ The leaflets of compound leaves can become modified into tendrils that act like grappling irons, twisting around the supporting plant to keep themselves upright.

leaflet

tendril

stipule

ⓒ When plants such as peas (*Pisum sativum*) grow close together, they become self-supporting as their tendrils become entwined.

This same function is common to many other climbing plants, and is very clearly seen in several *Clematis* species. Here, the leaf petioles, and sometimes the petioles of leaflets in compound leaves, twist around any cylindrical object they come into contact with. It is assumed that the mechanical/physical contact on one side of the petiole leads to uneven growth, resulting in the petiole curling around the twigs of the host plant. It is also assumed that there is a plant growth hormone involved in this response. Yet again, Charles Darwin investigated this experimentally.

ⓢ Gardeners will know the tenacity with which the leaf petioles of *Clematis* species relentlessly cling to plant supports.

Leaves that are stems

Ⓛeaves are not defined by their function, shape or colour. However, we all know that generally speaking leaves are photosynthetic organs that are oval-shaped and green. This means that when we see a flat, leaf-shaped green photosynthetic structure on a plant, we naturally assume it is a leaf – although sometimes we will be wrong. The way to avoid this mistake is to check carefully whether there is anything amiss, such as flowers and fruits growing out of the middle or along the side of the 'leaf'.

ⓒ Leaf or stem? The presence of flowers on this green 'leaf' of *Phyllanthus angustifolius* betrays the actual nature of the structure, which is a stem.

Ⓥ Less immediately obvious than those of its relative *Phyllanthus angustifolius*, the green structures of *P. pulcher* are still stems and not leaves.

A bit on the side

The genus *Phyllanthus* is a large group of perhaps 1,000 or more species. At first sight, the plants look quite normal, with long simple leaves (e.g. *P. angustifolius*) or pinnate leaves (e.g. *P. pulcher*). However, for part of the year the leaves look odd. This is because the edges of what were assumed to be leaves are decorated with flowers. This raises two questions. First, these cannot be leaves, but what are they? And second, what has the plant done with its leaves? The answer to the first question is that these 'leaves' are actually flattened photosynthetic stems known as cladodes. And the answer to the second is that the leaves are where they should be, that is they are appendages on the shoots that terminate in flowers. The difficulty arises from the fact that these shoots are very small and the leaves are very, very small.

Sweeping aside objections

Butcher's broom (*Ruscus aculeatus*) is a common shrub native to the woodlands on the northern Mediterranean shore, and a popular ornamental plant in temperate gardens because it has relatively large red fruits. These fruits are situated right in the middle of what any casual observer would have identified as leaves. However, as in *Phyllanthus* species these are cladodes, although their similarity to leaves is quite unnerving.

A BORN MIMIC

On the edges of the temperate rainforest of the Canary Islands – its iconic laurel forest – is a plant whose stems twine up the trees growing here. This plant, *Semele androgyna*, has what appear to be large pinnate leaves, right down to the fact that the leaflets have petioles. Again, however, this is a deception, because the leaflets are cladodes. The deception collapses when the plant blooms, when little shoots with tiny brown leaves grow from the leaf margin and terminate in small white flowers. The flowers subsequently develop into fleshy green fruits.

∨ Finding fruit on the 'leaf' margin of *Semele androgyna* helps to identify the structure as a stem.

▷ Without the flowers of *Semele androgyna*, it would be easy to mistake the pinnate 'leaves' for true leaves.

◁ Butcher's broom (*Ruscus aculeatus*) is another plant with stems masquerading as leaves. The position of the fruit on the 'leaf' is the key feature that identifies the structure as a stem.

Stems that are leaves

We have already seen that some plants – especially members of the legume and cactus (Cactaceae) families – have leaves that are modified into short, pointed spines to help defend the plant from herbivory. Plants in the legume family are true masters of inventiveness when it comes to modifying leaves into stem-like structures, the most extraordinary example of which even keeps giraffes at bay.

The acacia, the ant and the giraffe

Acacia trees (*Acacia* spp.) are one of the iconic plants of the Kenyan savannah. As such, they represent an attractive food supply for giraffes. However, young giraffes in particular are very sensitive to the bite of *Crematogaster* ants. In order to recruit these ants to protect it from grazing by giraffes, the whistling thorn (*Acacia drepanolobium*) provides the insects with lodging and provender. The lodgings come in the form of swellings at the base of the thorns, and the provender is provided by extra-floral nectaries. The trees that employ the ants as their minders have thorns that are shorter than those on the ant-free trees, which suggests that the cost of keeping the ants can be offset by having to produce fewer physical defences. This is another example of plants carrying out cost–benefit analysis.

What you are looking at?

When we try to classify plants and group them with their closest relations (from an evolutionary point of view), we compare the structure of the two organisms. To do this successfully, we must be sure that we are comparing like with like. This is generally difficult, but some families of plants have modified their appearance in such a way as to confuse early taxonomists. The legume

⌃ ⌄ The whistling thorn (*Acacia drepanolobium*) enlists the support of ants to see off browsing giraffes by providing housing for its army of helpers.

EXTRA-FLORAL NECTARIES

The leaves of acacias are well known (in botanical circles at least) for the production of extra-floral nectaries (see main text). Nectar is a common attractant and/or reward employed by flowers that are pollinated by insects or other animals. However, plants exploit animals in other ways, too, and nectar is a reward for all purposes.

The northern catalpa (*Catalpa speciosa*) has extra-floral nectaries on its leaves that it uses to attract ants, ladybirds and other predacious insects. This benefits the tree because the predators eat the eggs and larvae of the *Ceratomia catalpae* hawkmoth, which is a major herbivore on the tree.

⊗ Northern catalpa (*Catalpa speciosa*) rewards its ant helpers by providing a sweet, nutritious treat for them, produced by extra-floral nectaries on the leaves.

family of plants has a particularly large number of these confusing modifications, and those of the genus *Acacia* could form the subject of many books on their own. Of relevance here is the fact that acacias often seem to have simple leaves, in a family that is well known for having compound leaves with well-developed stipules. However, it turns out that these 'leaves' are in fact phyllodes, which are flattened petioles that function like the leaf blade of a true leaf. But that is not the end of the trickery, because some acacias have leaf-like structures that are in fact cladodes or flattened photosynthetic stems. The moral here is: never judge a plant part by its function.

⊗ Leaf, stem or leaf petiole? Members of the legume family such as *Acacia holosericea* have 'leaves' that are deceptive, with both stems and leaf petioles resembling true leaves.

Scented leaves

(L)eaves bring many different scents to our gardens, from the beautiful lemon verbena (*Aloysia citrodora*) to the repugnant giant honey flower (*Melianthus major*). The reasons why these plants go to the expense of producing odiferous molecules are many and varied, and some have nothing to do with the scent we smell.

⌃ Although an attractive foliage plant, the leaves of the giant honey flower (*Melianthus major*) have an unpleasant smell when handled, resulting in one of its common names being touch-me-not.

Some like it hot

Leaves that smell often do so because they contain aromatic oils. These oils are highly flammable, and some plants appear to have become extra flammable in order to wipe out any competition when there is a wildfire. These plants get away with this seemingly superficial suicide strategy only if they have a robust mechanism for surviving a hot fire. They do this by regrowing from buds on underground storage organs called lignotubers, as seen in *Eucalyptus* species. The same plants may use the same oils to lower their temperature for photosynthesis, by evaporating the oils rather than precious water. The latent heat of evaporation of oils is higher than that of water and so more heat is lost, thus reducing the heat of the plant further. An oily coating can also be waterproof, and so these oils can help to reduce water loss in this way. Furthermore, the oils can act as a sunscreen, helping to reduce damage to the plant from ultraviolet rays.

⌄ Bushfires trigger epicormic regrowth from the base of a eucalyptus tree that enables it to survive these seemingly extreme but entirely natural events.

Protection

Just like animals, plants are attacked by pathogens and predators, but as they are sedentary they do not have the option to run and hide. This means that they must either keep their defences up at all times, or have a way of sensing when their neighbours are being attacked so that they can pull up their chemical drawbridge quickly. Some plants (e.g. common gum cistus, *Cistus ladanifer*) use their aromatic oils as a deterrent to herbivores such as molluscs. Other oils are powerful antibiotics, which the plants (e.g. Greek oregano, *Origanum heracleoticum*) use to keep fungal infections at bay – something we humans exploit. When the leaves of some plants are damaged by herbivores, they may emit chemical signals such as jasmonates (e.g. Spanish jasmine, *Jasminum grandiflorum*). It has been shown that other plants use these signals as a cue to produce repellent chemicals in order to protect themselves from similar attacks. The final way plants use the chemicals in their leaves to protect themselves is by making their leaves toxic to the germinating seedlings of other species. An infamous example of this is the common rhododendron (*Rhododendron ponticum*), whose leaf litter is so rich in tannins that almost nothing can grow through it.

The eastern cyclamen (*Cyclamen coum*) is one of the few plants that is able to withstand the tannins in common rhododendron (*Rhododendron ponticum*) leaf litter that prevent the growth of most other potentially competitive plants.

ATTRACTANT ODOURS

It is suspected but not yet proven that the scents produced by the leaves of some plants are part of the mechanism for attracting pollinators. The logic of this can be seen in rosemary (*Rosmarinus officinalis*), whose flowers are quite insignificant and yet the plant is normally smothered by bees.

Plant reproduction

All living organisms reproduce. The earliest cells, like most modern bacteria, simply split into two, with each new individual receiving a copy of the parent cell's DNA. Many plants, and some simple animals, can also reproduce asexually by splitting in two or budding off new individuals.

Asexual reproduction can rapidly increase the number of individuals in a population – an advantage in some situations, but such genetically identical populations are ill-prepared to adapt to new threats over the long run. Genetically diverse populations, on the other hand, are more likely to harbour individuals that can survive a new disease or other environmental challenge.

Genetic novelty arises through mutation, and organisms in which mutations can be rapidly shared, mixed and combined with other mutations in a population have an advantage over organisms that cannot. Sexual reproduction achieves such mixing.

Plants reproduce sexually through the fusion of sperm and egg, as in animals, but with the major complication that plants cannot move. The evolutionary history of plants has largely centred around the varied and sophisticated ways that have emerged to move sperm from one individual to another, culminating in the spectacular and varied phenomenon of pollination in flowers.

Asexual reproduction

Suppose you could take one of your fingers, provide it with nutrients, and have it grow into a copy of yourself. This common science fiction theme is approaching the realm of reality, although it is clearly not something that higher animals normally do. Splitting or budding is practised only by very simple invertebrates, such as rotifers, sea anemones and corals. For many plants, however, it is a routine part of life. Such asexual, or vegetative, reproduction allows a population to increase rapidly in size, but all individuals are genetically identical.

The floating pennywort (*Hydrocotyle ranunculoides*) forms invasive colonies, spreading across water or damp soil via long internodes. The nodes consist of a leaf, a cluster of roots and a bud for further branching.

This bamboo grove in Kyoto, Japan, is part of a clone that originated from a single seed. It has spread outward over many years via branching underground rhizomes that periodically turn upward to form the green culms.

Strawberry plants (*Fragaria* spp.) may form extensive clonal colonies via specialized elongate runners (very long internodes with terminal buds), which produce new plants where they touch down.

Plant growth units

Plant growth by nature involves the production of repetitive units consisting of short lengths of stem (internodes) with attached leaf and bud (at nodes). Such units may accumulate to form a massive tree, but for plants that stay close to the ground, they also produce roots and form extensive colonies. This is the routine growth pattern in most monocots, including grasses, lilies, irises, gingers and orchids, as well as herbaceous eudicots such as the water pennyworts (*Hydrocotyle* spp.) and strawberries. A colony of bamboo, a giant member of the grass family, may consist of thousands of genetically identical stems produced from an extensive underground rhizome system. In a genetic sense, such a colony is one gigantic individual, but when these colonies become fragmented, new 'individuals' emerge. It is easy for horticulturists to exploit these natural growth patterns to multiply such plants rapidly.

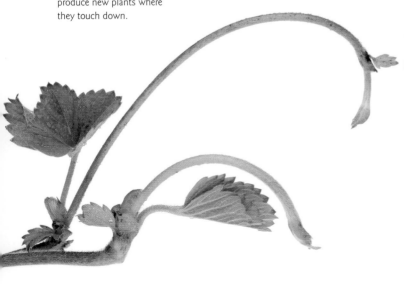

Adventitious buds

The quaking aspens (*Populus tremuloides*) of the Rocky Mountains produce multiple trunks, not from an underground stem system, but from roots growing horizontally just below the soil surface. Buds that develop on roots, the lower parts of tree trunks or other unusual locations are called adventitious buds. Aspens are unusual, however. Most woody plants are not naturally clonal, but with the application of the right hormones they can be induced to root from cuttings. Modern tissue-culture processes can induce even more reluctant plants to propagate vegetatively.

New plants from leaves

Other plants produce embryo-like bodies in more unusual ways. The miracle leaf (*Bryophyllum pinnatum*) produces 'babies' from adventitious buds along the edges of its leaf. Some other succulent plants, such as African violets (*Saintpaulia* spp.), begonias (*Begonia* spp.) and *Sansevieria*, can also be induced to produce new plantlets from cut leaves. Daylilies (*Hemerocallis* spp.) and some members of the ginger family (Zingiberaceae) will occasionally produce little plantlets in their inflorescences after the flowers have faded. Mosses and liverworts frequently produce clusters of embryo-like bodies called gemmae, which can separate from the parent and produce new plants.

Plants from unfertilized flowers

Some plants can produce new individuals from the reproductive organs of flower parts without fertilization, typically from tissues around the egg, rather than from the egg itself. This is known as agamospermy or apomixis. The common dandelion (*Taraxacum officinale*) usually reproduces this way, making it a highly successful cosmopolitan weed.

⌃ The individual dry fruits of the dandelion (*Taraxacum officinale*) are equipped with a parachute-like tuft of hairs that enables wind dispersal. Each dry fruit contains a single seed that developed without fertilization.

SELF-FERTILIZATION

Although not strictly asexual, some plants employ self-fertilization to produce large numbers of individuals, particularly to colonize disturbed or unstable habitats. Some species of sundew (*Drosera* spp.) that inhabit fluctuating pond edges do this. Most of these plants will, however, occasionally outcross, introducing new genetic material into the population.

⟨ The succulent leaves of the miracle leaf (*Bryophyllum pinnatum*) bear tiny plantlets along their edge that can detach and form new individual plants. The adventitious plantlets form directly from vegetative tissues of the leaf.

Sexual reproduction in green algae

The ancestors of land plants were green algae (Chlorophyta). Like the majority of aquatic organisms, most green algae reproduce sexually by releasing sperm cells directly into the water. Union with a nearby egg, held on its parent plant, results in a diploid zygote (see page 24). The zygote undergoes meiosis to release special dispersal cells called zoospores. Zoospores are also motile, resembling sperm cells, but are programmed to form new individual plants upon arrival at an appropriate location, rather than seek out eggs.

Sperm cells follow chemical gradients

Algal sperm cells are equipped with flagella – commonly just two at the leading end of the cell – and contain chloroplasts and mitochondria to fuel their locomotion. Eggs are held by the parent plant, often within specialized cells or simple chambers, and emit chemical attractants. Sperm cells swim toward eggs, following a concentration gradient of the attractant.

Zoospores and long-distance dispersal

The fusion of sperm and egg results in a diploid zygote. In the green algae, the zygote may remain dormant for a short time, then undergo meiosis, resulting in haploid cells (see diagram). These haploid cells take the form of zoospores, which often resemble sperm cells but are adapted to swim greater distances, and without regard to the chemicals emitted by eggs. When arriving at a suitable habitat, the zoospore forms a new multicellular individual. These adults then grow, photosynthesize and eventually produce gametes (sperm and egg cells) for another round of sexual reproduction. If zoospores from genetically different parents arrive and develop within close proximity, sperm cells are likely to fertilize genetically different eggs, mixing genes and maintaining genetic diversity within the population.

Meiosis creates haploid cells

In most green algae, cells of vegetative ('adult') individuals are haploid – they each contain a single set of chromosomes. When sperm and egg combine, the resulting zygote is diploid – it contains two sets of chromosomes, one from each gamete. Meiosis is the special form of nuclear division that separates the two sets of chromosomes back into single sets again. In doing so, however, the chromosomes brought in by the two gametes are also mixed and re-sorted, resulting in zoospores with different combinations of parental chromosomes. Meiosis typically consists of two nuclear divisions, resulting in four cells.

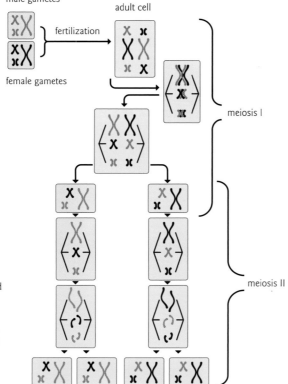

∧ Green algae, reproducing sexually with sperm and egg, were the precursors of the land plants.

> During fertilization, chromosomes from the gametes are combined into a diploid set. In meiosis I, chromosomes that code for the same sets of genes pair up. They then separate in random mixes to the two sides of the cell, and again form haploid nuclei. Chromosomes in the haploid cells are a mix from the two original gametes. In the second division (meiosis II), the two identical chromatids in each chromosome separate, resulting in four haploid cells.

male gametes

female gametes

diploid adult cell

fertilization

meiosis I

meiosis II

spores

V O L V O X R E P R O D U C T I O N

Volvox species are found in freshwater habitats and form hollow, spherical colonies, consisting of flagellate cells connected by thin strands of glycoprotein. Internal objects are new colonies produced either sexually or asexually. Clusters of sperm cells and eggs are formed within separate, unicellular chambers. Sperm cells emerge from their chambers and swim toward eggs, uniting within the egg chamber. The zygote then forms a thick-walled, desiccation-resistant spore, which can survive the drying of the pond. It will undergo meiosis upon germination.

⑦ In this dark-field image, the young colonies of *Volvox aureus* that have formed within an older one seem to glow.

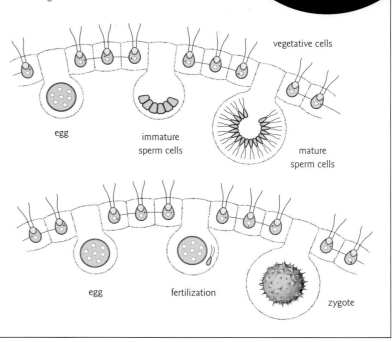

vegetative cells

egg

immature sperm cells

mature sperm cells

egg

fertilization

zygote

Fertilization is analogous to combining a red deck of cards with a blue one, in order. In this analogy, meiosis is the process of separating the combined deck into two complete decks again, but with the red or blue cards being allocated randomly to each. Each new deck then has an ace, two, three, etc. of hearts, for example, but each a random mix of red and blue cards.

RED ALGAE – AN ALTERNATE STRATEGY

Red algae (Rhodophyta) are ancient cousins of the green algae. Instead of producing a relatively small number of sperm cells equipped for active swimming, they produce a much larger number of tiny, inert spermatia, which are dispersed passively by water currents. Similar strategies are seen time and time again among plants, such as those that depend on wind to disperse pollen randomly, rather than on elaborate mechanisms to induce insects to do the job.

Life cycles

(T)he series of events in sexual reproduction – including meiosis, the creation of gametes, fertilization and the development of new individuals – is called a life cycle. The development of a new individual from embryo to adulthood often involves many distinct phases, as does the production of gametes and movement of sperm to the egg. Many kinds of organisms also have special dispersal phases, involving spores, pollen or seeds. Life cycles are therefore varied and can be quite complex.

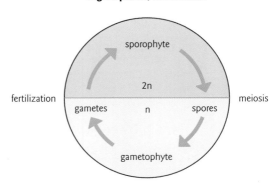

Many protists

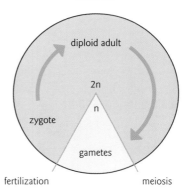

Some protists and all higher plants, not animals

Animals

> The cyclical patterns of growth and reproduction for different kinds of organisms are represented in these life-cycle diagrams. The diploid phase always begins with union of sperm and egg, while the haploid phase always begins with meiosis. 'Adult' activities of growth and energy-gathering took place in the haploid phase in simple organisms, but have shifted to the diploid phase as organisms became larger and more complex.

Fertilization and meiosis

What all life cycles have in common are the two nuclear events – fertilization and meiosis – that separate diploid and haploid phases and result in a new generation of functioning individuals. In most green algae, the diploid phase is minimal: a zygote that undergoes meiosis after a short period of dormancy. Such life cycles are said to be haploid-dominant. In animals, including ourselves, the opposite is the case. Our bodies are diploid and the haploid phase is minimal, consisting only of the gametes. So in green algae, meiosis occurs after fertilization and results in new haploid individuals, while in animals, it occurs in diploid individuals and results in gametes.

Haploid and diploid phases

True plants are the descendants of green algae that colonized the land. They share a common ancestor that evolved characteristic multicellular chambers in which gametes were produced, and include mosses and liverworts, as well as ferns, gymnosperms and flowering plants. The first plants were haploid like their algal ancestors. Haploid plants produce gametes to initiate sexual reproduction, and so are called gametophytes. The first sperm cells on land still had to swim through water

(in the soil) to reach the eggs. Long-distance dispersal by zoospores was not possible, however, and a secondary diploid plant body evolved to produce and disperse dry, inert spores. This diploid individual, arising directly from the zygote, is called a sporophyte. Organisms with life cycles that include both multicellular haploid and diploid individuals are said to exhibit alternation of generations (see box).

ALTERNATION OF GENERATIONS

Ulva is an unusual marine genus of green algae commonly known as sea lettuces, in which multicellular diploid individuals (sporophytes) alternate with identical-looking haploid individuals (gametophytes). On the sporophytes, specialized cells undergo meiosis to produce zoospores, which then travel some distance before settling down and developing into new haploid individuals. Specialized cells on this haploid form produce gametes, which are both motile, although the 'egg' is slightly larger. Such gametes mingle in the water to fuse. The resulting zygote will then produce a new diploid sporophyte plant. Alternation of generations is a central theme in plant evolution. In land plants, the two generations are always morphologically distinct.

In the life cycle of green algae in the genus *Ulva*, the diploid and haploid phases are identical in size and appearance.

Sea lettuce (*Ulva lactuca*) and its relatives are among the relatively few green algae that inhabit salt water, forming massive colonies on rocks near the seashore.

Individual plants (thalli) of *Ulva lactuca* resemble sheets of lettuce or cabbage, hence the species' common name of sea lettuce.

The circle of life

Life cycles are typically illustrated as circles, with the end leading back to the beginning, but it must be remembered that with each fertilization event, a genetically distinct new individual is created. Likewise, when spores are produced, new plants may be established far from the parents. All life cycles include fertilization and meiosis as dividing points between haploid and diploid phases, but what happens between these two events varies widely among the different groups of plants.

Reproduction in early land plants

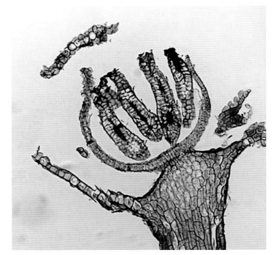

(T)he ancestors of land plants were green algae, which were haploid (like most modern green algae) and had short-lived zygotes that underwent meiosis to give rise to haploid zoospores. Early land plants were distinguished from such algal ancestors by the evolution of distinctive multicellular chambers for sperm and egg, and by the evolution of a diploid sporophyte that produced and dispersed dry spores. Modern bryophytes have probably changed relatively little from their early land plant ancestors, and hence provide models of structure and sexual reproduction in those plants.

(⁊) A colony of the simple land plant *Conocephalum conicum*, consisting of flat, forking thalli similar to many green algae.

(∨) In *Porella* liverworts, sperm cells are produced within round chambers called antheridia, each nestled at the base of a specialized leaf.

(⏀) Eggs in *Porella* are solitary and each is produced at the bottom of a narrow vase-like chamber called an archegonium.

Antheridia and sperm production

In bryophytes, sperm cells are produced in small globose to sausage-shaped chambers called antheridia, with sperm-generating tissue enclosed by a multicellular jacket. The sperm cells are flagellate, and often of bizarre shapes, and follow concentration gradients to swim through water on the soil surface or between plants toward egg cells on nearby plants. Bryophytes and their early land plant ancestors were thus confined to habitats that remained moist long enough for sexual reproduction to take place, and they had to grow close to the ground or flat against other substrates.

Many bryophytes, however, are adapted to survive long periods of desiccation in between – such plants might be considered the 'amphibians' of the plant world. Only with the later advent of seeds did plants escape from this paradigm.

Archegonia and egg production

The egg-producing chambers in bryophytes are called archegonia. They are vase-shaped and each contains just one egg. An opening forms at the tip when the egg is mature to allow the entry of a sperm cell, which is attracted by chemicals emitted by cells in the archegonium. After fertilization, the zygote

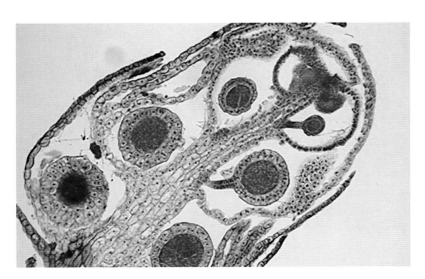

begins its development into a sporophyte individual within the archegonium. To avoid self-fertilization, sperm cells may be released prior to the egg being ready for fertilization on the same plant, or sperm and egg may be borne on separate plants (termed dioecious). Both strategies can be seen among mosses, liverworts, ferns and other seedless plants today. Archegonia and antheridia are located on the upper surface of thallose liverworts and hornworts, and in the axils of leaves or the tips of leafy stems in mosses and leafy liverworts. In the *Marchantia* genus of 'giant' liverworts, archegonia and antheridia are located on umbrella-like stalks (see box below).

SPECIALIZED REPRODUCTIVE STRUCTURES OF *MARCHANTIA*

In the highly specialized liverwort genus *Marchantia*, antheridia are embedded in umbrella-like structures called antheridiophores (right), opening to the upper surface to release sperm cells; archegonia hang upside down below the spokes of similar, but open, structures called archegoniophores. Because of their large size, *Marchantia* species are often used to represent liverworts in classrooms, even though they are quite atypical of the group. Whereas *Marchantia* thalli are differentiated internally, with an upper layer of specialized air chambers, most liverworts are more uniform in cross section. *Marchantia* species also reproduce asexually, producing small discs of tissue, known as gemmae, in cup-like structures. When washed out of the cups by rainwater, the gemmae can grow directly into new gametophytes.

⑦ ▷ In the specialized common liverwort (*Marchantia polymorpha*), the archegonia hang upside down, hidden by the finger-like projections of the umbrella-like archegoniophores (above). The sperm-producing antheridia are embedded within the umbrella-like antheridiophore, and emerge through pores on the upper surface (below).

Sporophytes and long-distance dispersal

For successful sexual reproduction to take place in seedless land plants, both extinct and living, haploid plants have to remain small and grow closely together in genetically mixed populations. Such mixed populations come about through long-distance dispersal of spores – as in the zoospores of algae, but with desiccation-resistant spores dispersed in the air in land plants. Spores in all living land plants are produced by diploid sporophyte individuals.

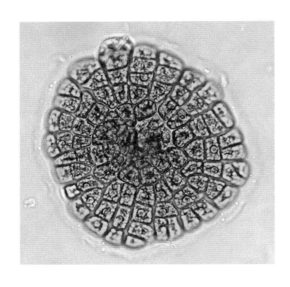

⊘ The haploid green algae in the genus *Coleochaete* live in fresh water on rocks and other surfaces. The plants develop outward from the centre to form a round disc. Sperm and egg are produced within specialized cells of the disc, and zygotes undergo meiosis to produce flagellate zoospores. *Coleochaete* species are the closest living relatives of land plants.

Spreading spores

The most critical evolutionary shift in early land plants occurred with the adaptation of spores for aerial dispersal. In *Coleochaete* and other green algae, the zygotes undergo meiosis to produce zoospores, which then swim some distance before establishing new haploid plants. Zoospores could not get very far on dry land, so the spores of early land plants lost their flagella, became much smaller and more numerous, and were encased in a hard wall. The hard wall was strengthened by a highly durable substance called sporopollenin. This substance is also found on the hard, desiccation-resistant zygospores (dormant zygotes) formed by some green algae. The earliest land plant spores were probably dispersed by physical forces such as

floods, or possibly on the feet of ancient animals, which were coming out onto land at about the same time. When spores landed in a suitable habitat along with spores from genetically different sporophytes, a mixed population of gametophytes resulted, followed by another round of sexual reproduction.

Sporophytes

Since dry spores are dispersed passively, and do not require apparatus for life functions or motility, they are smaller and produced in great numbers (a strategy similar to the production of passive, inert spermatia by red algae – see box on page 183). So in land plants fertilized eggs do not immediately undergo meiosis, but instead divide multiple times to produce a mass of diploid tissue, most of which eventually undergoes meiosis to produce a large number haploid spores. The multiplying mass of diploid cells constitutes the sporophyte, a genetically distinct individual plant. Sporophytes begin development within the archegonia, and remain dependent on the gametophyte for nutrients and water. Land plants thus exhibit alternation of haploid and diploid generations, although not in identical forms as in the unrelated *Ulva* (see box on page 185). In early land plants, and all modern bryophytes, the haploid gametophyte generation remains large and long-lived, and the sporophyte is small and ephemeral.

RICCIA
REPRODUCTION

Species in the modern liverwort genus *Riccia*
provide a model of what early land plants might
have looked like, as they appear to be adapted, or
readapted, for a simple means of spore dispersal.
As in *Coleochaete*, the *Riccia* gametophyte
grows as a flat thallus, and the sporophyte is
very simple, lacking the stalk that elevates the
spore chambers (sporangia) for wind dispersal in
most modern sporophytes. The sporangium also
lacks the elaters (worm-like cells that twist to push
spores out) found in many other liverworts. Sperm and
eggs are produced in antheridia and archegonia on the upper
surface of the gametophyte. When the eggs are fertilized, the zygote
develops into a roundish sporophyte that remains embedded within the gametophyte body.
The sporophyte consists of a simple jacket of cells, and tissue within that undergoes meiosis to
produce 100–200 spores. The spores
are typically not dispersed until the
parent disintegrates.

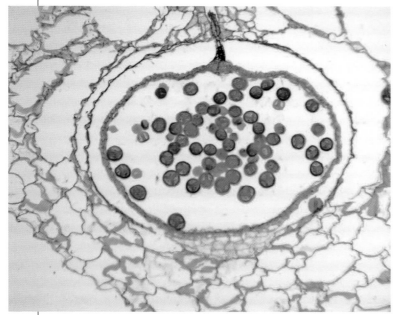

⊘ This simple embedded
sporophyte of *Riccia nigrella*
has produced many spores
through meiosis of its internal
tissue. These will not be
released until the parent
gametophyte disintegrates,
and will be dispersed by
water currents or on the
feet of animals.

⋀ The modern-day
haploid liverwort, *Riccia
natans* forms flat colonies
on wet soil, spreading
outward from a central
starting point as a
forking thallus.

(⌐) *Sphagnum* moss grows in dense bogs, its decay-resistant remains building up thick deposits over time. These peat deposits are dug up for fuel or horticultural use.

(∧) This *Polytrichum* moss forms thick tufts of upright stems. Other mosses creep along the surface of tree trunks, exposed roots or rocks. Dense spongy moss colonies absorb and hold water needed for survival and reproduction.

Dispersal of spores in bryophytes

(L) iverworts, hornworts and mosses are collectively referred to as bryophytes. Thousands of bryophytes live successfully and abundantly around the world, despite lacking vascular tissues for internal transport of water and materials. They also lack roots for deep absorption of water and have minimal capacity to store water, so most species live intermittently and many are found in moist habitats. This limited functionality adds to the challenge of sexual reproduction. The vegetative plants that we see, in a vast *Sphagnum* moss bog, for example, are haploid gametophytes, and require water for sperm movement.

(∧) The spore capsules of many mosses have one or two sets of teeth around the opening, which move in and out with changing humidity to loosen spores for dispersal.

(⟩) As in most mosses, the spore capsules of *Pohlia nutans* are elevated on slender stalks, positioning them better for wind dispersal of the tiny, lightweight spores.

Sporophytes are small

The diploid sporophytes of liverworts, hornworts and mosses are all relatively small, produce only one spore capsule (sporangium), and remain attached and dependent on the gametophytes for their short life. The typical sporophyte consists of a foot, which absorbs nutrients and water from the gametophyte, a stalk and the single spore capsule. The stalk develops from a relatively small number of cells between the foot and the developing spore capsule, which elongate rapidly as the spores mature. Once spores are shed, the sporophytes shrivel up and disintegrate. The gametophytes continue, however, sometimes for many years, and will undergo sexual reproduction repeatedly as conditions and seasons permit.

Raising the sporangia

Bryophytes have evolved several ways to raise their spore capsules for efficient launch of the spores into passing breezes. *Sphagnum* mosses and a few others form a stalk from the parent gametophyte tissue instead of within the sporophyte itself. In the giant *Marchantia* liverworts, sporophytes develop hanging upside down from the umbrella-like structures that bore the archegonia (see box on page 187). The sporophytes of hornworts in the genus *Anthoceros* are of a completely different form. They grow as a horn-like structure that elongates through division of cells at its base (known as intercalary growth). Meiosis occurs within interior cells, starting at the more mature top and continuing as new cells are produced at the bottom. Spores likewise mature from the top of the sporangium downward, and are released as the sporangium wall splits to reveal them.

Releasing the spores

Bryophyte spores are mostly dispersed by wind, the most powerful way to move tiny objects great distances. Many moss capsules have an elaborate structure, including a small cap, or operculum, that pops off when the spores are mature. They often also have a ring of teeth around the mouth of the capsule that move in and out of the spore chamber with changing humidity, which helps to dig out the spores and expose them to air currents. Many liverworts instead produce worm-shaped structures within the sporangium called elaters, which twist with changing humidity to loosen the spores and push them out.

⋀ Like many liverworts, hornwort plants (*Anthoceros* spp.) are flat thalluses. The sporophytes arise as slender rods, growing from the base. The older tips seen here have already split to release spores, while new spores are being produced in succession from the bottom.

INSECT DISPERSAL

Dispersal of spores by animals is uncommon in bryophytes. In some bryophytes that do not elevate their spore capsules, like *Riccia*, spores may be transported on the feet of large animals, but in members of the family Splachnaceae, spores are more specifically adapted for dispersal by insects. Capsules are coloured and emit foul-smelling odours to attract flies that feed on dung or carrion. The spores are sticky and attach to the bodies of visiting insects. Such mimicry is common among flowering plants that attract carrion-feeding insects for pollination, but highly unusual among bryophytes.

⋁ In the unusual moss *Tayloria mirabilis*, spores are sticky and produced on colourful, smelly sporangia. Flies seeking dung are tricked into landing, and carry spores away.

Reproduction in seedless vascular plants

(E) arly vascular plants, including several lines that continue to flourish today, do not produce seeds but instead achieve dispersal by spores. They include the pteridophytes, comprising ferns, related plants like whisk ferns (*Psilotum* spp.) and horsetails (*Equisetum* spp.), and a separate lineage, the lycopods (including the clubmosses). These all have small, short-lived gametophytes and large, independent, long-lived sporophytes, reversing the pattern seen in bryophytes. However, in common with bryophytes sperm cells must still swim through water on the surface of the soil to unite with eggs.

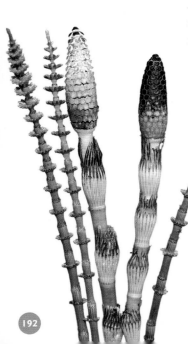

⊙ In the common horsetail (*Equisetum arvense*), the sporophyte consists of rapidly growing upright shoots, some of which bear cones of sporangia at their tips. Leaves are arranged in circles (whorls) at the nodes, and in all extant species are reduced to short scales.

Ephemeral gametophytes

The fern gametophyte has roughly the form of a small liverwort, but the archegonia and antheridia are located on the lower rather than the upper surface. Also unlike the liverworts, the fern gametophyte is short-lived, surviving just long enough for the fertilization of an egg and the early development of the new sporophyte embryo. Gametophytes of clubmosses and horsetails are different, in that they are more like tubers, living partially or completely underground. They do not photosynthesize but instead depend on symbiotic fungi to provide them with nutrients.

Long-lived sporophytes

As gametophytes became smaller and more short-lived, the sporophytes became large, independent and long-lived. What we perceive as a fern plant, a clubmoss or a horsetail is the sporophyte generation. While bryophytes are gametophyte dominant, the seedless vascular plants, and the seed plants that eventually emerged from them, are sporophyte dominant. The sporophytes are independent of the gametophytes, can grow and branch indefinitely, and live for many years. They can grow tall, and ultimately produce billions of spores. Vascular plants also produce root systems, and so they

can be active for longer periods of time in drier habitats than bryophytes. Moisture on the soil surface is required only for the short existence of the gametophytes.

Sporangia in ferns

In ferns, sporangia are produced in large numbers directly on the leaves, typically grouped into clusters called sori. Such clusters may be covered by a flap of leaf tissue called an indusium. Meiosis occurs within the sporangia, resulting in hundreds of spores in each. In the primary clade of modern ferns, the sporangia have a special mechanism for opening. As a sporangium matures, its wall begins to dry out. In a special ridge of cells around the sporangium called an annulus, tension builds up during this drying process until the sporangium wall rips and the spores are flung out into the air. The tiny spores are then transported by the wind, sometimes for thousands of miles.

Sporangia in the clubmosses, horsetails and whisk ferns

In the clubmosses, sporangia are produced in simple cone-like strobili, each sporangium nestled at the base of a bract. In the horsetails, in contrast, sporangia are produced in complex cones at the ends of regular or specialized shoots. And in the specialized whisk ferns, naked sporangia are produced in clusters of two or three.

THE LIFE OF A FERN

In the fern life cycle, the durations of the sporophyte and gametophyte generations are reversed compared with bryophytes. The main fern plant is the sporophyte generation, which may live for many, even hundreds, of years, and may produce millions of spores each year. Spores are produced through meiosis inside the sporangia. They germinate to produce the small haploid gametophytes, which bear archegonia and antheridia. When an egg is fertilized by a swimming sperm cell, a new diploid sporophyte plant is formed.

ⓥ The main fern plant is the long-lived sporophyte generation.

adult sporophyte

young sporophyte

sorus (sporangia clusters)

Fern life cycle

gametophyte

spores released from sporangia

Heterosporous plants

Some seedless vascular plants, including species of aquatic ferns and some clubmosses and spikemosses, have evolved seed-like structures through modification of their gametophytes. Like seeds, these structures contain a diploid embryo within a mass of stored food, but unlike true seeds they lack a distinctive seed coat. The sperm cells must still swim through a watery medium – unlike those of seed plants, in which sperm cells travel within pollen grains. The best-known example of this type of life cycle is the spikemoss genus *Selaginella*.

Megaspores and microspores

In *Selaginella* species, gametophytes are specialized to bear either sperm cells or eggs, and they are quite different in size and appearance. Spores are produced in sporangia, as in ferns and other seedless plants, but the spores (called megaspores) that develop into egg-producing gametophytes (megagametophytes) are about 20 times larger in diameter than those (microspores) destined to produce sperm-bearing gametophytes (microgametophytes). The production of megaspores and microspores is called heterospory, and most plants that practise this sort of reproduction are aquatic or semi-aquatic. The notable exception is *Selaginella* species, which occur in a variety of dry to damp habitats.

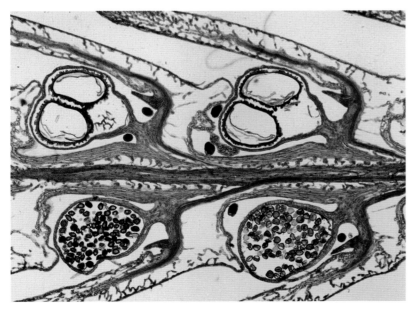

⌐ In *Selaginella ornithopo-dioides*, leaves are tiny and scale-like, but borne on branching stems so as to create a fern-like assemblage. Sporangia are borne in some of the upper branches.

< Two distinct kinds of spores are produced in *Selaginella* species, a phenomenon known as heterospory. On the upper side of this shoot, each sporangium bears two to three large megaspores, which will develop into specialized female gametophytes. On the lower side, sporangia bear a large number of microspores, which will develop into male gametophytes.

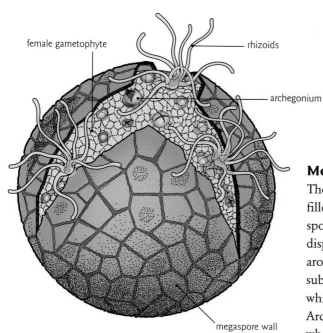

female gametophyte — rhizoids

archegonium

sperm cells

microspore wall

sperm cells

megaspore wall

Megagametophytes

The megaspores are quite large, and are already filled with food reserves as they ripen in the sporangium. These spores are too heavy to be dispersed by wind, but some may be moved around by animals or water currents. They subdivide to form an internal gametophyte, which eventually ruptures the megaspore wall. Archegonia form at the upward-facing end where the spore wall has opened up, usually after the megaspore has fallen onto the soil.

Seed-like progeny

Eggs are fertilized in the archegonia by swimming sperm cells. The diploid zygote, the beginning of the sporophyte generation, develops within the megaspore, drawing nutrients from the gametophyte tissue. At this point, the megaspore and its enclosed embryo may be further dispersed by water currents or animals, although very little is known about this. The young sporophyte continues to grow and eventually becomes independent of the gametophyte.

△ The egg-bearing megagametophyte develops within the wall of the megaspore, eventually partially rupturing it to expose the archegonia. The microgametophyte develops and subdivides into a number of sperm cells within the microspore wall.

Microgametophytes

The microspores of *Selaginella* species themselves are larger than typical fern spores, but little is known about how they are dispersed. They do not develop independent gametophytes, but subdivide internally, essentially producing one large antheridium inside the spore. The spore wall breaks open to release mature sperm cells, and these follow chemical gradients emitted by the female gametophytes to swim toward receptive eggs.

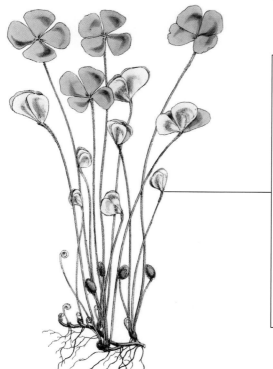

OTHER HETEROSPOROUS PLANTS

Besides *Selaginella*, heterospory is also found in the related quillworts (*Isoetes* spp.), lycophytes that inhabit wet bogs and marshes. Their sporangia are located at or below soil level, at the base of the grass-like leaves. Heterospory was also present in some ancient tree-like lycophytes, such as *Sigillaria* and *Lepidodendron* species, which lived in swampy forests from the Middle Devonian through the Carboniferous periods. The life cycle is also found in some aquatic ferns, including those in the genera *Marsilea*, *Salvinia* and *Azolla*.

Ovules and pollen

rue seeds contain an embryo and stored food – like the matured megagametophytes of *Selaginella* (see page 195), but with some significant differences. The embryo and food-storage tissue of a seed are enclosed by the jacket-like integument, not the wall of the megaspore. A young seed is called an ovule, and is the site of gametophyte development, fertilization and embryo development. The pollen grain is a spore with a highly reduced gametophyte within. In the next few sections, we look at the development of seeds and pollen in gymnosperms.

The ovule

In gymnosperms, a young ovule is typically egg-shaped, bound by a single integument, with a short stalk at the base and a narrow opening at the top called the micropyle. The interior of the young ovule is technically a megasporangium, but one that produces just a single megaspore. Meiosis occurs in a spore mother cell, resulting initially in four haploid cells, but three of these cells disintegrate, leaving just the one. Unlike in earlier plants, the megaspore is tiny and no food reserves are stored within. Instead, a food-storage tissue called the nucellus surrounds the megaspore. The megaspore never sees the light of day. It does not develop a hard, watertight wall, and it is not dispersed like other spores. Instead, it develops directly into a megagametophyte within the ovule, which remains attached to the parent plant until it matures into a seed. The developing megagametophyte absorbs most of the food reserves from the nucellus, and comes to occupy the interior of the ovule. Two eggs are typically produced, but normally only one will be fertilized.

Pollen grains

As the first ovules evolved from specialized sporangia, pollen grains evolved from microspores. Pollen grains are smaller and lighter than the microspores of *Selaginella*, and so are better adapted to dispersal by wind or animals. The sperm-bearing gametophyte develops within the pollen grain, before it is shed from the parent plant, and is extremely tiny. It consists of just two functional cells, one of which will form an outgrowth called the pollen tube, and the other of which, the

∧ A very young pine cone is hardly recognizable as such. It will take two-and-a-half years to reach full size, during which time it will be pollinated, the eggs will be fertilized and the ovules will become seeds.

⊘ This image shows the ovule (oval shape), within which is the gametophyte tissue that contains two archegonia (dark blue), each with an egg. The pollen grain lands at the upper end, and the pollen tube penetrates the thick gametophyte tissue, absorbing nutrients from it and slowly making its way toward the eggs. This alone may take many months. Usually, only one ovule develops into a seed on each scale.

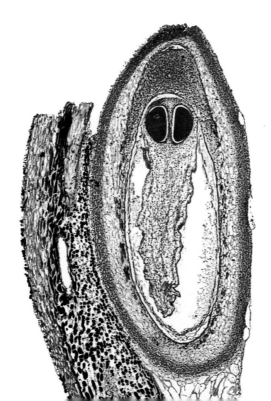

generative cell, will divide to form two sperm cells. In ancient seed plants, and still in archaic gymnosperms like cycads and the ginkgo (*Ginkgo biloba*), the sperm cells are flagellate, even though they are never released to swim through soil water as in earlier plants. More advanced seed plants, like conifers and flowering plants, have non-motile sperm, which are little more than naked nuclei.

Fertilization of the egg

Pollen grains are carried through the air, randomly by the wind or more directly by animals, to the ovules held on another plant. Upon arrival at the ovule, the pollen grain is drawn through the micropyle into the outer pollen chamber by a drop of fluid. It then develops the pollen tube, which penetrates the food-storage tissue of the megagametophyte and extends toward the egg. The sperm cells are released into an interior chamber and either swim or are drawn to the eggs.

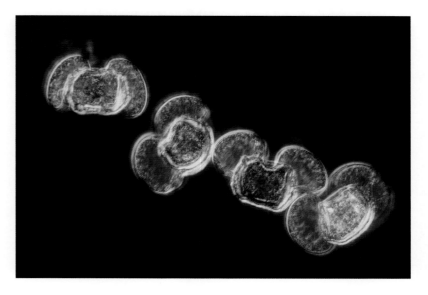

⑦ In this dark-field image of Scots pine (*Pinus sylvestris*) pollen, the dense, sperm-generating gametophyte tissue is stained red, while the pollen grain wall and the puffy 'wings' are stained blue. The wings lower the overall density of the pollen, allowing for effective wind dispersal (magnification ×750).

⊘ Because pine pollen is dispersed randomly by the wind, large quantities must be produced to ensure some will land on the female cones of nearby trees. Such pollen is abundant in the fossil record.

COATING THE SEED

As the ovule matures into a seed, the integument may develop multiple specialized layers that protect the young embryo and food reserves within. In some cases, the integument becomes fleshy and/or brightly coloured to attract animals that will aid in dispersal, as in this snow totara (*Podocarpus nivalis*). Others, such as those of the pine family (Pinaceae), are borne in cones and have thinner integuments.

Early seed plants

(T)he first plants to produce true seeds and pollen were the Palaeozoic seed ferns that appeared at the end of the Devonian period. They resembled modern ferns in that they had large compound leaves on trunks that were simple or sparsely branched. Unlike ferns, however, they had true secondary growth, with layers of secondary xylem (wood) and secondary phloem. They were most likely relatives of the seedless progymnosperms, but with simpler stems. Ovules and microsporangia were borne directly on the leaves.

(V) Palaeozoic seed ferns, such as this *Lyginopteris*, had large compound leaves (A). Seeds and microsporangia were borne directly on leaf segments, as in this *Sphenopteris* (B). The earliest ovules, as in these *Genomosperma* ovules, had incomplete integuments that were made up of leaf segments (C).

The first ovules
Ovules evolved as ancient megasporangia that came to be surrounded by leaf segments, which at first may have formed a cage-like barrier. The segments gradually fused together to form a continuous integument, leaving just a small opening at the tip for the entry of sperm cells. The integument was most likely an adaptation that protected the nutrient-rich nucellus from ancient herbivorous insects, as well as from desiccation. Leaves and leaf-like structures that bear only ovules are called megasporophylls, and those that bear only pollen are called microsporophylls.

The first pollen
Pollen grains were likewise produced in simple microsporangia borne directly on leaf segments. Early pollen may have been dispersed by the wind, but also quite possibly by the same herbivorous insects that sought to consume the young ovules. These insects likely carried pollen grains stuck to their bodies, which rubbed off on the sticky tips of the ovules. If seed ferns were indeed ancestors of all other seed plants, then both ovule- and pollen-bearing structures in modern gymnosperms and angiosperms can be viewed as modified leaves or leaf segments.

Seed plants in the Mesozoic
Gymnosperms that cohabited the Earth with early reptiles and then dinosaurs had more complex pollen- and ovule-bearing structures, often in cones or highly modified sporophylls that were distinct and separate from the vegetative leaves. None perhaps was as elaborate as those of the extinct Bennettitales. In the complex cones of these plants, ovules were borne on a central dome or column that was surrounded by elaborately branched pollen-bearing structures and tough outer bracts. The entire bisexual structure was analogous to the flower of the angiosperms, but so different in structural details that it could not have been a direct ancestor to true flowers, as was often thought in the past. The massive structures perhaps provided protection against the more aggressive herbivores of the time.

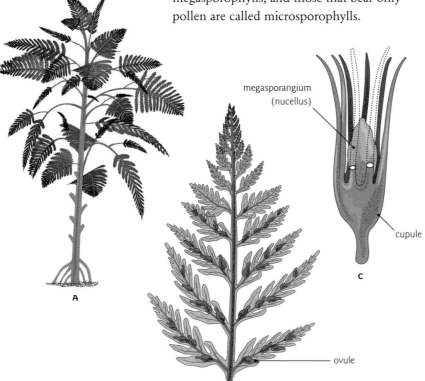

megasporangium
(nucellus)

cupule

C

ovule

A

B

microsporophyll with microsporangia

ovules

bisexual strobilus (cone)

bract, with hairs

⊘ Left: *Cycadeoidea*, an extinct member of the Mesozoic Bennettitales, with bisexual cones on its trunk. Right: The bisexual strobilus of *Cycadeoidea*, with large compound microsporophylls, and ovules borne on the central axis.

THE CASE OF CYCADS

Cycads were another Mesozoic group of seed plants that descended from ancient seed ferns, but unlike the Bennettitales, a number of species have survived to the present day. They retain simpler, leaf-like ovule- and pollen-bearing structures. In the genus *Cycas*, the megasporophylls are clearly leaf-like, with a compound blade and ovules carried on the stalk below. They are borne loosely around the terminal bud of the plant. In more advanced genera, however, the megasporophylls are simpler and smaller, and are aggregated into cones. Microsporophylls of all cycads are small and aggregated into cones. Like seed ferns and Bennettitales, cycads have large compound leaves. Most of these are simply compound and resemble modern palm leaves, but in one genus of cycads, *Bowenia*, the leaves are doubly compound as in seed ferns. Ancestors of other modern seed plants, like conifers and ginkgos, were also present in the Mesozoic, but had even more condensed reproductive structures and simpler leaves.

⊙ Like *Cycadeoidea*, cycads also arose in the Mesozoic era, and while they also superficially resemble the bennettitaleans, their reproductive structures are much simpler.

⊘ In members of the genus *Cycas*, such as this *C. ophiolitica* from Australia, ovules are borne along edges of very leaf-like megasporo-phylls produced in a loose cluster at the shoot apex.

Sexual reproduction in cycads and ginkgo

Cycads and the ginkgo tree have retained some aspects of ancient reproduction. They produce pollen grains and ovules like gymnosperms and flowering plants, but the sperm cells have many tiny flagella called cilia, a holdover from life cycles involving free-swimming sperm. Their fruits are adapted for animal dispersal.

Ginkgo reproduction

Ginkgo is the sole survivor of an ancient lineage of seed plants characterized by fan-shaped leaves with forking veins. Its existence as a wild plant in China is still debated, as it has been cultivated there for thousands of years. Pollen and ovules are borne on separate trees (i.e. plants are dioecious), a strategy to prevent self-fertilization. Pollen is borne in loose, flexible strobili called catkins, and is adapted for wind dispersal. The ovule-bearing structure consists of, usually, two naked ovules (sometimes one) at the end of a short stalk.

⌃ Ginkgo or the maidenhair tree is considered a living fossil among seed plants. It has fan-shaped leaves resembling the leaflets of some ferns. This hardy tree is much planted in cities in temperate climates, as it is tolerant of air pollution.

⌄ The sperm cells of ginkgo (*Ginkgo biloba*) are rather large, and bear many flagella on the top side, as do those of the cycads. Once inside the ovule, they still must swim a short distance to where the egg sits.

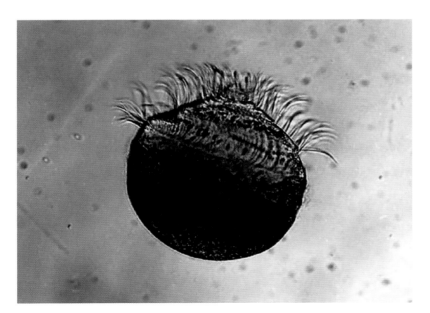

When receptive, the ovules produce a sticky droplet at their tips, which catches pollen grains randomly swirling around in the wind. At the end of the day, the droplet retracts, pulling the pollen grain into a small chamber at the tip of the ovule, where the pollen tube emerges and develops into a tiny microgametophyte. It takes up to five months to produce its two sperm cells, which are motile by way of thousands of tiny flagella arranged in a spiral band in the upper two-thirds of the top-shaped cells. The sperm cells then swim the very short distance to the waiting egg cells, of which only one is usually fertilized.

While the microgametophyte is developing, the megagametophyte (nucellus) is expanding and filling with food reserves. A variable time after fertilization, the ovules drop off the tree, and the embryo embedded within the nucellus continues its development. A fleshy layer, emitting a peculiar foetid odour, develops from the integument. It is believed that this developed to attract carrion-feeding animals, which swallow the ovules and disperse them.

Cycad reproduction

Cycads consist of many living species, although their ancestors arose in the Permian period, as did the ginkgos. The development of sperm cells and fertilization in cycads are similar to those in ginkgo, but the seeds become coloured and remain attached to the megasporophylls until mature. Plants are also dioecious, but both pollen and ovules are borne in large strobili, except in some species of *Cycas*, in which megasporophylls resemble small leaves and bear up to 12 ovules along the midrib. In that genus, the megasporophylls are produced loosely around the apical bud of the trunk. This apical bud resumes leaf production after reproduction, pushing the mature megasporophylls to the side.

(V) In the genus *Cycas*, ovules are borne on leaf-like megasporophylls, arranged in a loose cluster, rather than in a closed cone.

CYCAD POLLINATION

While traditionally thought to be wind pollinated like ginkgos and other gymnosperms, most cycads are now known to be pollinated by tiny insects such as weevils or thrips, which feed on pollen in the male cones. The insects are lured to female cones by colours and odours that resemble those of the male cones. Typically, some pollen grains are stuck to their bodies as the insects enter the female cones, and these are rubbed off on the ovules inside. There is nothing there for them to eat, so after a while they leave. This is an example of deception in attracting pollinators, a phenomenon found extensively among angiosperms.

(Λ) In most cycads, such as *Bowenia spectabilis* (top) and *Macrozamia lucida* (above), the cones remain tightly closed until pollination. Here, insects have been feeding in a male cone, but leave when the supply of pollen runs low. They may then be lured to a female cone.

Reproduction in conifers

C onifers are the most widespread, abundant and diverse of all the living
gymnosperms. Their ovule-bearing strobili are typically large woody cones, but
sometimes they are small and fruit-like. The largest and oldest trees – the sequoias
and redwoods – are conifers. In conifers, the leaves remain small and simple,
most taking the form of needles or scales, although a few are flat and blade-like.
They constitute a separate lineage from cycads, ginkgos and angiosperms,
all of which have more complex leaves.

> Conifer cones, such as these of a fir, develop very slowly. Pollination occurs early, but pollen tube growth and seed development take many months.

⌄ In this longitudinal microscope image through a male pine cone, the developing pollen grains can clearly be seen within the microsporangia.

Pollen-bearing catkins

Conifers are wind pollinated. Their pollen
is borne in slender, flexible strobili called
catkins (similar to those of ginkgo), which
release lightweight pollen grains in visible
clouds (see photographs, pages 178 and 223).
In some conifers, notably pines (*Pinus* spp.),
pollen grains are equipped with two air sacs
that increase their buoyancy in the air.
Within the pollen grain, the tube nucleus
forms the pollen tube, while the generative
cell subdivides to form two non-motile
sperm cells that drift to the tip of the tube.

Ovule-bearing cones

Ovules are in borne in cones that are typically
large and woody, but with some interesting
exceptions. In the pine family, cones are
complex structures, with thin bract scales
initially forming in a spiral arrangement
around the cone axis. Seed scales then develop
in the axils of each bract scale, and as they
mature they become the dominant structures
that both bear the ovules and form the tough
closing scales of the cone. This structure
is interpreted as a modified branching
cone structure found in ancestral conifers.
Other families of conifers have even more
reduced structures, sometimes with the bract
and seed scales completely fused together.
Development of pine cones is extremely
slow, taking up to two-and-a-half years from
initiation to seed release.

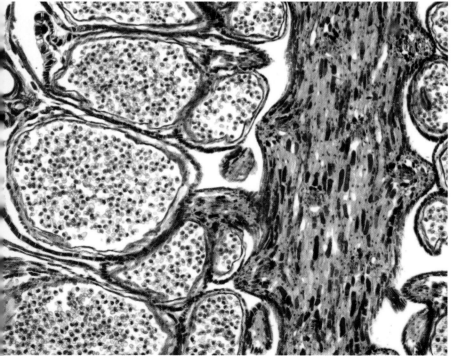

Fruit-like cones

In the genus *Podocarpus*, the strobili are small, with a few sterile cone scales at the base fused together to form a brightly coloured receptacle. Fertile ovules (most commonly just one) emerge from the top of the receptacle, and are covered by another modified cone scale called an epimatium. Among the many species of *Podocarpus*, the colour of the receptacle is blue, purple, red or yellow, and the seeds are typically green or blue, again in modification for animal dispersal. In junipers (*Juniperus* spp.), strobili are small, fleshy and blue, an adaptation for animal pollination. A third example of coloured, animal-dispersed units is found in the genus *Taxus*, the yews. In their reduced cones, a single scale at the base is modified into a bright red fleshy structure called an aril, which surrounds the single seed like a doughnut. Such adaptations are relatively recent in the history of conifers, and constitute an example of co-evolution with fruit-eating birds (see page 211).

⊓ The cones of Yew (*Taxus* spp.) are fruit-like, with a brightly coloured cupule that develops from a single bract. This is an adaptation for dispersal by birds.

⊃ Juniper (*Juniperus* spp.) 'berries' are actually small cones in which the scales have become fleshy and fuse tightly together when ripe. They are also dispersed by birds.

PINE AND SPRUCE SEED DISPERSAL

In pine (*Pinus* spp.) and spruce (*Picea* spp.) seeds are often winged for dispersal in breezes, although in some, such as the pinyon pines of North America and the Swiss stone pine (*Pinus cembra*) of the European Alps, they are gathered by animals and buried. Unused seeds may then germinate and form new trees.

◁ *Gnetum gnemon* grows in tropical forests, and has large, flat leaves that resemble those of many flowering plants. Here, a cluster of young fruits has begun to develop at the base of one of the leaves.

▽ Some species of *Ephedra*, such as this *E. monosperma*, have brightly coloured seed coats to attract birds. The birds eat the coat but drop the seed away from the plant, thereby acting as dispersers.

Reproduction in gnetophytes

The gnetophytes are an ancient and enigmatic group of gymnosperms, of which only three genera exist today: *Gnetum*, comprising tropical vines or trees with opposite leaves; *Ephedra*, desert shrubs with whorled leaves reduced to tiny bracts; and *Welwitschia*, containing a single bizarre species found only in the Namib Desert of southwestern Africa. Some gnetophyte reproductive features superficially resemble those of the angiosperms, leading some botanists to suggest that the two groups are closely related. However, DNA-based phylogenetic studies indicate that the gnetophytes are most likely a specialized offshoot of the conifers, but with more complex strobili.

▷ *Ephedra foeminea.*
Left: Male cone with nectar drops produced by sterile ovules at the distal centre of the cone. Right: Female cones produce similar pollination drops exposed at the micropylar openings.

Pollen strobili

All living gnetophytes are dioecious (with male and female strobili on separate plants). Pollen is borne in small but complex strobili, with microsporophylls superficially resembling the stamens of flowering plants. Some species of *Ephedra* are pollinated by insects, as are *Gnetum* and *Welwitschia* species. Other species of *Ephedra* are pollinated by the wind, which is thought to be a specialization within the gnetophytes.

THE DIVERSE USES OF *EPHEDRA*

Many people are aware of the former use of extracts of *Ephedra* as an aid in weight loss. The alkaloid drug ephedrine, found primarily in the Chinese species ma huang (*E. sinica*), was eventually discovered to have serious side effects and has been banned in the United States and other countries. American species, lacking ephedrine, were used by Native Americans and European settlers to make a tea (sometimes called Mormon tea).

Ovule-bearing strobili

Ovule-bearing strobili are complex, but typically produce only a single seed. The ovule has the same structure as those in other gymnosperms: straight, with a single integument, but with an unusual protruding tube that produces the pollination drop. In some, the megagametophyte exhibits modifications such that it resembles that of angiosperms, but with a reduced number of cells. In *Gnetum* and *Welwitschia* species, there are no more archegonia. Instead, cells in a specific location serve as eggs.

Ripe seeds

In *Gnetum*, seeds have a brightly coloured, fleshy integument adapted for animal dispersal, while seeds of *Ephedra* are surrounded by brightly coloured cone scales for the same

⟨ Seeds of a *Gnetum gnemon* have a fleshy seed coat that is attractive to birds. Fruit-like seeds in gymno-sperms presumably evolved relatively recently, perhaps in the Cretaceous, as fruit-eating birds evolved.

purpose. In contrast, the seeds of welwitschia (*Welwitschia mirabilis*) are adapted for wind dispersal, being equipped with a papery wing similar to that seen on the seeds of many pine species. It may be that the dry habitat of this species does not support the type of birds that might disperse fruit-like seeds.

WEIRD AND WONDERFUL WELWITSCHIA

In welwitschia, only two leaves are produced, which last for the lifetime of the plant. The leaves lengthen continually from a meristem at the base, reaching up to 4 m (13 ft) long, a feature otherwise known primarily from the monocots. In older plants, the leaves split lengthwise, creating the appearance of many leaves around the top of the stem. The Namib Desert has virtually no rainfall, so welwitschia appears to be dependent on ground water seeping through the flat habitat from regions further east, and has an extensive root system. Some water may also be absorbed from condensing fog from the ocean, which is the primary source of water for insects and other animals that live here. Most welwitschia plants are aged at about 500 years, but carbon dating of some individuals has put their age at an astonishing 2,000 years.

⟨ In welwitschia (*Welwitschia mirabilis*), from the desert of south-western Africa, there are only two leaves, which continue to grow from the base of the plant throughout its life.

∧ The seeds of welwitschia, unlike those of most *Ephedra* and *Gnetum* species, are dry and flat. They are dispersed by the wind. Here, they can be seen being released by the cone at the upper left.

The angiosperm flower

The angiosperms evolved a unique bisexual strobilus that we call a flower (see more in Chapter 6). Female reproductive organs called carpels are located at the tip of the strobilus (the centre in a more flattened flower), with male, pollen-bearing organs called stamens located below or outside them. In typical flowers, coloured or protective organs called tepals surround the sexual organs. Parts are connected to a platform called the receptacle, and elevated on a stalk called the pedicel. Tepals are most commonly in two series: the outer, often green, sepals and the inner coloured petals.

∧ *Fuchsia* flowers represent one of many specialized flower types, but are based on standard sepals, petals, stamens and pistil. Only agile pollinators like hummingbirds can access the nectar at the back of the flower, by hanging upside down.

⊤ In tulip (*Tulipa* spp.) flowers, the three carpels are united into a common pistil, which rises in the centre of the flower and is surrounded by the six stamens. Monocot flower parts are typically arranged in threes, as here.

< The flowering rush (*Butomus umbellatus*) is also a monocot with flower parts arranged in threes, but represents the more archaic condition of apocarpy, whereby the six carpels remain separate from one another.

The case for carpels

Carpels consist of an ovule-containing chamber called an ovary, a neck-like section of varied length called a style, and a pollen-receptive surface at the tip called a stigma. The evolutionary origin of carpels is still a mystery. They may be modified megasporophylls, stem-like structures or a combination of the two. In more archaic angiosperms, carpels remain separate from one another, but in the majority, carpels are fused together into a single structure called a pistil, with shared or closely placed stigmas. Pollination is presumably more efficient when carpels are joined like this, as a single brief visit by a pollinating animal can provide pollen to all carpels at once. Protection of developing ovules may also be enhanced in a multi-carpellate pistil, by sharing thin inner walls and a common fortified outer wall.

A TYPICAL FLOWER

The typical angiosperm flower consists of sepals, petals, stamens and pistil, always in that order, attached to a base called the receptacle on a stalk called a pedicel. Arrangement of floral organs into such a bisexual strobilus is considered a key feature in the evolution of angiosperms. In more specialized flowers, some sets of organs, such as petals, may be lacking. Unisexual flowers are also common, especially in wind-pollinated plants, or where flowers are condensed into compound flower-like inflorescences. In such cases, stamens are lacking or non-functional in female flowers, likewise for carpels and pistils in male flowers.

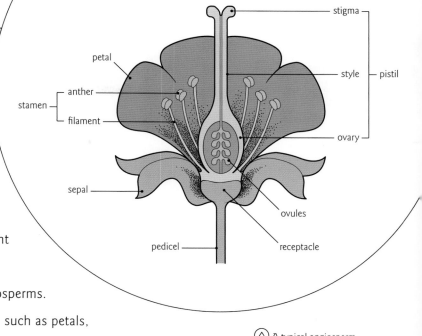

⋀ A typical angiosperm flower, illustrating the standard parts. The pistil contains a number of ovules, although specialized flowers may have only one. One or more sets of parts may also be lacking in some flowers.

Stamens

Typical angiosperm stamens consist of four elongate pollen sacs (microsporangia) joined into a common structure called an anther. The anther is elevated on a slender stalk called a filament. In some more archaic angiosperms, the stamen is flattened and more or less leaf-like, with four parallel pollen sacs on the upper surface. Specialized stamens may have fewer pollen sacs. Pollen is varied in size, shape and ornamentation, so can often be used to identify plant material in geological strata, archaeological remains or forensic evidence.

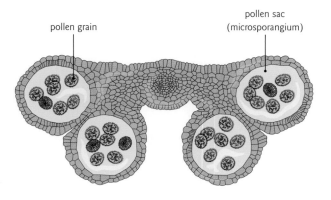

Tepals: sepals and petals

Tepals are generally leaf-like in shape, and are thought to have evolved from leaves surrounding the reproductive organs. In more archaic angiosperms, series of tepals may grade from green and leaf-like around the outside, to more colourful and petal-like nearest the stamens and carpels. More commonly, tepals are in two distinct series: the outer, leaf-like sepals, and the inner coloured petals. As the primary organs that attract pollinators, petals may have highly modified shapes, colour patterns and fragrances.

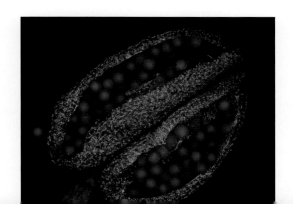

⟨ Left: Fluorescence micrograph of a thale cress (*Arabidopsis thaliana*) anther. Above: In cross section, an anther can be seen to consist of four parallel pollen sacs, although the two on either side may join together and appear as one.

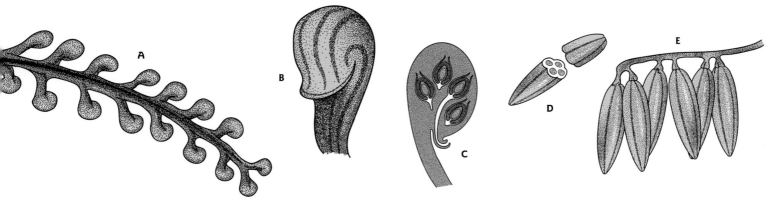

Origin of angiosperm carpels and ovules

In the fossil genus *Caytonia*, cupules were produced on an elongate, simply branched axis (A). Each cupule was curved so that its opening faced back toward its base (B). Within each cupule, ovules were lined up along its 'backbone', with their openings facing the opening of the cupule (C). Pollen sacs were grouped in fours (D) and borne on branching structures (E).

The origins of the angiosperms lie in how their distinctive carpel, stamen and ovule came about through modification of earlier gymnosperm structures. In some organisms, the fossil record is highly useful for tracing their evolutionary history. In the case of flowering plants, however, the fossil record is meagre. The known fossils most closely related to the angiosperms appear to be the Caytoniales, but there are considerable gaps between the two groups. More distantly related fossil plants can also provide clues on the nature of ancestors that are as yet unknown.

Cupule conundrums

In many Mesozoic gymnosperms, ovules were protected in structures called cupules. In many ways these functioned like carpels, most likely protecting ovules from desiccation and/or herbivorous insects. As in the earlier evolution of the integument around ancient megasporangia (see page 198), cupules appear to have evolved from leaf segments that wrapped around clusters of ovules. In all known fossil cupules, the ovules are straight and have only one integument, as in other gymnosperms. The characteristic ovule of angiosperms, however, is bent backwards on its axis (anatropous) and has two integuments. Ovules, in turn, are surrounded by the carpel. So carpels most likely did not evolve directly from the cupules of any known gymnosperms, otherwise angiosperm ovules would still be straight and missing the second integument (see box opposite).

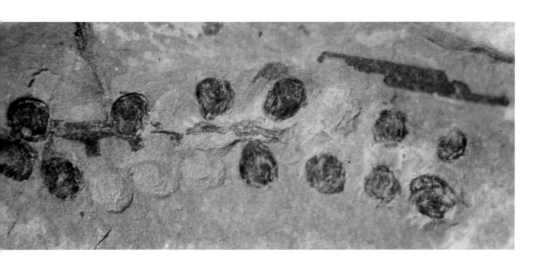

In this fossil of *Caytonia* sp., ovule-bearing cupules are lined up on the two sides of a specialized shoot.

THE SECOND INTEGUMENT

According to one persuasive theory, the cupule in some ancient plant similar to *Caytonia* (in the family Caytoniales) became, not the carpel, but the second integument of the angiosperm ovule. These cupules were bent so that the opening faced back to the base. Ovules were lined up along the back of the cupule, not along the edges as in most angiosperm carpels. The theory posits that the number of ovules in the cupule was reduced to one, and that the cupule came to close tightly around the single ovule, becoming the second integument. As the original cupule was bent backwards, the angiosperm ovule took on that configuration.

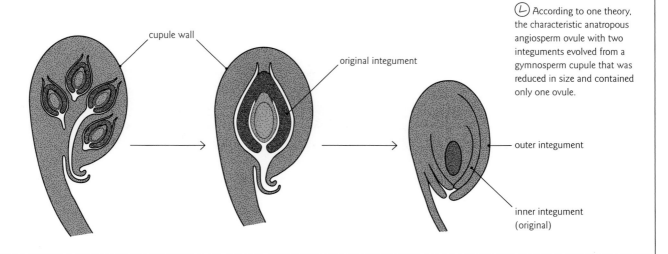

cupule wall

original integument

(L) According to one theory, the characteristic anatropous angiosperm ovule with two integuments evolved from a gymnosperm cupule that was reduced in size and contained only one ovule.

outer integument

inner integument (original)

Microsporangiophylls to stamens

In the Caytoniales, pollen was borne in highly branched structures, but the pollen sacs (microsporangia) were elongate and joined together in parallel groups of (usually) four, as in the much simpler anthers of the angiosperms. If something like this was the ancestor of the angiosperm stamen, it is unclear how it became reduced to a single four-sac unit, or what the relationship was to the flattened leaf-like stamens seen in some archaic angiosperm families.

Megasporophyll to carpel

As to how the first carpels wrapped around the ovules, we have even less evidence. There is a void in the Mesozoic fossil record for possible ancestral structures. In fact, there are two different ideas on what the first carpels were like. In the traditional view, the ancestors of the carpels were simple leaf-like megasporophylls with ovules lined up along the edges. These folded together along the midrib to bring the ovules inside. This is the form found most widely in angiosperms. In another view, the original ovules were unfolded, urn-shaped (asciidiate) structures with just one or two ovules inside. The most archaic angiosperms had asciidiate carpels, suggesting that the first angiosperms had the same, but there are no models among fossil gymnosperms for what ancient structures evolved into such carpels while, at the same time, accounting for the two integuments and anatropous shape of the ovules.

Archaic angiosperms

The earliest branches of the angiosperm family tree are referred to as the ANA grade, from the orders Amborellales, Nymphaeales and Austrobailyales. Presently, Amborellales, consisting of a single species, is considered the first branch, and hence the most archaic, but followed closely by Nymphaeales and Austrobailyales. The ANA grade and the magnolids constitute the basal angiosperms, and apparent archaic characteristics are scattered among them.

Apocarpy – separate carpels

Early flowers were condensed shoots containing leaf-like tepals and condensed ovule- and pollen-bearing structures that may have been leaf-like or stem-like. The parts were all separate from one another at first, but became joined together in many varied ways in adaptation for more efficient pollination by animals and other forces. Carpels were separate from one another (apocarpous), arranged spirally or in whorls (circles) around a central axis. Separate carpels predominate in the ANA grade and magnolids, and also in basal monocots and eudicots. Transitions to united carpels (syncarpy) occurred in parallel within all these major groups. Stamens, petals and sepals can also be united in various ways.

Ascidiate or plicate carpels

The traditional model of carpel evolution viewed them as folded (plicate) leaf-like structures, bearing ovules along the infolded margins, something like a modern pea pod. The margins of the carpels were sealed, but loosely so in the region of the stigma. This is the basic form of carpel in modern magnolids, monocots and eudicots. However, the discovery that a very different form of carpel is basic in the ANA grade, whose ancestors pre-dated all other angiosperms, seems to contradict that traditional model. In these archaic angiosperms, carpels are urn-shaped, unsealed at the top, and contain only one or a few ovules lined up along the back. These carpels appear more like tubes pulled up over the ovule(s), although the stigma does show signs of folding in *Amborella trichopda* and other ANA species. Unfortunately, fossils that might clarify the relationship between the two types are lacking.

Amborella trichopoda is considered the most archaic angiosperm. Clusters of small, inconspicuous male flowers can be seen below. A female flower (below right) consists of a series of tepals, not distinguished as sepals and petals, and a number of separate, ascidiate carpels. Each carpel contains a single seed.

Drupe fruits

The ascidiate carpels of *Amborella trichopoda* and some other ANA grade angiosperms mature as single-seeded red fruits called drupes. This sort of fruit is adapted for dispersal by birds, with the red colour being particularly conspicuous against the dim greenness of a tropical forest. Such fruits parallel the development of fruit-like structures in gymnosperms like *Gnetum*, *Taxus* and *Podocarpus* (see pages 203 and 205). The seamless ascidiate carpel of *A. trichopoda* allows development of such fruits in archaic angiosperms without sealed carpels, although similar fruits evolved in the plicate carpels of many eudicots (e.g. *Prunus* – plums and cherries), magnolids (Lauraceae) and monocots (palms). Thus the evolution of small red drupes was closely tied to the evolution of fruit-eating birds.

pollen sacs

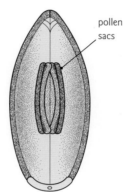  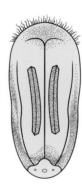 

Austrobaileya   *Himantandra*   *Degeneria*   *Magnolia*

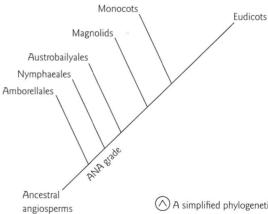

THE MISSING LINK

Stamens in basal angiosperms, including magnolids, are often flat and blade-like, suggesting an ancestor that was leaf-like. This contrasts with the highly branched pollen-bearing structures of the Caytoniales. Fossils bridging the gap between angiosperm stamens and earlier gymnosperm structures are lacking, as are fossils that might explain the origin of the angiosperm carpel.

Flower colour, nectar and scent

As adaptations to the visual capabilities of various animals, flower colours are quite varied and of every hue. Nectar serves as an additional incentive for some animals, with nectaries in various parts of the flower, most commonly on the carpels or petals. Nectar may accumulate in the base of the flower or in specialized tube-like extensions of the petals called spurs. Scents are produced in special glands called osmophores.

(>) These two very different looking flowers are the female (top) and male (bottom) flowers of the tropical palm genus *Ptychosperma*. Both, however, produce drops of nectar, the common denominator that gets insects to move from one flower to another.

Flower colour

Flower colour is closely tied to the visual abilities of pollinating animals, and floral pigments have evolved to create every conceivable hue. Red coloration, for example, is particularly likely to appear in flowers adapted for bird pollination. Red, however, can be created by unrelated compounds, indicating much parallel evolution of flower pigments. In many plants, red to blue pigments belong to the chemical class of anthocyanins, but in carnation (*Dianthus* spp.) and cactus flowers, as well as in the taproots of common beet (*Beta vulgaris*), the pigments are in the class of betalains. Yellow flowers, such as those in the daffodils and buttercups (*Ranunculus* spp.), however, are coloured by carotenoids, which

include the orange pigments of carrots. Flowers even exploit wavelengths of light that are invisible to humans, such as ultraviolet. For bees, a patch of ultraviolet pigment in a flower can create a strongly contrasting dark pattern (see box on page 253).

Nectar

Nectar is a particularly nutritious version of the plant sap that runs through the phloem, and may have begun as a 'leakage' of such sap inside the flower. In modern plants, nectars are produced in specialized glands. They are composed primarily of sugar, but numerous other substances unique to particular species give nectars, and the honey derived from them, distinct flavours. The production of nectar by flowering plants is a huge source of nutrition

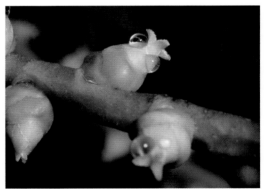

SOMETHING OTHER THAN NECTAR

In the South African *Diascia longicornis*, a member
of the figwort family (Scrophulariaceae), long spurs are
filled with a nutritious oil rather than nectar, which is
consumed by certain bees. The oil has a higher calorific
value than nectar, and is therefore attractive to the
specific bees that are adapted to retrieve it.
Such bees are in the genus *Rediviva*,
and females have modified hairy
forelegs, with which they retrieve the
oil from the narrow flower spurs.

⊘ *Diascia denticulata*
produces an alternate
food reward for insects,
a nutritious oil located
in the long spurs.

for a great variety of animals that have co-
evolved with them. Hummingbirds, butterflies,
moths, bats, and many kinds of bees and flies
are partially or totally dependent on them.
In turn, the honey made by bees feeds
many wild animals, including, famously,
bears and humans.

Scent

The scents produced by flowers are also hugely
varied. They are commonly sweet or musky,
attracting a wide range of animals, particularly
at night when colours are useless for this
purpose. The pleasant smells of roses and many
other species, essential to attract pollinators and
achieve sexual reproduction in plants, have
ironically been exploited in the
manufacture of perfumes and
other fragrance products used to
enhance human efforts to attract
mates. Not all flower fragrances are so
pleasant, however. Some resemble the
odour of rotting flesh (see page 81),
attracting more ghoulish pollinators, and
some mimic the sex pheromones of insects,
luring males into false sexual encounters.

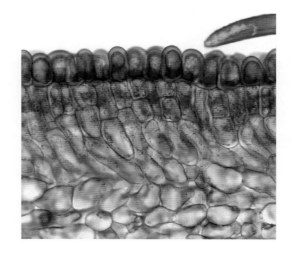

◁ In this *Ditassa gracilis*,
the egg-shaped epidermal
cells and the reddish cells
below them form an
osmophore, a scent-
producing gland.

▽ Hummingbirds feed
almost exclusively on nectar,
which is hidden away in
narrow tubes. Bird-pollinated
flowers are most frequently
red, as birds have good
sensitivity to this colour.

Inflorescences and superflowers

(A) grouping of flowers on a specialized shoot or branching set of shoots is called an inflorescence. In many plants, small flowers are grouped compactly into an inflorescence that resembles or functions like a single large flower.

(∨) The flowers of the foxglove (*Digitalis purpurea*) form a simple inflorescence called a raceme. Flowers mature one at a time for an extended time, enticing pollinators to come day after day. This maximizes cross-pollination as the visitors must feed on many different plants each day.

The inflorescence

In many plants, such as *Hibiscus* and *Magnolia* species, flowers appear singly along ordinary leafy shoots or at the tips of such shoots. An inflorescence is a more compact and segregated cluster of flowers, and leaves within it are modified into bracts. The bracts are most commonly smaller than ordinary leaves, and may be green or the same colour as the inflorescence branches. Some, however, are large and brightly coloured, taking on the function of attracting pollinators. The flowers within an inflorescence most commonly open one or a few at a time, in sequence. This establishes a pattern of repeated visitation by pollinators, which may return to the same inflorescence day after day.

Superflowers

In other inflorescences, the opening of flowers may be synchronized so as to create a mass pollination event over a much shorter period of time. This occurs in many species of palms, in which unisexual male and female flowers occur on the same inflorescence (i.e. they are monoecious), but with their opening, or anthesis, offset. In the genus *Bactris* of the new world tropics, and in the genus *Hydriastele* in New Guinea,

it has been shown that the female flowers open as soon as the large bract enclosing the inflorescence opens. A massive migration of small insects, attracted by a distinctive fragrance, brings pollen from a previously opened inflorescence to the new one. These insects feed on tissues and pollen of the unopened male flowers, departing when the male flowers open, after which the remaining pollen is shed. These inflorescences function like single flowers, in which the release of pollen and the receptivity of the stigmas is staggered to avoid self-pollination.

(>) In this *Hydriastele* inflorescence, the small female flowers are all receptive at once, although largely hidden beneath the larger male flowers. The latter will open 24 hours later.

⟨ The common poinsettia (*Euphorbia pulcherrima*) is a compound flower, with large coloured bracts serving to attract pollinators to the small clusters of flowers in the centre.

False or compound flowers

A compound flower is an inflorescence that not only functions like a single flower, but is also modified to resemble one. The most famous of such 'false flowers' are common to the daisy family (Asteraceae), but are also the rule in the arum family (Araceae), poinsettias (*Euphorbia pulcherrima*) and dogwoods (*Cornus* spp.). In each case, bracts or specialized flowers around the edge of the inflorescence are modified to look like the petals of a single flower. In sunflowers (*Helianthus annuus*) and poinsettias, the small central flowers open in sequence, as in a simpler inflorescence, but in aroids, male and female flowers are typically synchronized for a massive pollination event like those described for *Bactris* and *Hydriastele* above.

⟨∨⟩ The musk thistle (*Carduus nutans*) belongs to the daisy family (Asteraceae), whose members all have very compact inflorescences that appear to be a single flower.

INSECT ATTRACTOR

In the family Araceae, flowers are all tiny and grouped into inflorescences that function like single large flowers. The flowers are packed tightly on the central spike, called a spadix. A large bract, called the spathe, protects the young flower spikes, and may be coloured to attract pollinators, or sometimes forms a chamber were pollinators may be temporarily trapped or spend the night. Pollination occurs within 24 hours. When the large spathe opens, the female flowers are receptive and the appendage emits an odour to attract flies and other insects. Later, the male flowers shed pollen, dusting the insects, which then fly off and repeat the cycle.

⟨∧⟩ The numerous tiny flowers of the titan arum (*Amorphophallus titanum*) are densely packed on the central spadix, and surrounded by a single large spathe.

Pollination and fertilization

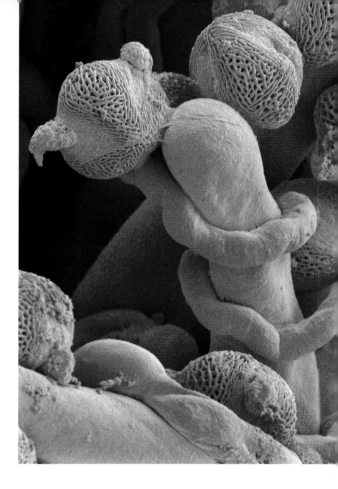

(A)ngiosperm pollen grains do not enter the ovules as they do in gymnosperms. Rather, they germinate on the stigmas, and their pollen tubes grow into the carpels to reach the ovules inside the ovary. Sperm are non-motile and are delivered directly to the egg cell at the tip of the ovule. A second sperm nucleus combines with two polar nuclei in the megagametophyte to form a triploid nutritive tissue called endosperm. This 'double fertilization' must occur before the ovule can grow or store food reserves, unlike the ovules of gymnosperms.

(⊤) This coloured scanning electron micrograph shows pollen that has germinated on the stigma of a prairie gentian flower (*Gentiana* sp.) and sent its narrow pollen tubes winding down toward the ovules in the ovary.

(∨) A typical angiosperm pollen grain that has germinated to form a pollen tube. Two sperm nuclei can be seen near the tip; one will unite with an egg and the other with the two polar nuclei to form the triploid endosperm.

Pollen germination

When a compatible pollen grain lands on the moist, receptive stigma of a flower, it germinates. A pollen tube emerges from the tube cell inside the pollen grain and grows through the tissues of the stigma and style, following a concentration gradient of the chemical gamma-aminobutyric acid (GABA), which is emitted by the ovules. In some species, the generative cell of the pollen grain divides to form two sperm cells as they mature in the anther. In others, the generative cell divides after it has entered the pollen tube.

Double fertilization

In angiosperms, a unique form of double fertilization occurs. One of the two sperm cells combines with the egg cell, located just inside the micropyle of the ovule. It then forms a zygote, which subsequently develops into a diploid embryo. The second sperm cell enters the ovule as well, but does not fertilize an egg, because there is only one. Instead, it migrates past the already fertilized egg and combines with the two polar nuclei that reside in the large central cell of the ovule megagametophyte. This forms a triploid nucleus, and the central cell begins to develop into a multinucleate or multicellular nutrient-storage tissue called endosperm. Thus in angiosperms, the formation of food-storage tissue does not begin until fertilization, rather than before as in other seed plants. In this strategy, energy is not wasted on ovules that do not get fertilized.

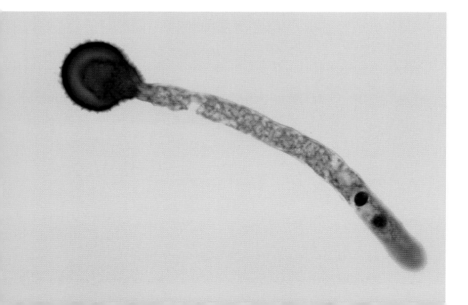

Self-incompatibility

Most angiosperms, like most organisms in fact, have mechanisms for avoiding self-fertilization. Some are based on physical impediments or differences in timing of female receptivity and pollen release. Many, however, employ mechanisms that cause stigmas to recognize pollen grains that belong to the same individual. Numerous forms of self-recognition have evolved independently in different groups of angiosperms. Commonly, these are based on the production of a protein coded in a particular gene (an s-gene). Many alleles (alternative forms) exist for these genes within a population, so different genetic individuals are likely to have different forms of the protein. It has been shown in populations of evening primrose (*Oenothera* spp.), for example, that up to 37 s-alleles may be present. If the proteins in the pollen grain are the same as those on the stigma, a set of reactions occurs that inhibits the growth of the pollen tube.

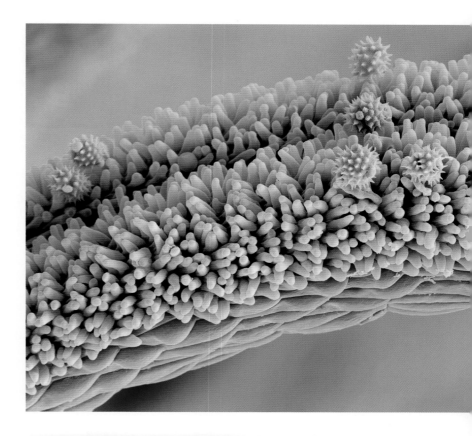

REJECTION OF FOREIGN POLLEN

Pollen from foreign species, should it land on a stigma, is also prevented from fertilizing the egg. Varied physical and chemical mechanisms result in failure of pollen grains to adhere to the surface of the stigma, to hydrate sufficiently for pollen tube growth to begin, to orient the pollen tube properly, or to be able to grow through the tissues of the pistil. In the genus *Brassica* (cabbage, mustard, broccoli, etc.), cell-to-cell recognition that initiates rejection of foreign pollen is part of the complex S-protein system also responsible for rejection of genetically identical pollen.

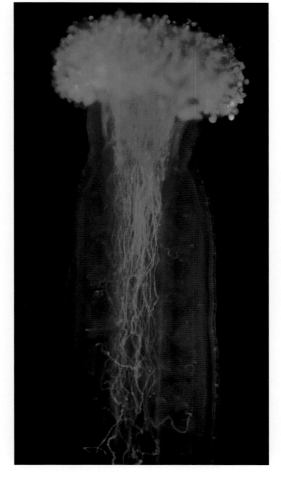

(∧) Some pollen grains (grey) can be seen stuck to the finger-like projections of the stigma (yellow) in this coloured scanning electron micrograph of an arnica (*Arnica montana*) flower. The pollen tube will undertake a lengthy journey through the stigma, style and ovary before reaching an ovule.

(<) This dark-field image of the entire pistil of a thale cress (*Arabidopsis thaliana*) reveals numerous pollen tubes extending from germinating pollen grains on the stigma down to the ovary, where each will connect to one of the ovules.

◁ Acorns are single-seeded fruits in which the dry, hard fruit wall adheres tightly to the seed. The caps are made up of small bracts.

▷ Blackberries are aggregate fruits formed from the separate carpels of a single flower. Each fruitlet contains one seed.

Fruit

$\widehat{A}$s seeds mature, the carpels that enclose them ripen into fruits. Fruits are variously adapted to aid in the dispersal of the seeds. Some dry up as they mature, and simply split open to release seeds equipped to disperse on their own, while others remain as the dispersal structures themselves. The fruit wall in those cases may be adapted for dispersal by wind, water, animals or mechanical means. Fleshy fruits are adapted to be eaten by animals, which disperse the seeds as they eat the soft fruit tissues. For more detail on seeds and fruit see Chapter 7.

▽ In the large fruit of the watermelon (*Citrullus lanatus*), a hard rind protects the numerous seeds that develop within the sweet, juicy pulp.

Pollination stimulates fruit development

As ovules begin to develop into seeds, they emit growth-regulating hormones, principally auxins, which stimulate the ovary wall to begin development into a fruit. Seedless varieties of fruit have appeared over time, by chance or by design. Seedless bananas and figs that develop without pollination occurred in ancient times, and have been propagated asexually ever since. The ancestors of the seedless banana were mutants that had three sets of chromosomes in each nucleus (they were triploid), which makes normal gamete production and fertilization difficult. In such plants, auxins are produced in the absence of ovule fertilization. The production of seedless watermelons mimics this condition, but since watermelon vines (*Citrullus lanatus*) are annual plants, triploid seeds must be created through hybridization each generation.

Ripening of fleshy fruits

In many fleshy fruits, a second hormone, ethylene, is emitted once the seeds are fully developed, which causes rapid changes in the fruit that make it more attractive to animal dispersers. Called climacteric changes, these include colour changes, softening, conversion of starch to sugar, and the generation of

fragrances. Fruits that undergo climacteric changes include apples, bananas, guavas, blueberries, apricots and tomatoes. Such fruits can be picked when full sized but still green, and will continue to ripen as ethylene is produced. Commercially grown bananas, tomatoes and pears are harvested and shipped this way, with ethylene gas often applied to complete ripening during or after shipping. Consumers may also put such fruits in paper bags to retain natural ethylene and accelerate ripening. Other fruits, such as strawberries, citrus fruits, grapes and pineapples, ripen more gradually, without such sudden change or stimulation by ethylene, and must remain on the plant.

THE FIRST FRUITS

Whether the first fruits were fleshy or dry is still unknown. If carpels evolved through the folding of a leaf-like megasporophyll (see page 209), then the early transitional fruits may have simply unfolded when the seeds were mature, like modern follicles or legumes. Fleshy fruits, which do not open and have additional developmental stages in ripening, would have come later. The most archaic angiosperms, such as *Amborella*, however, mostly bear small fleshy drupes. If these were the first type of fruits, then a completely different, and still unknown, form of ancestral structure would have to be sought.

⊘ The ascidiate carpels of *Amborella trichpoda* ripen into single-seeded red drupes visible to fruit-eating birds of the tropical forests of New Caledonia.

Dispersal of dry fruits

When the fruit wall becomes hard and dry as it matures, it plays different roles in seed dispersal. Capsules open to release their seeds, but in nuts, grains and achenes, the dry fruit wall remains as a protective layer around a single seed. Many of these dry fruits are gathered and buried by animals, but in the common dandelion (*Taraxacum officianale*), a parachute-like appendage forms on top of the single-seeded achene (more technically a cypsela), adapting it for wind dispersal.

⊘ This sycamore (*Acer pseudoplatanus*) has formed the dry winged fruits (samaras) that are characteristic of this large genus.

6

Cones and flowers

Cones and flowers are organs that produce seeds, which act as survival capsules for the young plant. However, before the seed can be put together with a coat, food supply and the embryo, there is a lot of work to be done. For example, when plants (and other organisms) reproduce, they generally take the opportunity to reshuffle their genes to see if there is a better combination for new conditions. This reshuffling is achieved by the creation of gametes (sperm and eggs), which are then mixed with other gametes, preferably from a different plant. And there is the problem.

Plants cannot move in the way that animals do when they spot another individual with which they would like to share gametes. Thus, cones and flowers must be able to facilitate not only the production but also the distribution of gametes, as well as be the midwife for the embryo. They achieve these functions in very distinct ways, because while cone- and flower-bearing plants share a common ancestor, one did not evolve directly from the other. The evolutionary origins of both cones and flowers are a bit of a mystery.

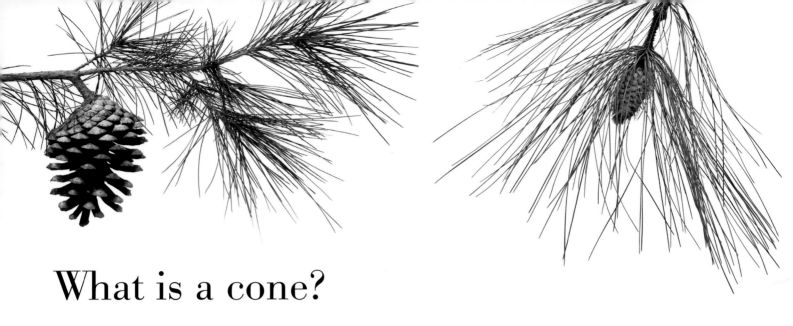

What is a cone?

Cones are normally woody structures, clearly visible to the naked eye and usually rather drab in colour. There are always two types of cones in a species: the smaller, more numerous male cones, which produce the pollen that delivers the sperm; and the larger female cones, which are the site of egg production and the seed-making machinery. Separating the production of sperm and eggs in this way helps to reduce the chance of self-fertilization, but some species have made this impossible by producing the two types of cones on different plants.

⋀ Pine trees have separate female (left) and male (right) cones that coincidentally help to reduce the risk of self-fertilization. The male cones are less numerous and smaller than the female cones.

The smaller male cones

The cones are made up of scales, which are modified leaves called sporophylls – literally meaning 'spore-bearing leaves'. In the smaller male cones, their correspondingly smaller sporophylls are called microsporophylls,

⋁ The female, or ovulate, cones of larch (*Laryx* spp.) are colourful in comparison to many other cones, although the reason for this is unclear.

and in the larger female cones they are known as megasporophylls. When you look closely at the male cones, you will see that the microsporophylls are arranged spirally, just like some true leaves, and that they are almost always brown and membranous. The microspores, which are produced in microsporangia on the microsporophylls, develop into the four-celled pollen grains. Each pollen grain eventually produces two sperm cells, but before that happens (usually after the pollen tube has started growing) it has to be released from the male cone and then land on a female cone on a tree of the same species. To aid its flight, the pollen of cone-bearing species often has two large air sacs attached on the outside. Conifer pollen has been recorded travelling up to 100 km (60 miles) in North America.

The larger female cones

At first sight, the female cones appear to be just a bigger version of the male cones, made up of thicker, woodier sporophylls. We now know that this is not the case, and that they are not made up of individual megasporophylls but are actually short, highly modified shoots, sometimes referred to as the seed-scale complex. These short branches are also arranged spirally and each has two parts: a sterile bract and, above this, a hard scale that produces the ovules. The ovules developed from the megaspores that were produced by the megasporangia on the ovuliferous scale. The megasporangium is protected by integuments, which go on to become the protective seed coat. The ovules are the location of the eggs that the pollen's sperm are looking for.

The moment of truth

When a pollen grain lands on the watery pollen drop of one of the female cones, it germinates and grows a pollen tube that extends towards the egg. The arrival of the pollen grain does not go unnoticed by the female reproductive apparatus, which starts to develop and produce eggs. When the tube reaches an egg, the sperm are released. If the sperm and egg are compatible, one sperm cell will fuse with the egg cell and hopefully an embryo will develop. This embryo is protected by the integuments that previously protected the megasporangium, and the tissue that provided nutritional support for the egg-making organs becomes the food supply for the embryo up to the time that the young seedling can support itself by photosynthesis. The whole process described here can take up to two-and-a-half years.

⋀ The young, immature green female cone of the Aleppo pine (*Pinus halepensis*), with the megasporophylls tightly shut.

⋀ When mature, the megasporophylls of the female cones turn brown and open up to release the seeds, only a small proportion of which may contain viable embryos.

CLOUDS OF POLLEN

Anyone who has left their car parked overnight below a conifer tree in spring will have noticed the yellow dust that covers the paintwork in the morning. This is pollen, and the dust will comprise millions of individual pollen grains. If you tap a male cone loaded with ripe pollen, you will be rewarded by a cloud of dust. It has been calculated that there is less than a one in a million chance that any individual conifer pollen grain will successfully deliver a sperm cell that leads to the production of viable seed.

⋁ Relying on wind-pollination is a risky business, and for there to be sufficient successful pollination large clouds of pollen are released into the air by conifers.

Unusual cones

Some of the cones produced by gymnosperms are very un-cone-like. The most extreme examples are those of *Ephedra* species, which look more like the flowers of angiosperms than the cones of gymnosperms. *Ephedra* also employs double fertilization, which is almost a defining feature of the angiosperms. This discovery led to the proposal that the genus was the 'missing link' between the angiosperms and gymnosperms. Sadly, all the evidence of the past 15 years does not support this theory, and mystery still surrounds the origins of flowering plants.

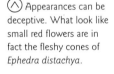

∧ Appearances can be deceptive. What look like small red flowers are in fact the fleshy cones of *Ephedra distachya*.

∨ Although it looks like a berry, the aromatic juniper (*Juniperus communis*) 'berry' is in fact a fleshy female cone.

⊐ The bright red fleshy arils of the European yew (*Taxus baccata*) are the only part of the tree that is not seriously toxic.

The berry in gin

The junipers (*Juniperus* spp.) are one of the better-known gymnosperms because of the role they play in cooking and in flavouring gin. The part of the plant used in cooking is often described as a berry, but 'berry' is one of those words that have both a precise botanical definition and a less precise colloquial meaning. In this case, the juniper 'berry' is not a berry in botanical terms because the fleshy part is derived from the fusion of the scales of the female cone.

A NEW USE FOR A BICYCLE PUMP

The ancient ancestry of cycads can pose a problem for conservation workers, because the pollinator of a particular species may have become extinct, leaving the plant without its go-between. In some cycad species, such as the African *Encephalartos ferox*, the pollen cannot penetrate the female cone far enough without the help of the probing snout of a weevil. In the absence of the weevils, which are now extinct, ingenious horticulturalists employed a bicycle pump filled with pollen and water to blast sperm into the cones so that they were close enough to swim to the eggs. Latterly, the bicycle pump has been replaced by the much more sophisticated turkey baster!

The yew berry

The Latin word for berry is *bacca*, which is the origin of the scientific species name for the European yew (*Taxus baccata*). The fleshy part of the yew fruit – the bright red aril – is derived from the end of the shoot on which the female reproductive parts were perched. The aril appears to function as a combined attraction and reward for vertebrates, especially birds, for taking the seeds away from the parent plant. It is just about the only non-toxic part of the tree, confirming that its function is in seed dispersal. Although it may seem like passage through the gut of a bird is dangerous to seeds, many survive the journey unharmed because birds have inefficient guts compared to those of mammals. Some species of plant further reduce the risk of their seeds being digested by including a laxative in the fleshy part of the fruit.

Beetlemania

Groups of organisms come and go over time, but one group that appears to be beyond its best-before date is the cycads. These gymnosperms were certainly around 285 million years ago and very diverse 135 million years ago, but like so many groups they seem to have taken a knock during the Earth's mass extinction events. They have some of the largest cones in the gymnosperms, and they exude a fluid containing sugars and amino acids to attract and reward animals that can act as pollinators. Some cones even warm themselves up to be especially attractive to

visiting beetles. The fact that beetles are used as pollinators is indicative of the ancient ancestry of these plants, because the beetles were a well-developed group long before more iconic pollinators such as birds and bees arrived on the scene. Today, weevil pollinators can still be seen feeding from the cones of the Western Australian zamia palm (*Macrozamia riedlei*).

∧ The Australian cycad *Macrozamia miquelii* is a member of an ancient group of gymnosperms that existed up to 285 million years ago, when one of the few insect pollinators were beetles.

⟨ Weevils were among the first animals to be recruited by gymnosperms for pollination services.

MEGA SPERM

The largest sperm in the world do not belong to whales or some other large vertebrate, but to the cycads. They can be 0.3 mm long and visible to the naked eye, and it is thought that this gigantism may be a result of the sperm having to swim further than most other sperm. This is one of the characteristics of cycads that points to the fact that they are an ancient group.

What is a flower?

$\textcircled{F}$lowers are the reason many people select particular species and varieties to grow in their gardens, and they have always inspired poets and playwrights, from Omar Khayyam to William Shakespeare. They are also important symbolically in many religions and cultures. The beautiful flowers of the sacred lotus (*Nelumbo nucifera*), for example, represent the purity of Buddhist enlightenment emerging from the muddy dukkha of our lives. But all this is nothing to the plant, which produces flowers for much more pragmatic reasons.

A bit of a mystery

$\vee$ Despite being undeniably beautiful and religiously significant in parts of the world, the most important function of the sacred lotus (*Nelumbo nucifera*) flowers is the production of seeds to perpetuate the parent plants.

Flowers are responsible for the production of a fruit containing a seed, or seeds, which hopefully contain viable embryos. They are found on the majority of plant species, more than 90 per cent of which are angiosperms, or flowering plants. Despite the ubiquity of flowers, their evolutionary origins are an unsolved riddle. Charles Darwin called it an abominable mystery, and 150 years later thus it remains. Unravelling the events that led to the construction and selection of structures that would evolve into flowers is inextricably entwined with the evolution of seeds. But here comes the twist in the tale, because while flowers are responsible for the production of seeds, seeds evolved long before flowers – 100 million years before, in fact.

PLANTS ARE NOT ANIMALS

The assertion that plants are not animals may seem like a superfluous truism, but understanding how they work is often hampered by the fact that schools teach far more animal biology than they do the biology of any other group. It is therefore often tempting to try to make all organisms conform to the way animals live, which leads to many problems. One example of the profound differences between plants and animals is that, while animals start their lives with their reproductive structures in place (although not mature), plants rebuild their equivalent structures every year.

The flower's job description

If you were a recruitment consultant for a plant and were looking to employ flowers, you would have to describe clearly to prospective applicants what was required. The essential duties of the post would be:

• Produce pollen that can be supplied to other flowers.
• Select appropriate pollen to provide sperm.
• Select the best of the appropriate pollen.
• Produce viable egg cells.
• Create the conditions for the fusion of sperm and egg.
• Facilitate the production of a food supply for the embryo.
• Put together seeds.
• Produce a fruit that facilitates the protection of the seeds.

Desirable duties would be:

• Facilitate the transfer of pollen to and from a flower on a different plant.
• Exploit a vector that both takes away and brings pollen from and to suitable plants, including possibly the attraction and reward of an animal.
• Produce a fruit that facilitates the dispersal of the seeds.

A job share would be acceptable, where one flower carries out the duties connected with the production and dispersal of pollen, and the other carries out the remainder. And if a job share was set up, the flowers may be on the same plant or different plants.

⌃ *Cosmos* is just one example of an annual plant that must create new reproductive structures every year.

⌃ Bees are just one of the many insect pollinators that enable flowers to fulfil the criteria set out in the job description for a flower.

Inflorescences

Only a few plants produce just one flower. In fact, it is difficult to name many species that use this strategy – except those producing the largest (*Rafflesia*) and the smallest (*Wolffia*) flowers (see pages 230–31). Some perennial plants – including *Fritillaria* species, orchids in the genus *Paphiopedilum* and the giant waterlily *Victoria amazonica* – bear just one flower per plant at a time, but they are very much the exceptions. Most plants instead put forth an inflorescence consisting of two or more flowers, suggesting that there is a definite advantage to this strategy.

Safety in numbers

Why is it better to produce more than one flower? First, such a strategy extends the time the plant is in reproductive mode, because flowers tend to be fragile and thus short-lived organs. This view is supported by the fact that plants producing one flower, such as *Paphiopedilum* species, have very robust, long-lived flowers that are fully functional for up to two months (this also makes them very good houseplants). Second, the longer the plant is in flower, the greater the chance of a visit from a pollinator bringing pollen from another plant, thus promoting outbreeding and reducing the risk of inbreeding and all that entails. However, there is a risk attached to producing many flowers of similar ages at the same time, which is that there is a chance a visiting pollinator will self-pollinate the plant. *Victoria amazonica* reduces this risk to zero by producing just one flower at a time, and being female on day one and male on day two.

HOW DO THEY DO THAT?

Botanists are interested in many areas, including what controls development in plants. The narrow range of potential patterns of inflorescences implies some developmental constraints in the tip of the shoot. This is not fully understood, but the different genes that control the production and position of flowers are gradually being identified. There is still a great deal of investigation to be carried out before we understand how plants produce their inflorescences.

▷ The talus fritillary (*Fritillaria falcata*) produces one flower per year, gambling on the chances of its successful pollination and fruit production.

▷ The inflorescences of hogweed (*Heracleum* spp.) minimize the risks of pollination by increasing the number of flowers for the pollinators.

▷ Stonecrops (*Sedum* spp.) are among many plants that produce numerous flowers over a period of several weeks or months.

Types of inflorescence

There are many different ways flowers can be arranged into an inflorescence. Some plants produce a determinate inflorescence, where the main axis ends in a flower, as opposed to an indeterminate inflorescence, where the oldest flowers are those produced at the base of the main axis. Some inflorescences are described as racemose, where there is a single main axis; or cymose, where there is no main axis; or paniculate, where the branches terminate in a flower. Some are described as either simple or compound, depending on whether there are any secondary branches. And some inflorescences have bracts at the base of each flower, while others do not. All four of these variables can be put together in every combination. However, despite the large variation there are many other patterns that plants could make but do not, suggesting that there is some developmental or genetic constraint on the structure of inflorescences of which we are unaware (see box).

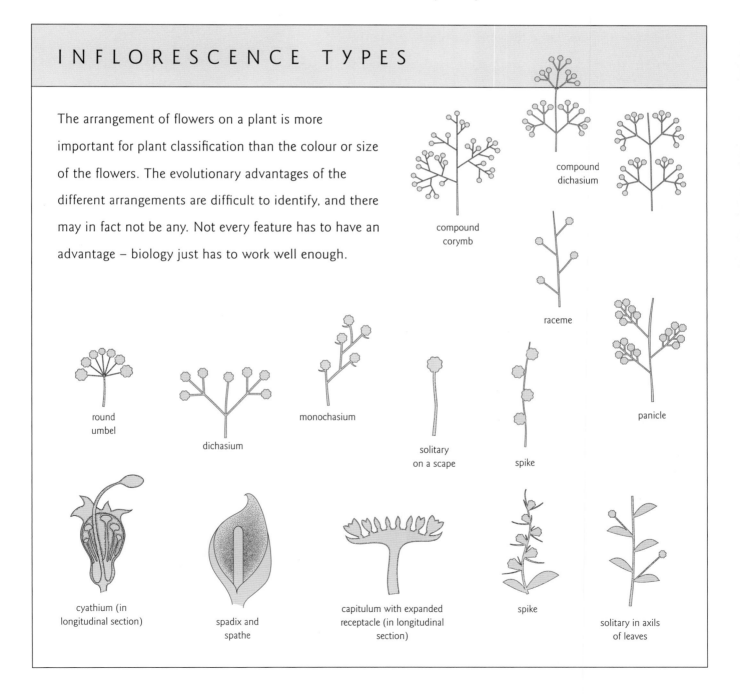

INFLORESCENCE TYPES

The arrangement of flowers on a plant is more important for plant classification than the colour or size of the flowers. The evolutionary advantages of the different arrangements are difficult to identify, and there may in fact not be any. Not every feature has to have an advantage – biology just has to work well enough.

compound dichasium

compound corymb

raceme

round umbel

dichasium

monochasium

solitary on a scape

spike

panicle

cyathium (in longitudinal section)

spadix and spathe

capitulum with expanded receptacle (in longitudinal section)

spike

solitary in axils of leaves

Flower diversity

There is no such thing in biology as a perfect organism or perfect species; there are just too many variables, with the result that organisms constantly have to make compromises. These compromises are frequently a balance between size and number, and can clearly be seen in reproduction, whereby a plant either produces many small flowers or a few large flowers. Similarly, a plant might produce many small seeds or a few large seeds – more on this later (see pages 286–87).

△ Although lacking the familiar decorative parts, the female flower of *Euphorbia goetzei* has all the essential parts for reproduction: the stigma, style and ovary.

▽ The biggest flower on Earth, the corpse flower (*Rafflesia arnoldii*), takes up to two years to develop fully and produce pollen

The largest flowers

The very largest flowers on Earth are produced by the corpse flower (*Rafflesia arnoldii*), a plant that is bizarre in many ways. First, it is parasitic on vines in the genus *Tetrastigma*, which transfer some genes to the corpse flower. Second, the parasitism is so complete that the two species form a chimera, in which the two different types of tissue are so closely attached that they appear to be one. Third, the flowers are so huge – 800 mm (31 in) in diameter and weighing 7 kg (15 lb) – that they take almost two years to reach anthesis, when the pollen is released (this is the point at which a plant is said to be 'in flower'). A fourth bizarre feature of the plant is that it smells like carrion to attract flies and other insect pollinators. Cultivating the corpse flower has proved very difficult, but a close relative, *R. keithii*, has been successfully propagated by Sabah Parks staff in East Malaysia.

The smallest flowers

Flowers can be very small, even to the point that their component parts are invisible to the naked eye. However, with just a ×10 magnification hand lens, a whole world opens up. Not surprisingly, the smallest flowers in the world are thought to belong to the smallest flowering plants, in the genus *Wolffia*. These plants can be as small as 0.6 mm by 0.33 mm, or the same size as a single cycad sperm. As well as being very small, the flowers are very simple, consisting of one stamen and one carpel. It could be claimed that the male flowers of species in the genus *Euphorbia* are just as small, because they comprise a single stamen and no carpel.

MINUTE *WOLFFIA*

It is inevitable that the smallest flowering plant in the world will have very small flowers. These flowers will be disproportionately large, and so flowering will take a disproportionate amount of the plant's energy. This

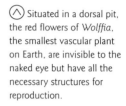

may be why the majority of the reproduction of *Wolffia* is asexual, with tiny individual leaves breaking off and drifting away from the parent plant.

(∧) Situated in a dorsal pit, the red flowers of *Wolffia*, the smallest vascular plant on Earth, are invisible to the naked eye but have all the necessary structures for reproduction.

(<) The individual plants of *Wolffia* are tiny, resembling little green dots with no identifiable features unless examined with a hand lens.

Too many of a good thing

The number of flowers produced by an individual plant is a function of its overall size and the size of the flowers. The record for the number of flowers produced on a single plant in one year is 60 million, held by a talipot palm (*Corypha umbraculifera*) growing in the Singapore Botanic Gardens. Eight months after it flowered, millions of fruit were produced. But there is a sad moral to this story, because the cost to the plant was so great that four months later it died. Despite the cost, species that produce single mass flowerings are not uncommon. Such monocarpic (literally 'one-fruiting') plants have life spans ranging from eight weeks in the case of thale cress (*Arabidopsis thaliana*), to 100 years in cultivated *Agave salmiana* var. *ferox* individuals.

(↻) The talipot palm (*Corypha umbraculifera*) holds the record for the most number of flowers produced on a single plant in one year.

The asexual parts of a flower

$\left(\text{O}\right)$ne of the 'desirable duties' of a flower's job (see page 227) is the ability to attract and reward an animal to remove pollen and transport it to a suitable beneficiary. More often than not, these duties are performed by two whorls of structures around the flower known collectively as the perianth (derived from the Latin prefix *peri-*, meaning 'around', and the Greek word *anthos*, meaning 'flower'). Sometimes, there is an extra whorl around the perianth, known as the epicalyx.

$\wedge$ The characteristic orange lanterns of the Chinese lantern (*Physalis alkekengi*) are in fact the fused segments of the calyx.

The calyx

The term calyx is derived from a Latin word meaning 'to cover' or 'conceal', and this hints at the most common function of this outer part of the flower: to cover and protect the young, developing, and perhaps fragile, flower. Generally, the calyx is made up of flat greenish structures called sepals, although this is not written in stone. In the Lenten rose (*Helleborus orientalis*), for example, the calyx is actually bright, colourful and petaloid (petal-like). And in the Chinese lantern (*Physalis alkekengi*), it is the vermilion-coloured 'lantern' that surrounds the fruit. The take-home message here is that you should not expect a calyx to be sepaloid.

The corolla

The inner whorl of the perianth is known as the corolla, which is derived from the Latin word meaning 'little garland'. This is not an unreasonable description of this whorl, because generally it is made up of the colourful, flat structures we call petals. However, again this is not written in stone, and in the Lenten rose

$\vee$ Although the decorative parts of hellebores or Lenten roses (*Helleborus* spp.) appear to be petals, they are in fact the calyx.

GLISTENING NECTARIES

Nectar is expensive to produce. The colourful parts of a flower are there to attract the pollinator and to advertise the presence of the food. However, if these parts glisten as well, there is an even better chance that the pollinator will see them.

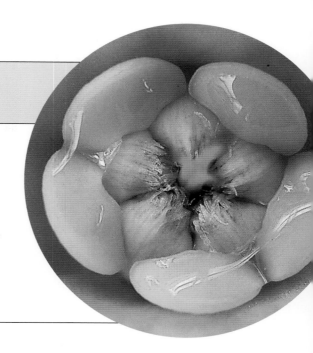

⊙ The glistening nectar in flowers such as those of euphorbias can make the flower more attractive to pollinators in the absence of colourful petals.

the corolla is actually made up of nectaries. In some garden varieties of hellebores the flowers do not have nectaries, but instead have petals where the nectaries should be. This reduces the chance of that flower being pollinated, but it is useful for demonstrating that the nectaries are where we would expect to see petals. Hellebores are in the buttercup family (Ranunculaceae), and if you look closely at the petals of buttercups (*Ranunculus* spp.) you will often see tiny nectaries at the base of the petals.

The sweetener

Nectaries are found on many different parts of plants but the majority are in the flowers, because nectar is one of the rewards given to visiting pollinators. The bigger the animal pollinator, the greater the volume of nectar that will be secreted and thus the greater the cost to the plant. Nectar is a ferociously sweet substance with a very high calorific value. If it is also scented, it can double up as the attractant for the pollinator.

ALL FOR ONE AND ONE FOR ALL

The production of scent and nectar is not restricted to any particular part of a flower. The calyx, corolla, anthers and pollen have all been recorded as producing scents, although often different scents are produced by the different parts (see pages 242–43). Likewise, the nectaries can be found on the sepals, petals, stamen and carpels, or there may be an extra part of the flower in the form of a nectary disc that serves this function. Extra-floral nectaries can be found anywhere else on the plant, although the leaves are the most common location (see box on page 175).

◁ Marsh marigold (*Caltha palustris*) demonstrates a 'typical' flower, with a colourful corolla or petals.

▷ Large Amazonian ants are rewarded for their services with nectar produced from extra-floral nectaries on the leaves of these plants.

The sexual parts of a flower

In the hypothetical flower job description given earlier (see page 227), the emphasis is very clearly on the role flowers play in sexual reproduction in angiosperms. The fact that about 50,000 species of plants engage in sexual reproduction without employing flowers means that they are not essential, but if popularity equates to success, then the most successful way to reproduce sexually is to use flowers. As we saw in Chapter 5, plants reproduce using cells called sperm and eggs, just like animals, although these cells are produced through a very different process.

The male parts

For sexual reproduction to take place, sperm must be produced. This is carried out inside the pollen grain once it is in position. However, before this process can be set in motion at the right moment, the pollen grain must first be produced, which takes place in the anthers. The anthers release their pollen in a number of ways, the most common being by simply splitting from top to bottom. This is very clearly seen in flowers like Madonna lilies (*Lilium candidum*). Alternatively, the anther may be built like a tube, with a pore opening at one end when the pollen is ready to be released. This is seen in members of the heather family (Ericaceae). Once the pollen has been released, the anthers often fall off.

The female parts

The female parts of a flower are more complicated than the male parts, mainly because they are left holding the embryo. The female organ, known as the carpel, has to oversee the selection of suitable pollen through the activities of the stigma, and then has to select the best pollen as the pollen tube grows down the style. Once the sperm have been released in the ovule, the carpel has to provide the physical resources necessary to build the seed that will be sent off into a hostile world. It may also produce a fruit to help the seed on its way.

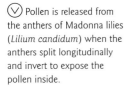
⌄ Pollen is released from the anthers of Madonna lilies (*Lilium candidum*) when the anthers split longitudinally and invert to expose the pollen inside.

⌄ The orange male parts of *Solanum hispidum* flowers contrast beautifully with the blue petals, making them conspicuous to passing pollinators.

THE FASTEST PLANT ON EARTH

A few species actively eject their pollen to increase its dispersal, including the Canadian bunchberry (*Cornus canadensis*). Its petals take just half a millisecond to fly open, which is the fastest plant movement on record. In the process, the filament and the anther at its tip are flung out of the flower, and the pollen in the anther is scattered onto the animal that has landed on the flower.

THE SIMPLEST FLOWERS IN THE WORLD

It has already been stated that some flowers carry out only male or only female activities. If a flower is male, it will produce only stamens. In fact, to fulfil its job description it need only contain one stamen. Furthermore, the job of providing attractions or rewards could be contracted out. This is the strategy adopted by *Euphorbia* species, whose male flowers consist of just one stamen. The female flowers are a bit more complicated, with three uniovulate carpels. In what might look like evolution being capricious and trying to confuse, these simplest of male and female flowers are arranged on the plant in the same way as a normal flower, with a whorl of male flowers around one female flower.

⊽ From the left: *Euphorbia septentrionalis, E. neohumbertii, E. cupularis* and *E. echinulata. Euphorbia* species have simple male and female flowers arranged in a group to resemble a 'normal' flower.

Shapes of flowers

(F) lower shapes fall into two broad categories. The first of these are regular flowers like roses (*Rosa* spp.), buttercups and tulips (*Tulipa* spp.). Also described as radially symmetrical or actinomorphic, they can be cut in half in two or more ways so that each half is a mirror image of the other. The other category contains the irregular flowers, like wisteria (*Wisteria* spp.), snapdragons (*Antirrhinum* spp.) and moth orchids (*Phalaenopsis* spp.). These are also described as bilaterally symmetrical or zygomorphic, and can be cut into two equal halves in only one way.

(∨) Orchid flowers are zygomorphic, or bilaterally symmetrical, meaning they have only one axis of symmetry.

Which came first?

Evolution is the one great law of biology. As a result, plant biologists are always looking for the reason why a particular feature gives one plant an advantage over another that has an alternative feature. There is also an underlying belief that life gets more complicated as it evolves. If this is true, then simpler structures were the original (or primitive) state and more complex structures are the later (or derived) state. Following this logic, regular flowers – in which the sepals, petals, stamens and carpels in each whorl are similar – should be closer to the first flowers. This certainly seems to be true, and in fact in some of the oldest fossilized flowers (*Magnolia* spp.), all the sepals and petals appear to be very similar.

However, even though we accept that regular flowers represent the original state, it is not unknown for lineages that have irregular flowers to revert to regularity.

(∧) *Porana oeningensis* is one of the few existing fossils of a primitive actinomorphic flower.

When is it better to be irregular?

Flower evolution is still one of the great mysteries, and there will be a huge boost to the reputation of the biologist who finally explains the origin of flowering plants. However, once the first flower emerged it is easy to see why irregular flowers might have been selected, and indeed Darwin used this as a model for how natural selection may work. Imagine a regular flower. Any animal of the right size could bring pollen to it from any other flower of any other species. This could plaster the stigma with unsuitable suitors and prevent access for the right sort of pollen. Any alteration in the structure of the flower that could select pollinators bringing the right pollen would therefore be advantageous. Irregularity seems to be frequently associated with increased pollinator specificity and thus efficiency. There may also be other reasons for irregularity, such that irregular flowers might be better at keeping the pollen dry.

(<) Actinomorphic, or radially symmetrical, flowers have numerous axes of symmetry and can be cut into equal halves many ways.

How do flowers become irregular?

Flowers that are irregular may have advantages, but how does a plant modify its flowers? What are the genetic changes that lead to changes in the flower structure? There seem to be several answers to this question, all of which may be true in some cases. First, there is a gene for symmetry that is linked to the genes that control the structure of the inflorescence. It seems that irregular flowers are most common on indeterminate racemose inflorescences, while regular flowers are found on both racemose and cymose structures. Irregularity in flowers appears to have evolved separately at least 25 times, and three times in just one family – the daisy family (Asteraceae). As mentioned above, irregularity has been reversed in some of these cases, although several of these involved the loss of parts, which is genetically more complicated and hence less likely to occur. That said, gardeners often report growing common foxgloves (*Digitalis purpurea*) that have just one regular flower at the top of the inflorescence, which has been shown to be caused by a mutation in a single gene (see box).

FUNNY FOXGLOVE

Foxgloves normally have an indeterminate inflorescence, which means that it goes on producing new flowers at its apex until it either dies or is perhaps told to stop by more mature flowers further down the stem. In the 'Monstrosa' mutant, a relatively large, radially symmetrical flower is produced at the apex and that is the end of flowering. This mutant was first recorded in 1869 and has been seen many times since.

⊘ Mutations in *Digitalis* species can result in a single actinomorphic terminal flower while the remaining flowers are still zygomorphic.

FINDING THE GENE

When researchers are trying to discover which genes are responsible for the development of a plant, they look for mutant plants where the gene in question is broken. For example, in a normal common foxglove flower, the four petals are fused together into a tube. However, there is a common foxglove cultivar in which the four petals are not fused to each other. In this plant, the gene that is responsible for fusing the petals together in the very young buds is broken or 'knocked out'.

◁ *Digitalis* 'Serendipity', a mutant plant, demonstrates that the foxglove's zygomorphic flowers are formed from the fusion of four petals.

 A *Colletes* bee collects bright orange pollen from an indigo bush (*Psorthamnus* sp.) flower, as a reward for pollinating the plant.

 *Hydnora africana* emits the odour of faeces, which serves to attract its natural pollinators, dung beetles.

The spadices of *Amorphophallus* species look and smell like rotting meat, making them particularly attractive to pollinating flies.

Pollination

P ollination is one of the most familiar plant–animal interactions after herbivory. In fact, in some cases it is seen as a form of herbivory, because the visiting animal is not there for the benefit of the plant *per se*, but for the 'reward' provided by the plant. Pollination is one of the most intimate associations between two organisms, and one where plants and animals meet on equal terms. Nowhere else in the natural world does one organism entrust another with the safe and accurate delivery of its sperm.

It all starts with attraction

Pollination may be a perfect example of a symbiosis between two organisms, but that is only after it has been established. At the start, one of the partners has to make the first move, and that is likely to be the plant, which has to make itself attractive. It might do this by exploiting an existing signal that the animal will already recognize and associate with something good. An unsavoury example might be the propensity of flies to gravitate towards faeces, so a flower that smells of faeces would also catch the attention of the flies. Colours as well as odours are regularly used attractants, and familiar shapes and patterns can be used in conjunction with these.

THE NEED TO OUTBREED

A plant is an organism that cannot move under its own steam in a coordinated and planned direction over long distances. One of the difficulties this creates is the challenge of bringing its eggs and sperm into contact with the sperm and eggs of a different organism. If sperm and eggs from the same, or even closely related, organisms are used in sexual reproduction, there is a risk of inbreeding depression. This is when the product of sexual reproduction between closely related organisms builds up an above-average proportion of suboptimal versions of genes. When the effects of all of these poor genes are added together, the organism becomes less able to breed successfully. This becomes critical in a small, isolated population because there may be no option other than accepting your own or a relative's genes. Inbreeding depression is something that conservation biologists work hard to avoid, although in very rare species this may be impossible.

Rewards are always welcome

Having attracted an animal's attention and got it interested, the plant has to make sure that it will return again and again, and visit other similar plants that advertise the same attractive come-hither signals. Being simple creatures, animals can easily be satisfied. As already noted, a common reward is sweet, energy-rich nectar for them to lap up, or they may be offered tissues rich in starch that they can nibble on. The reward may also be something slightly tangential, like the increased chance of finding a mate or even be somewhere to lay eggs, as is the case in the figs (*Ficus* spp.) (see page 302).

◁ *Ceratosolen galili* wasps are rewarded for their pollination services with a ready-made nursery in the fig fruit.

How close should you be?

In a relationship like the one between the plant and its pollinator, there is a need for fidelity and reliability; one might almost say trust. This can result in some relationships becoming very exclusive, where just one animal species and one plant species are involved. While this guarantees that the pollen delivered by the animal is suitable and that it is always put in the right place, thus reducing wastage, it also leads to great vulnerability. If one of the partners changes their behaviour or disappears, for example, the other can be in trouble. Organisms can change their behaviour when the climate changes, or they can even disappear due to extinction events or as a result of migrations. These very close relationships may not appear to be such a good idea right now, as climate change and its associated effects accelerate.

Flower colour

The prevailing colour of plants is green, and any variation on this is for one of two reasons. First, it may be coincidental – for example, the colour of bark is a by-product of the molecules it contains. Or second, the colour has a function and bestows a selective advantage that compensates for the cost of producing it. Flowers that are wind pollinated are uniformly drab, whereas those pollinated by animals are coloured. This is very strong evidence that the colours of flowers are signals that are interpreted by animals.

Blue flowers such as those of forget-me-not (*Myosotis* spp.) are particularly attractive to bees.

Flowers that are pollinated by birds, such as this Little Spiderhunter (*Arachnothera longirostra*), are frequently red or bright yellow.

How to make colours

The parts of flowers are coloured as a result of pigments (chemistry) and cell shape and structure (physics). There are four large groups of molecules involved in colouring plants. The first – the tetrapyrroles – include chlorophyll and are involved in photosynthesis and other light-sensing activity. The other three – the carotenoids, flavonoids (including the anthocyanins) and betalains – are all found in flower parts, including pollen and nectar. There are hundreds of different pigments within these three groups, and they are put together in a plethora of combinations to create the dazzling array of colours we see in flowers. The pigments are either stored in the cell vacuole, or in plastids in the cytoplasm or in the vacuole (e.g. anthocyanoplasts). Colours can be altered by temperature and cellular pH, the presence of metals and sugars, and cell shape. The cells of the epidermis may be conical, flat or pointed, and in the case of conical cells, their shape makes colours appear more saturated.

Petal colours

It is universally acknowledged that petal colours are central to attracting a variety of pollinators. Darwin introduced the concept of pollination syndromes, where well-defined combinations of characters are always linked to specific guilds of pollinators. While this can work, it is now clear that there are no hard and fast rules. Some scientists have found it very difficult to believe that Darwin may not have been right, but there are some broad generalizations that link specific colours with specific pollinators. Generally speaking, bees are attracted to blue and yellow, beetles to white or dull colours, moths to whites and yellows at twilight, butterflies to anything bright, birds to reds and bright yellows, flies to whites and browns, and bats to large, dull-coloured flowers.

COLOURED POLLEN

The bright colours of pollen, often contrasting with the petals of flowers, are well known to gardeners. However, this may be the exception, because ideally a plant does not want its pollen to be eaten by robbers. When many plants are examined under ultraviolet light – a part of the light spectrum visible to reptiles, birds and many insects, but not to humans – the pollen is the same colour as the centre of the flower, as if the plant is trying to hide it from robbers.

Coloured nectar

The fact that nectar is sometimes coloured and/or scented is a classic example of a plant feature having more than one function. Coloured nectar appears to have evolved at least 15 times independently across the evolutionary tree in variety of habitats. One study of a small group of just 67 taxa found that coloured nectar was correlated with three situations: vertebrate pollinators, including lizards, on Mauritius; insular habitats; and high altitudes. When it comes to the function of the colour in nectar, it might be a non-functional trait, it might be an attracting signal to the pollinator, it might be a repelling signal to robbers, or it could be that the pigment has antibiotic properties to ensure the nectar remains safe to drink.

⌃ Bees are not only attracted to blue flowers but also to blue pollen. This bee is collecting blue pollen from the red flowers of *Echium wildpretii*.

FLOWERS OF THE JADE VINE

Occasionally, nature produces colours that look unreal, one example being flowers of the jade vine (*Strongylodon macrobotrys*). These are pollinated by bats, which is unexpected, given their colour. The petals are covered in cone-shaped cells, which may contribute to their extraordinary 'unnatural' colour, which is strangely reminiscent of Cambridge blue, the greenish hue worn by Cambridge University sports teams.

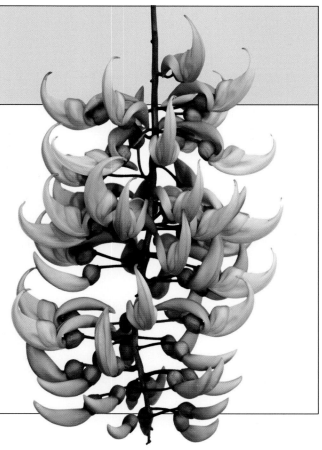

⌄ The jade vine (*Strongylodon macrobotrys*), with its curious blue-green flowers, is pollinated by bats and drips with nectar.

Flower scents

(M)any people cannot resist sniffing a scented rose, especially an old-fashioned bloom. The smell it gives off is the result of the synthesis of small, simple, volatile organic compounds that can cross cell membranes easily, evaporate quickly and travel a reasonable distance. Within this specification are about 1,700 known compounds and more waiting to be identified, which plants mix into bouquets to create countless aroma combinations. While the majority of flowers use 20–60 compounds to create their scent, a few use just one and some use as many as 150.

Making scent

There are three classes of compounds that plants synthesize to make up the majority of odours: terpenoids (including 556 compounds), aliphatics (528), and benzenoids and phenylpropanoids (329). Within these groups are specific scents that are named after the plant from which they were first extracted, such as geraniol, camphor, menthol, vanillin, jasmine and so on. Despite their names, it is important to remember that these compounds are not restricted to these plants, or even most

common in them. For example, geraniols and citronellol are more common in roses than in *Geranium* and *Citrus* flowers, respectively. For those of us who are not chemists, odours are also classified more simply as flowery, dung or carrion, and pheromonal.

Location of scent production

Most plant parts can produce odours. Roots, stems, leaves and fruits are all smelly in some plants, but the most commonly scented part of a plant is undoubtedly the flower. Within the flower, no part has unique responsibility for scent production. For example, in the flowers of nicotianas (*Nicotiana* spp.), the petals

STOP AND SMELL THE ROSES

Sensor cells enable animals to detect odours. Bees have up to 170 different sensor cells in their antennae, whereas fruit flies have only 62, so bees can detect a wider variety of compounds. One study claimed that individual bees could detect and remember up to 1,729 different scent combinations out of 1,816 pairs they were exposed to, and in total more than 7,000 floral bouquets are recognized by bees and associated with combinations of floral characters. There is still controversy among ornithologists as to how good birds are at smelling, but it is clear that bird-pollinated flowers are often weakly scented. When a pollinator is attracted by scent, it will approach the flower along a zigzag route, constantly testing where the greatest concentration of scent is located. In contrast, if the pollinator is attracted by a visual signal, it approaches in a straight line.

(∨) The sweet scent of *Nicotiana* species is produced predominantly by the petals.

∧ Bees are most active in the morning; rhododendrons capitalise on this by emitting their scent at this time of day.

⅂ Perfoliate honeysuckle (*Lonicera caprifolium*) is pollinated by moths at night-time and is most fragrant as the sun goes down.

produce the majority of the scent, but the stamens, the style and stigma, and the sepals also produce scent, to lessening extents. Within the flowers of roses, the sepals, petals, anthers and pollen play equal parts in scent production, and the chemicals produced vary. The majority of the scent produced by the sepals and petals are fatty acid derivatives, while the anthers synthesize benzenoids, and the pollen uses terpenoids.

Timing of scent production

As with colour, there are some generalizations that can be made to match types of scent with type of animal. However, the timing of scent production is important and related to the pollinator. For example, flowers pollinated by moths, such as nicotianas, honeysuckle (*Lonicera* spp.) and night-flowering cacti, are scented only in the late evening and early night-time. In contrast, many orchids, rhododendrons (*Rhododendron* spp.) and snapdragons are scented in the early morning, when their bee pollinators are most active.

GREEN FLOWERS

If a plant has green flowers, it must rely on scent to attract a pollinator. One such plant is the Mexican *Deherainia smaragdina*, whose flowers have stamens that spring apart after they have released their pollen, only then exposing the stigma to incoming pollen. The pollinator will be attracted by the peculiar scent, which very closely resembles the inside of a teenager's trainers.

∨ With its inconspicuous green flowers, *Deherainia smaragdina* relies on its revolting scent to attract its pollinators.

Hot flowers

(F) lowers and floral structures that are intentionally heated from the inside (thermogenesis) are found in hundreds of species in a dozen plant families. These families are predominantly among the more ancient lineages at the base of the angiosperm evolutionary tree. There is also one recorded example of thermogenesis in a cycad in the gymnosperms. Thus, this is a feature that has evolved several times for perhaps four different reasons. Often it seems to be associated with beetle pollination, protogyny (where the female parts mature first) and floral gigantism.

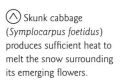

⌄ Skunk cabbage (*Symplocarpus foetidus*) produces sufficient heat to melt the snow surrounding its emerging flowers.

> Sacred lotus (*Nelumbo nucifera*) maintains its flowers at a constant temperature regardless of the ambient conditions.

Why have centrally heated flowers?

The first, and perhaps most obvious, reason to heat your flowers is to make the scent go further. This is a common behaviour in humans, who use perfume more effectively by putting it on pulse points – areas of the body where the blood vessels are near the skin surface. However, the plants that heat their flowers often produce scents that are revolting to us. The second reason for warming the flowers is to protect them from frost damage. This is suggested for the skunk cabbage (*Symplocarpus foetidus*), whose flowers can appear through snow. Third, the heat is proposed to be an attraction/reward to beetles. Finally, and most recently, it has been shown that pollen germination and pollen tube growth is better at higher temperatures. It has also been found that some subantarctic mega herbs can raise their internal temperature as much as 10 °C (18 °F) above the ambient temperature, not by burning stored fuel, but with a combination of hairs, colour and shape that allows increased absorption of the sun's heat.

How do they do that?

The heat is produced in up to three different parts of the flower. For example, in the flowers of the scared lotus, half of the heat is produced in the receptacle (the structure upon which the flower parts sit), one-quarter by the petals and the rest by the stamens. This enables the flower to use the heat in different ways at different times during flowering. The mechanism for producing the heat seems to be the same in all thermogenic plants: by burning fuel in their mitochondria. This combustion is controlled by a cyanide-resistant alternative oxidase enzyme (AOX), and the enzyme is controlled at both the level of the gene and after it has been synthesized.

Keeping the customer satisfied

The production of the heat not only coincides with flowering, but it can also vary through the day. For example, heat production in the giant Amazonian water lily (*Victoria amazonica*) is bimodal, being lower in the daytime and peaking at sunset, which is when the pollinating beetles are seen entering and leaving the flowers. Why are beetles closely associated with many of these warm flowers? The answer seems to be that because beetles have a very high energy requirement for their activity, they try to get help with keeping warm wherever they can. *Philodendron solimoesense*, the record holder for floral temperature (46 °C, or 115 °F), is pollinated by beetles that mate within its floral chamber, and it has been demonstrated that the beetles need 2–4.8 times the amount of energy to mate in the open compared with mating in the flower. By promoting the reproduction of the beetle, the plant is helping to ensure that its pollinator survives.

⌃ The Amazonian water lily (*Victoria amazonica*) creates a warm environment for visiting beetles.

⌄ *Philodendron solimoesense*'s ability to raise its temperature provides an ideal mating environment for its pollinating beetles.

PLANTS ARE NOT ANIMALS, BUT...

Having stated several times that plants are not animals and that it is often futile to draw parallels between the two groups, thermogenesis is one area where there are some similarities. Thermogenetic plants that heat their flowers can also regulate that heat within a narrow range, independent of the ambient temperature. The most spectacular example is the skunk cabbage, which maintains its flower temperatures within a 3.5 °C (6.3 °F) range (22.7–26.2 °C, or 72.9–79.2 °F) over a 37.4 °C (67.3 °F) range of ambient temperatures (−10.3–27.1 °C, or 13.5–80.8 °F). This is achieved due to the fact that the activity of the AOX enzyme is inversely proportional to the ambient temperature.

< The flowers of the skunk cabbage (*Symplocarpus foetidus*) can regulate their temperature within a precise range.

Nectaries and spurs

(T)he petal spur holds a lofty position in plant mythology thanks to Darwin's orchid (*Angraecum sesquipedale*), a fragrant white flower found in Madagascar that has an impressive 450 mm-long (18 in) spur with a nectary at its base. The eponymous naturalist looked at this flower and deduced that it must be pollinated by a moth with a proboscis long enough to reach the bottom of the tube. Twenty years later, he was proved correct, when Morgan's sphinx moth (*Xanthopan morganii praedicta*) was discovered pollinating the flowers.

(>) The crimson columbine (*Aquilegia formosa*) of North America is pollinated by bees and hummingbirds.

(>) The Sierra columbine (*Aquilegia pubescens*), from the same area as the crimson columbine, is pollinated by hawkmoths and so avoids hybridization with its relative.

(∧) Darwin's orchid (*Angraecum sesquipedale*) has an extremely long petal spur that is 450 mm (18 in) long.

The expensive amber nectar

Nectaries can be found in many places on a flower, and indeed in many places on plants, including their leaves. However, a spur is a common location for a nectary. A floral spur is a tubular pocket with nectar-secreting glands at the distal, or far, end. Nectaries vary in size, shape and structure. The nectar they produce is not just leakage from the phloem, but is deliberately made. This means that there is a cost to its production, and where there is a cost there must be a compensating benefit. The amount of nectar produced is proportional to the size of the nectary, and if the nectar is not consumed by the pollinator (or a robber) then it will be reabsorbed by the flower and at least some of the costs will be recuperated.

How difficult is it to make a tubular pocket?

The answer to this question seems to be not very. There are many different spur-bearing flowering plants. It is a character that has evolved, and been lost, many times. It also seems that it is a character that can evolve very quickly, which means that it might involve very few (perhaps just one or two) genetic changes or mutations. Detailed cell-level studies of the flowers of columbines (*Aquilegia* spp.) have shown that it is asymmetrical expansion of the cells, rather than cell division, that creates the characteristic spur morphology. Furthermore, it has been found that spur length is not proportional to the number of the cells but to the length of individual cells. This means that making a longer tube is very straightforward.

Which came first, the long tube or the long tongue?

The floral nectar spurs of columbines, snapdragons, delphiniums (*Delphinium* spp.) and aconites (*Aconitum* spp.) have all been studied extensively, both at the genetic level and at the ecological level of pollination and speciation. It has been found that short-spurred species are pollinated by bees and bumblebees, medium-length species by hummingbirds and long-

(<) Morgan's sphinx moth (*Xanthopan morganii praedicta*) has a proboscis that is the perfect length to pollinate Darwin's orchid.

The flowers of the beautiful Mediterranean stavesacre (*Delphinium staphisagria*) start out looking like those of other members of the buttercup family, being radially symmetrical and with nectaries on the petals. However, nearly halfway through their development, spurs grow and the flowers become zygomorphic, and with this change the number of possible pollinators falls dramatically.

⌃ Depending on their stage of development, stavesacre (*Delphinium staphisagria*) flowers are either actinomorphic or zygomorphic.

spurred species by hawkmoths. It was thought that the flowers and the pollinators co-evolved, changing together over time. However, it is now believed that only the flowers have changed and the animals have remained the same, simply adopting new flowers to pollinate as they appear. Spur length is just one of the variables in flowers of the 70 columbine species, which also vary greatly in colour, odour and orientation. Changes in any or all of these factors can lead to a change in pollinator, which leads to reproductive isolation, which leads to speciation – in other words, evolution. The process can happen so quickly that the plant species remain inter-fertile if crossed

by hand. The crimson columbine (*Aquilegia formosa*) and Sierra columbine (*A. pubescens*) readily hybridize but remain separate species solely due to the different animals that pollinate them – hummingbirds in the case of the crimson columbine and hawkmoths in the Sierra columbine.

TOXIC NECTAR – WHY?

The natural world constantly surprises and contradicts some of the simplest of assumptions. Nectar is a reward or payment for services rendered, so it would not make sense for it to be toxic, and yet this is widespread. We do not yet know why nectar is toxic to some organisms, but there are four hypotheses. First, toxic nectar would deter nectar robbers – assuming the proper pollinator could tolerate the toxin. Second, toxic nectar could select specific pollinators, thus reducing the amount of useless pollen brought to the flower. Third, the nectar may be toxic only to microbes that would otherwise live in the nectar. Finally, it might be there 'by mistake', having leaked in from the surrounding tissues – if it does no harm to the pollinators, then this is not a problem. There is clearly work to be done here.

Composite 'flowers'

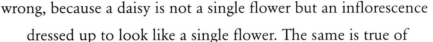

F lowers are sometimes more complicated than they appear. Taking the common daisy (*Bellis perennis*) as an example, if you ask someone who is familiar with this plant to describe the symmetry of its flowers, they will almost always say that they are regular, radial or actinomorphic. Whichever one they choose, they stand a good chance of being both right and wrong, because a daisy is not a single flower but an inflorescence dressed up to look like a single flower. The same is true of flowers in other families and genera, the most common being the floral structure of *Euphorbia* species.

⑦ The common daisy (*Bellis perennis*) is actually composed of two types of flowers, the central yellow actinomorphic disc florets and the outer white zygomorphic ray florets.

◁ Dandelions (*Taraxacum* spp.) lack central disc florets and are made up solely of yellow ray florets.

Stamp on a foxglove and you get a daisy

Earlier (see box on page 237), mention was made of the widespread common foxglove cultivar *Digitalis purpurea* 'Monstrosa', which has one large, regular, terminal flower atop a 'normal' raceme or spike of irregular tubular flowers. Now imagine that you could stamp down perfectly on the tip of the foxglove inflorescence, squashing the stem from 1.8 m (6 ft) to 2.5 cm (1 in) without damaging any of the flowers. You would end up with a short stem, now known as a receptacle, on which sits one regular flower in the middle surrounded by many irregular flowers. Each flower would still have a bract at its base, and the whole inflorescence would be surrounded by many bracts, known as the involucre. In fact, you would have something that is not so very different from a daisy. While this clearly doesn't happen, it illustrates the important fact that, while a daisy flower appears odd at first sight, it is simply a variation on a theme we have encountered before; it is called a capitulum.

VARIABLE CALYCES

The calyces of Asteraceae florets are not as conventional as the corollas. In some members of the family the calyx is absent altogether, or if present it may consist of a ring of membranous tissue or lots of simple hairs. In some cases, it can be made up of compound hairs that look like tiny combs when viewed through a good hand lens. The 24,000 species in the daisy family have a well-earned reputation for being difficult to identify, but it cannot be denied that they are beautiful.

Division of labour

The daisy family (Asteraceae) is the second largest plant family in the world (after the orchid family, Orchidaceae), and the diversity of its members – in terms of both vegetative and floral characters – is large. However, many of them do have an inflorescence like an aster (*Aster* spp.) or a common daisy, in which there is both a ring of irregular flowers with long, strappy petals radiating out around the edge, and a centre disc of smaller regular flowers, which may be a different colour. Often the flowers around the edge – called ray florets – are female and lack any male parts. Their function is to attract the pollinators. The disc florets are normally bisexual, and they produce the pollen and the majority of the seeds.

Taking the plunger

The disc florets have an intriguing way of presenting their pollen. The five stamens are in a ring around the carpel, but the anthers are fused into a tube. The pollen is released into the centre of the tube, and the style and stigma then grow up the centre of that tube, pushing the pollen out like a plunger. To prevent self-pollination, the characteristic two-pronged, Y-shaped stigma is firmly closed as it moves up the tube, opening up only when the style is clear of the anthers.

Aberrant asters

Other members of the daisy family include the thistles. Their flowers are all the same, all bisexual and all regular, with a long tubular corolla made up of five fused petals. Thistles often have spectacularly beautiful but spiny bracts surrounding the capitulum, which are often very useful in plant identification. The dandelions (*Taraxacum* spp.) and the hawkweeds (*Hieracium* spp.) also have just one type of floret, but these are all ray or ligulate florets. The most bizarre inflorescences in the group are those on the globe thistles (*Echinops* spp.), in which the inflorescence is an inflorescence of inflorescences: a collection of single-flowered capitula, each with an elaborate collection of involucral bracts.

⟩ The purple 'petals' of the alpine aster (*Aster alpinus*) is, in fact, a mass of tiny separate flowers, typical of the daisy family, Asteraceae.

⋀ Examining a daisy 'flower' with a ×10 hand lens reveals a plethora of tiny flowers most of which will all have the prerequisite sexual parts.

⋁ Globe thistles (*Echinops* spp.) are a complex amalgamation of inflorescences of inflorescences.

Flowers for vertebrates

(F) lowers are mostly pollinated by animals that fly, and most animals that fly are pollinators. The major exception are the dragonflies and their relatives, which are carnivorous and are more likely to eat pollinators than nectar. Among the large flying animals that pollinate flowers are bats and birds, but non-flying vertebrates are also important players, including lizards and possums. As you would expect, the flowers that feed these animals are very different from each other and from those pollinated by much smaller invertebrates.

(⌄) Birds with long bills, such as this white-necked jacobin (*Florisuga mellivora*), are superbly suited to retrieving nectar from long tubular flowers.

(⌐) Bats, such as this lesser long-nosed bat (*Leptonycteris yerbabuenae*) covered in pollen, are clumsy pollinators that require easily accessible, robust flowers that produce copious amounts of nectar as a reward.

Bats do it

Bats are a well-researched group of pollinators. Because they are predominantly nocturnal and have limited visual perception, the flowers they pollinate are not normally brightly coloured, a famous exception being the jade vine (see box on page 241). Bats can also be a bit clumsy, and so the flowers have to be robust, and although the pollinators are amazing navigators, the flowers are generally on the very outside of the plant. Furthermore, bats are very energy-expensive animals, partly because of their small size and partly because of their lifestyle, so bat-pollinated flowers must produce large volumes of nectar as a reward. A typical bat-pollinated plant is the Mexican *Agave salmiana* var. *ferox*, which produces such large quantities of nectar that it drips from the flowers, accumulating in pools on the leaves below the inflorescence. One of the most remarkable bat-pollinated plants is *Marcgravia evenia*, with a concave leaf above the inflorescence that serves as a sounding beacon for echolocation of the flowers.

Birds do it

Flowers that are pollinated by birds also produce copious volumes of nectar – the cockspur coral tree (*Erythrina crista-galli*), for example, produces 1 ml of nectar per flower per night. The flowers of this member of the legume family (Fabaceae) revolve through 180 degrees to ensure that the pollen is placed on a suitable part of the bird. Flowers pollinated by birds tend to be tubular and red, and they also often demonstrate approach

herkogamy, in which the stigma is held out above the anthers. The consequence of this arrangement is that the approaching bird deposits pollen from a previously visited flower before it penetrates further into the flower to drink the nectar and collect pollen to take to the next flower. Unlike bat-pollinated flowers, those pollinated by birds can be fragile, because many species – notably hummingbirds – can be very careful and considerate as they hover over them.

Even lizards do it

Reptiles in general and lizards in particular are not obvious candidates for pollinators, because most lizards are carnivorous and those that do eat plants normally consume large quantities of fruit, along with flowers, pollen and nectar. Consuming flowers, complete with nectar and pollen, does not result in pollination! However, on several oceanic islands – including New Zealand, the Balearics, Mauritius and Madeira – reptiles have been seen drinking from the flowers of both native and non-native plants.

Why flowers on islands are more likely to be pollinated by lizards than those on the mainland remains to be clarified, but possible reasons include the absence of sufficient suitable insects or mammals to act as pollinators; the fact that only big lizards tend to reach islands, and big lizards are generally herbivores; and finally the possibility that the lizards can come out into the open without the fear of predation.

⌃ The arrangement of the stamens and stigma in cockspur coral tree (*Erythrina crista-galli*) flowers reduce the likelihood of self-pollination.

⌄ Although dominated by winged animals, pollination is also achieved when other vertebrates such as this Hawaiin gecko visit flowers for a nectar treat. Reptilian pollinators are almost exclusively restricted to oceanic islands.

WHAT QUALIFIES A VERTEBRATE AS A POLLINATOR?

For a vertebrate to be defined as a pollinator, it has to fulfil a number of obligations:

• Regular visits to flowers.

• Non-destructive feeding.

• Derive a significant amount of its diet from pollen and/or nectar.

• Collect and transport pollen.

• Deposit pollen on stigma of plants of the correct species.

• Promote viable seed production in the plants visited.

If these obligations are not fulfilled, then the animal is a visitor, not a pollinator.

Flowers for bees and wasps

Flowers and bees are almost synonymous with pollination, and yet flowers emerged at least 25 million years before bees. Bees evolved from wasps, and among the prey for these early wasps were the beetles that were already carrying out pollination. It may be that the wasps preferred the taste of the pollen on the beetles to the taste of the beetles themselves, and so started going straight to the flowers. It is believed that the rapid increase in diversity of the bees and flowers went hand in hand – in other words, diversity begets diversity.

⌃ It is not only social bees that pollinate flowers; solitary animals such as the common carder bee (*Bombus pascuorum*) also get involved.

⌃ Despite being wind pollinated, the flowers of willows are often visited by hungry bees, which snack on the pollen and might accidentally elicit pollination.

< *Lithurgopsis*, like all Megachilidae bees, have their pollen-collecting hairs (scopa) on the underside of the abdomen. This *Lithurgopsis* has filled her scopa with the large pollen grains of the cactus flower.

> This *Andrena* bee has to pass by anthers full of pollen, and a stigma (hidden among the anthers in this photo), before reaching sweet nectar. In this way, flowers ensure that visiting bees pollinate them.

A match made by evolution

Bees are very efficient transporters of pollen thanks to the hairs that cover their bodies. Social bees are very good communicators and thus can work together to move pollen between populations of plants of the same species. There is competition between plants for the attentions of pollinators and they will stop at nothing to win the affection of their bees. Included in the dirty tricks plants use is lacing their nectar with caffeine, nicotine and other addictive substances, to give the bee a reason for returning other than the energy-rich nectar.

THE BEST BEE POLLINATORS

Some bees are particularly good pollinators, among them the solitary mason bees (*Osmia* spp.), which build mud structures to house their eggs. Mason bees do not make honey, which means that the pollen is of more interest to them than the nectar. However, there are other reasons why they are a popular commercial pollinator. For a start, they visit more flowers than honeybees (*Apis* spp.). Second, they do not have the pollen baskets of other bees, which means that they are messy and the pollen falls off their bodies as they fly around. Finally, they are active at lower temperatures than other bees, and so are mobile for more months of the year and for a longer period during the day.

Designing a flower for a bee

Many different flowers are visited by bees and wasps, especially in our gardens, which are full of exotic species the bees will not have encountered before. As a result, it is difficult to be dogmatic about what flower designs are most commonly associated with bees. However, there are some types of flowers that are pollinated regularly by bees, including the pea-like flowers of members of the legume family. In these, the stamens and stigma are concealed and protected by two fused petals, and exposed to the hairy body of the bee only when it pushes into the flower to get at the nectaries deep inside. This type of design is also common to the bee-pollinated snapdragons.

ULTRAVIOLET REVELATIONS

Bees do not see the world as we humans do. The spectrum of colours visible to most human eyes runs from violet to red, while ultraviolet and infrared remain invisible to us. Bees, on the other hand, can see the ultraviolet colours. They pay a price for this at the other end of the spectrum, though, where yellow and orange become a messy yellow-green, and red (and infrared) appears as black. Flowers exploit this by including guidelines on their sepals and petals, visible only to their preferred pollinators and drawing them in.

⊙ Evening primrose (*Oenothera* sp.) in visible light (left) and false-colour reflected ultraviolet (UV) light (right). In UV light, the flowers' guidelines, intended to help their pollinators find their target, are visible.

Flowers for moths and butterflies

It is easy to assume that moths and butterflies are similar, but in fact the two groups are as different as bats and birds when it comes to their pollination behaviour. For a start, butterflies are active during the day, while most moths are nocturnal. Second, butterflies do not have a very well-developed sense of smell, while moths do. And third, butterflies are good at perceiving shapes and colours, while moths are not, on account of being nocturnal. The flowers they pollinate reflect these and other differences.

∧ Indian gooseberry (*Phyllanthus emblica*) flowers have short corolla tubes that are atypical for a moth-pollinated flower.

⊃ Despite being most commonly associated with butterflies, the butterfly bush (*Buddleja davidii*) flowers are also visited by day-flying moths, which may take the nectar without effecting pollination.

Specialist moths

We have already seen how Darwin was correct when he predicted that Darwin's orchid is pollinated by a moth with a very long proboscis (see page 246), and later we will also see how *Yucca* species have formed very close associations with different species of moths (see page 259). Another very close relationship exists between moths and plant species in the genus *Phyllanthus*. In this case, successful pollination is essential for the future of both plant and moth, because the larvae of the moth feed exclusively on the seeds of the *Phyllanthus*. This is a delicately balanced relationship, requiring the failure of some moth eggs to hatch or larvae to survive to ensure that some seeds are not eaten. Moth flowers are often solitary and hang down because moths feed without landing, preferring to hover in front of the flowers.

Generalist butterflies

In addition to the differences between moth- and butterfly-pollinated flowers listed above, flowers pollinated by butterflies tend to be in flat-topped clusters, such as those found on the butterfly-bush (*Buddleja davidii*) and lantana (*Lantana camara*). The flowers of these and other butterfly-pollinated species have a narrow rim with anthers that protrude beyond it, allowing pollen to be deposited on the face of the insect. These two plants are so popular with butterflies that they are used on butterfly farms to feed species that never see them in the wild.

One genus, many pollinators

Plants in the genus *Gladiolus* employ a remarkable diversity of pollinators, including bees, birds, flies, beetles, moths and butterflies. True to predictions, flowers of the butterfly-pollinated species open during the day, close at night, are red with white markings and produce relatively weak nectar. The moth-pollinated species, on the other hand, have richly scented, pale-coloured flowers that are only fully open at night and produce concentrated nectar. When you look at *Gladiolus* on the evolutionary tree, it becomes clear that the flowers have become adapted for moth pollination six times since the genus first appeared, and for butterfly pollination three times, and that these events have all been independent of each other. This shows that plants and their flowers are very labile and able to exploit the arrival of a new pollinator.

(∧) The swallowtail butterfly (*Papilio machaon*) and this *Gladiolus* species have co-evolved a close mutualistic partnership with the flower structure being well suited to butterfly pollination.

(L) Because of the copious amounts of nectar produced by the butter-fly-bush (*Buddleja davidii*), it is often used to encourage butterfly visitors such as the gulf fritillary or passion butterfly (*Agraulis vanilla*).

(∨) Lantana (*Lantana camara*) is another nectar-rich plant that can be used to feed butterfly populations, although it is now an invasive weed species around the world.

THE ROLE OF NON-NATIVE SPECIES

A highly controversial area of biology is whether non-native species should be tolerated. The butterfly-bush, originally from China and Japan, is now a common plant in many other temperate countries, but is considered to be an invasive pest in several of these, including the United Kingdom and New Zealand. Despite its unwelcome status in the UK, it is still widely planted to promote butterfly species that might otherwise be declining.

Flowers that deceive

To date, deception of a pollinator by a plant has been recorded in more than 7,500 species of flowering plants, two-thirds of them orchids. This is a disproportionately high ratio, even taking into account the size of the family. Some researchers go further and claim that one-third of all orchids (8,000 species) give no reward despite advertising that they do so. Floral deceit takes two forms: food deception and sexual deception.

⟨↑⟩ The orchid *Anacamptis israelitica* employs mimicry to deceive its bee pollinator into thinking that it is a Roman squill (*Bellevalia flexuosa*).

⟨>⟩ A female solitary bee (*Anthophora* sp.) visits a Roman squill flower and is rewarded with nectar. The rewards given out by the Roman squill are sufficiently frequent to make bee pollinators overlook the disappointment of visiting a barren *Anacamptis israelitica* orchid.

⟨↓⟩ The Chinese orchid *Dendrobium sinense* deceives its wasp pollinator by mimicking the alarm pheromones released by its normal honeybee prey.

Food deception in orchids

This is a relatively simple idea. An abundant plant produces a flower of a specific colour that emits a specific scent, and it produces nectar that can be accessed by an animal that in the process of drinking it transfers pollen between conspecific plants. A less abundant plant produces a flower of the same specific colour that emits the same specific scent, but it produces no nectar. As long as the second plant is in a small enough proportion of the population, its cheating will work. In other words, the pollinator will be rewarded sufficient times by the nectar-producing plant that it will always return to similar flowers even when it is not rewarded some of the time. This appears to be the successful strategy employed by a number of orchids, including *Anacamptis israelitica*, which mimics the common Roman squill (*Bellevalia flexuosa*).

Exploiting fear

When it comes to false promises, the Chinese orchid *Dendrobium sinense* has developed an ingenious ploy. This species is pollinated by the black shield wasp (*Vespa bicolor*), a species that captures honeybees and feeds them to its larvae. The hornet has therefore evolved to be very sensitive to the alarm pheromones given off by honeybees when they are being attacked, equating the pheromone with food for its

SEXUAL DECEPTION BY ORCHIDS

The flowers of the bee orchids (*Ophrys* spp.) are rightly infamous for the deception they play on young male bees, but other genera such as the hammer orchids (*Drakaea* spp.) and spider orchids (*Caladenia* spp.) use hormone-mimicking odours to attract wasps under similarly false pretences. The *Ophrys* deception is visual and olfactory. The labellum of the flower looks like a female bee, and inexperienced male bees are fooled and mate with the imposter. The flower has an advantage over the female bee in that it can put resources into accentuating those parts of the female that the males find most attractive, including pheromones, without having to allocate resources to wings and other expensive body parts. The poor male bees do not stand a chance, and the female bees simply cannot compete.

(▽) The flower of the sawfly orchid (*Ophrys tenthredinifera*) resembles a female bee and confuses the male bee pollinators further by emitting pheromones they cannot resist.

(◁) The dancing spider orchid (*Caladenia discoidea*) is another species that employs sexual deception to entice its pollinator to visit its flower.

larvae. The flower gives off the same chemical, (Z)-11-eicosen-1-ol, and when the hornet turns up looking for the bee in distress, it sits on the labellum of the orchid and removes the pollinia (see pages 261–63).

Another orchid, the eastern marsh helleborine (*Epipactis veratrifolia*), pulls a similar stunt by imitating the alarm pheromones of aphids, attracting the hoverflies that eat the aphids, and in the process duping them into pollinating the flower. The hoverfly may have the last laugh in this relationship, however, because it lays its eggs in the flower while on the orchid.

If it looks like faeces and smells like faeces…

Flowers that smell like faeces inevitably attract animals that lay their eggs in dung, in what is a simple yet effective exploitation of an existing behaviour. In South Africa, *Stapelia* species such as the carrion flower (*S. gigantea*) fool carrion flies into laying their eggs in the stinking, purple-brown fleshy flowers. This is an odd plant because it has pollinia like those found in orchid flowers. These are attached to the legs of the egg-laying flies, which transport them to the stigma on the flowers of another plant.

(∧) The carrion flower (*Stapelia gigantean*) has a distinctive faecal odour that encourages its pollinator to lay its eggs on the flower, pollinating the flower in the process.

(◁) The maggots of the pollinator emerge on to the hairy surface of the carrion flower, only to die of starvation.

Odd rewards

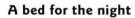

(T)he most common reward a plant gives a pollinator is nectar, which itself is not exclusively linked to flowers. However, the inducements flowers can offer go far beyond nectar and other simple food rewards, and they are not always just taken by the intended recipient. Robber bees are a common sight on snapdragons and runner beans (*Phaseolus coccineus*), biting into the back of flower buds and removing the nectar before the flower has opened.

edible pollen for the bees, but those that sleep in the flowers are 3 °C (5 °F) warmer in the morning than bees that sleep in the open. This means that the bees that slept in the flowers can be active much quicker than the others.

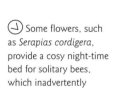

(↗) A buff-tailed bumble bee (*Bombus terrestris*) has worked out how to take the nectar reward from a *Penstemon* flower without providing its pollination services.

(↙) Some flowers, such as *Serapias cordigera*, provide a cosy night-time bed for solitary bees, which inadvertently pollinate the flower.

(↘) It is a common sight to see bees still sleeping on the 'mattress' provided by plume thistle (*Cirsium rivulare*) flowers in the early morning.

A bed for the night

Sleeping bees are a common sight on the flowers of the plume thistle (*Cirsium rivulare*), both in the evening and the following morning, despite being very exposed to predators and bad weather. However, some plant species go further to provide overnight accommodation. One such genus is the tongue orchids (*Serapias*), whose flowers resemble the nest holes of solitary bees. Male solitary bees are often unable to find a nest hole and will therefore sleep in the flowers. The flowers provide no nectar or

Who is rewarding who?

The flowers of the genus *Yucca* are icons of suburban front gardens and the Mojave Desert. The relationship that each species of *Yucca* has with a moth species is thought to be around 40 million years old. The yucca relies on the moth for pollination, and in turn the moth relies on the flower for somewhere to lay its eggs successfully. What is particularly remarkable about this pollination relationship is that the moth deliberately places pollen from another flower on the flower's stigma, thereby

BENEFITS FOR POLLINATORS

Nectar is not the only reward plants use to attract animal pollinators. Other incentives include:

• Food (starch structures, pollen).
• Warmth.
• Shelter for the night.
• Somewhere to meet a mate.
• Somewhere to lay eggs.

making successful cross-pollination more likely. However, mathematical biologists claim that obligate mutualisms like this should not exist for so long, because sooner rather than later one member of the party is going to start cheating, and in this case, it is more likely to be the moth. However, it appears that yuccas have a retaliation mechanism in place. The more eggs that are laid and the less pollen that is deposited, the more likely the flower is to drop off and not develop, thereby selecting against the greedy moth. The moths are learning this, because they now avoid, or even mark, flowers that have been visited to reduce 'over-laying'. Remarkably, although the yuccas and moths have evolved only once, their relationship has emerged independently many times.

The crunchy bits in a fig

As we shall see in the next chapter (see box on page 309), the fig 'fruit' we buy in the market is actually a collection of tiny fruits inside a fleshy sphere called a syconium, in which there is one opening (the ostiole). Inside the young syconium are tiny unisexual flowers, either male or female. The females come in two forms, some with long styles and some with short styles, and are pollinated by a wasp that enters the syconium through the ostiole in search of

somewhere to lay its eggs. It deposits its eggs in the ovaries of the female flowers, but only in those with short styles. The wasp cannot reach the ovary if the style is long, in which case it incidentally transfers the pollen on its body – picked up previously from visits to other syconia – to the stigma. When you eat wild figs, those crunchy bits are not all seeds! Like the yuccas and their moths, different species of fig (*Ficus* spp.) are pollinated by different species of wasp.

The yucca moth (*Tegeticula yuccasella*) has an obligate relationship with *Yucca torreyi*. The moth relies on the yucca flower as a home for its eggs, and in turn the flower relies on the moth to pollinate it.

Scarlet runner bean (*Phaseolus coccineus*) flowers can often be found with puncture holes at their base. These are tell-tale signs that robber bees have extracted the nectar before the flower opened properly.

Orchids

The orchid family is the largest family of flowering plants, with more than 25,000 species and countless thousands of hybrids and cultivated varieties. Orchid flowers are very different to those of any other family, and some taxonomists have even suggested raising the family to the rank of order and creating several new, smaller families. As a group, orchids have infatuated plant collectors to the extent that many are now endangered. While this is clearly unacceptable, there are many reasons to be fascinated with orchids, starting with the flowers.

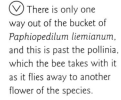

The asexual parts of orchid flowers

Orchids are monocots, which have floral parts in multiples of three. They are quite normal in this respect, with three calyx segments (sepals) and three corolla segments (petals). Things are not completely straightforward, however, because it is common for two of the petals to look like the sepals, all of which look like petals. However, the real departure from the norm is in the third petal, also known as the labellum or lip. The diversity of the size, colour and shape of the labella in the Orchidaceae is eye-watering and it is always related to pollination. As a direct consequence

(∨) There is only one way out of the bucket of *Paphiopedilum liemianum*, and this is past the pollinia, which the bee takes with it as it flies away to another flower of the species.

(<) *Paphiopedilum primulinum* flowers have a labellum that resembles a bucket or slipper, into which an unsuspecting pollinator will fall.

(⊓) The diversity of orchid flowers is mind-blowing. In particular, the labellum is capable of extraordinary modification as is the case in this bearded grass pink (*Calopogon barbatus*).

of the flowers having one petal that is different from the rest, the vast majority of orchid flowers are bilaterally symmetrical, or zygomorphic.

The sexual parts of orchid flowers

Following the monocot rule, the flowers of orchids 'should' have stamens and carpels in multiples of three. Generally, there are three carpels, which are fused together – this is easy to see when the fruits are mature. However, when you start to look for the stamens, something seems to be wrong. Growing from the top of the ovaries is a structure known as the column, and near the top of this and to one side is a sticky, glistening area. This looks like it could be the stigmatic surface, even though it is not on the customary terminal position. The remaining challenge is to find the stamens. The simple answer is that there are none. In the flower of an orchid, there is just one stamen and it is fused to the carpel. The column is therefore a combination of the stamen and the style and stigma.

Upside down and back to front

The flowers of orchids are quite literally upside down. When the flower is just a young bud, the labellum is at the top of the flower. However, as the flower develops and grows, the bud twists though 180 degrees (known as resupination), with the result that the labellum is at the bottom of the flower and can now act as a landing pad for the pollinating animal. The other area in which orchids turn normality on its head is in their pollen, which is not the dust-like substance characteristic of most other seed plants. In the orchids, the pollen sticks together into up to eight lumps called pollinia. These are 1 mm (1/32 in) or more in diameter and are attached to the column by threads called caudicles. The pollinia are often removed by the pollinator in a single visit, sticking to the animal by virtue of a sticky pad on the end of the caudicle called a viscidium.

One of the most peculiar features of orchids is the column. This unique structure is the result of the fusion of the male parts (the stamens) and the female parts (the carpel) of the flower. It works perfectly well, so why not? It also makes the identification of an orchid very simple, because if a flower has a column then it is a member of the orchid family.

⌃ The most obvious diagnostic feature of the orchid family is the column in the centre of the flowers, which is a result of the fusion of the male and female parts.

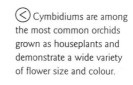

⌵ Cymbidiums are among the most common orchids grown as houseplants and demonstrate a wide variety of flower size and colour.

THE BUCKET ORCHIDS

The epiphytic bucket orchids (*Coryanthes* spp.) of tropical South and Central America have a labellum that is the shape of a steep-sided bowl or bucket. The pollinating male bee in the genus *Eulaema* (and it is only the males) is attracted to the flower by a powerful scent that appears to be species-specific. Like a change in the structure of the labellum, a change in scent can instantly isolate a plant, leading to the evolution of a new species through changed pollinator behaviour. The bee drinks in the scent and becomes intoxicated by it, falling into the bucket and becoming trapped there by the non-stick waxy sides. The only exit is past the end of the column, on which the pollinia are housed and where the stigma is located. If the bee has imbibed at another flower before, it may transfer the pollinia it collected from that plant to the stigma before taking another pair of pollinia to another flower.

⌃ The epiphytic orchid *Coryanthes verrucolineata* has taken the modification of the labellum to an extreme, forming a large bucket into which its specific pollinating insect is enticed and trapped until it can find its way out via the secret exit.

◡ This Lulworth skipper butterfly (*Thymelicus action*) has collected pollinia on its proboscis on an earlier visit to another pyramidal orchid (*Anacamptis pyramidalis*). On its visit to the second flower, the pollinia will be released onto the stigmatic surface.

◡ The orange pollinia are particularly prominent at the end of the column in this Bertolini's bee orchid (*Ophrys bertolonii*).

On the move

When a pollinator leaves an orchid flower, it may have two pollinia stuck to its body. When it lands on the labellum of the next flower it visits, those pollinia will not be positioned opposite the stigma of this flower but instead opposite its pollinia. The pollinator may just acquire another pair of pollinia rather than depositing the original pair on the stigma, which is located below the pollinia and separated from them by the rostellum. Orchids have arrived at an elegant solution to this potential problem. When the viscidium attaches itself to the pollinator,

the pollinium is drawn out from under the anther cap on the caudicle. As the pollinator flies away, the caudicle bends downwards through 90 degrees, such that the pollinium is relatively lower on the body of the pollinator and low enough to hit the stigma of the second flower. You can see this happen if you draw the pollinia out on the end of your finger. You will also discover how strong is the glue on the viscidium.

Orchid speciation

As mentioned above, orchids have a zygomorphic flower with a prominent petal called the labellum that has been modified in countless ways to attract many different pollinating animals. The pollination of orchids is an extraordinary business and can be influenced by a single mutation. If the labellum is altered enough, a different pollinator may step in. This means that

the new variety is reproductively isolated from all its former conspecifics and is able to evolve along its own trajectory. It will eventually be recognized as a new species, and will have evolved without being genetically incompatible with its sister species. This is one of the reasons why you can hybridize different species in the orchid family and produce fertile offspring, and also why there are so many cultivars.

⌃ The lip of the moth orchid (*Phalaenopsis* sp.) is the lowest of the three petals in an open flower but this is only because the bud rotates through 180 degrees before opening, a process know as resupination.

FOSSILS IN AMBER

Orchids do not generally make good fossils because they are not woody. However, their pollinating insects have a hard exoskeleton that does preserve very well in amber, and orchid pollinia are so large that they can be seen if attached to a preserved animal. In the past decade, there has been a run of discoveries of pollinators with pollinia attached beautifully preserved in amber. The current record for the oldest of these is a fungal gnat carrying the anther cap with pollinia and caudicle threads from a previously unknown species of orchid that has been named *Succinanthera baltica*, dated at between 40 and 55 million years old. The orchid family itself must therefore be much older than this, because it would have taken many millions of years for species to develop as far as the fossilized example.

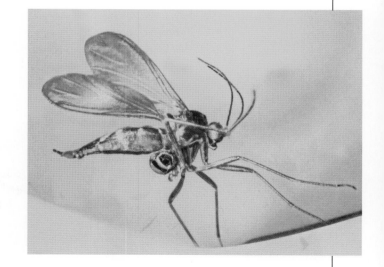

⌃ Proof that orchids have been around for more than 55 million years can be derived from the presence of pollinia attached to insects trapped in amber.

Wind pollination

(W)ind pollination is very common – nearly all gymnosperms do it, all grasses do it and most northern hemisphere timber trees do it. Before the evolution of seed plants, the vast majority of species used wind to move their spores around – after all, wind was already there when plants moved onto dry land. However, it would be wrong to assume that wind pollination is primitive just because it appears to be simple. The strategy is a derived characteristic and is such a successful one that many evolutionary branches have reverted to it.

(∧) Hazel (*Corylus avellana*) disperses its pollen to the wind from catkins, often in such quantities that the clouds of pollen are easily visible.

(↻) Grasses are wind pollinated, and their almost ubiquitous pollen is the cause of misery for many hayfever sufferers.

The answer is blowing in the wind

It has been stated before that plants have only a finite amount of resources to allocate to any one activity in their life cycle, and the amount of energy they allocate to flowering is one of those 'decisions' they make. Big, colourful flowers that offer a decent reward are expensive, so is there a cheaper mode of transport for pollen? There is: wind, the renewable energy for distributing your gametes. There are savings to be made here, because the wind is blind, so no colourful tissue is required. It is also self-powered, so no energy-rich reward is needed, and it turns up whether you want it to or not. It does, however, have some drawbacks: it is not there every hour of every day, it is not everywhere in equal strength and it will not give you any choice as to the direction in which it goes.

Wind power

Because wind is a cheaper option, it is widely and repeatedly used by plants for pollination. There are, however, a number of structural alterations that have to be made to a typical flower for wind pollination to be successful. First, unlike animals, the wind is blind, so there is no need for plants to synthesize colourful pigments for their flowers. Second, there is no need to synthesize energy-rich rewards because the wind works for naught. Third, the pollen has to be produced in large quantities because the wind distributes it indiscriminately, and it has to be small and smooth to increase its flight time (grass pollen is typically 0.03–0.05 mm in diameter). And fourth, the female stigmatic surface would do well to have a large feathery structure to increase its chances of being in the way of a passing pollen grain and hanging onto it. In some cases, the pollen has to fit snugly into a cup in order to be accepted and allowed to germinate.

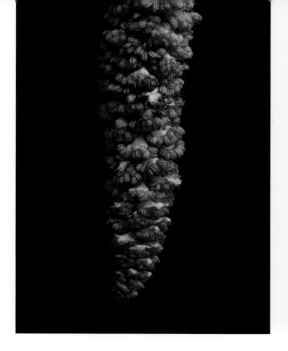

⟨ Although they are highly conspicuous, these willow (*Salix* sp.) flowers are pollinated by wind, which is oblivious to their bright colour.

∨ A scanning electron micrograph of a cluster of tiny grass pollen grains. Each species will have its own distinctive surface, from which it can be identified (magnification ×1,525).

∧ In early spring, poplar (*Populus* spp.) trees are covered with characteristic orange tassels. Upon close examination, these are seen to be tiny male flowers that are primarily just anthers.

Disproportionately common

While 95 per cent of plant species use an animal vector for their pollen, wind pollination remains hugely important because all the world's major arable crops, with the exception of the potato (*Solanum tuberosum*), employ wind pollination, including rice (*Oryza sativa*), wheat (*Triticum aestivum*), maize (*Zea mays*), sorghum (*Sorghum bicolor*), pearl millet (*Pennisetum glaucum*), oats (*Avena sativa*), rye (*Secale cereale*) and barley (*Hordeum vulgare*). Furthermore, most of the major commercial timber species are pollinated by wind.

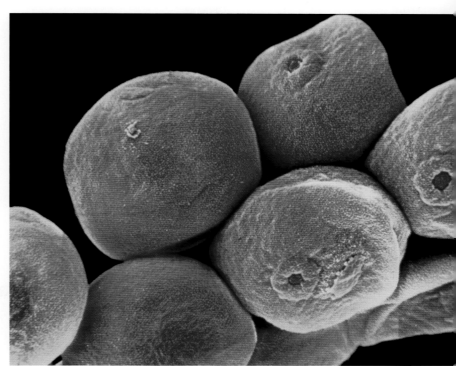

TRUTH IS STRANGER THAN FICTION

Sir Terry Pratchett regularly incorporated science in his *Discworld* novels, and with Ian Stewart and Jack Cohen he wrote a book loosely about biology as part of the series, titled *The Science of Discworld III: Darwin's Watch* (2005). In the earlier *Discworld* novel *Mort* (1987), Pratchett explains that 'Scientists have calculated that the chances of something so patently absurd actually existing are millions to one. But magicians have calculated that million-to-one chances crop up nine times out of ten.' He could quite easily have been writing about wind pollination here; absurd, but used by all the major crops.

Water pollination

ind pollination is well accepted and understood, but plants that use water as the vector are less well studied. Before going further, it would be helpful to review the stages of pollination: transfer of the pollen to the vector; movement of the pollen by the vector; transfer of the pollen by the vector onto a stigma; and fertilization and embryo formation. There are three different scenarios in which water fulfils the role of vector, and thus there are plants that can be said to be pollinated by water.

In tape grass (*Vallisneria spiralis*), the male flowers become boat-like vehicles, relying on water currents and surface tension to deliver the pollen to the female flower.

The strap-like leaves of tape grass are submerged, but pollination relies on the pollen being kept dry, with water acting as the transport vector for the flowers.

The difficulties of water pollination

Species from 33 genera in 11 families are pollinated by water, a strategy that has evolved independently many times. The distribution of these species is cosmopolitan, from northern Europe to southern South America, and they are found in both freshwater and marine habitats. Despite this, there are serious problems with using water as a pollination vector. First, the pollen will absorb water, causing it to explode and sink, and second, the water will wash off the adhesives found on terrestrial stigmatic surfaces. Fifty per cent of water-pollinated plants are dioecious (compared with 3 per cent on land), which can make synchronizing the production of male and female flowers very difficult. The fact that water pollination still occurs shows that plants have adapted to get around these problems, which they do in one of the three ways explored below.

Pollen remains above the water surface

Vallisneria species are the classic example of this strategy. The tiny male flowers break off the plant and float across the water with their anthers held above the surface. Two sterile anthers act as sails. The much bigger female flowers, which remain attached to the parent plants, create a small dip in the water surface big enough for the male flower to fall into, thus allowing it to deposit the pollen onto the stigma. In *Hydrilla verticillata*, the male flowers float to the surface in bud. They then open and fire their pollen into the nearby open female flowers, which are the only part of the plant that grows to the surface.

Pollen is transported on the water surface

Amphibolis and *Ruppia* are two genera whose members release their pollen onto the surface of the water, where it drifts in the hope of bumping into the stigma of a female plant. The pollen of *Ruppia* is released from the anther in an air bubble and floats to the surface, where it coalesces into a floating raft of many pollen grains. This greatly increases the chance of at least one pollen grain hitting a stigma. *Amphibolis antarctica* spends its entire life in sea water. Its pollen is huge – 6 mm ($1/4$ in) long – and lacks the exine (tough outer wall) found on all terrestrial pollen. The grains hook together in rope-like noodles that float and swirl across the surface of the water, eventually

hitting the feathery stigma on a female flower. In what is a remarkable adaptation, the stigma surface has a waterproof adhesive that is not affected by the salt water. Sometimes the embryo will grow in situ without the formation of a seed; in this case the plant is viviparous.

Pollen is released and transported underwater

Common eelgrass (*Zostera marina*) is one of the sea grasses that are so important as a fish spawning habitat. This species can live in the intertidal zone, where its pollen is transported on the water surface. However, it can also grow below the water to depths of 50 m (165 ft). In this situation, the pollen is released in clouds of grains that stick together. The density of the pollen is the same as the water, preventing it from either floating to the surface or sinking to the seabed, and allowing it to drift along until it hopefully hits the stigma of another flower. Search theory predicts that these clouds of joined pollen stand a much better chance of finding a stigma together than they would if they were released separately.

⋀ In *Amphobolis antarctica*, water is the transport vector for the pollen grains, which drift on the water surface until they collide with a female flower.

⋁ The pollen of common eelgrass (*Zostera marina*) is released underwater in large clouds, which drift with the currents they are until captured by a suitable stigmatic surface.

Asexual reproduction

Sexual reproduction is so ubiquitous that its superiority over asexual reproduction goes unquestioned. There are some clear advantages of sexual reproduction in the long term, most obviously the repeated combination and reassortment of the genes from two different individuals, which constantly tests whether there is a better version of the species. This enables populations of plants to adapt to changing conditions. But what if you are the best plant in the region and very well suited to the prevalent conditions? Why would you want to break up a winning team of genes?

The success of apomicty in dandelions (*Taraxacum* spp.) is evident in spring, when fields turn yellow with the vast numbers of flowers, all of them clones.

Evolution has no foresight

The theory of evolution and the origin of new species as proposed by Darwin and many others has no foresight. Natural selection filters out any organisms that cannot survive today, even if tomorrow they may have been just what is required. If a plant survives today and nothing changes, it will survive tomorrow. There is a process in plant reproduction called apomixis, which results in the formation of an embryo in a seed without the formation of eggs and sperm, and thus without sex

(i.e. asexually). This seed will give rise to a plant that is genetically identical to its parent. It will be a clone, but it may be a very successful clone.

Apomixis frequency

Apomixis has been recorded in more than 400 species, but about 300 of these are found mainly in just three botanical families – the daisy, rose and grass (Poaceae) families – and rarely in the orchids, the largest plant family. It appears, without any explanation yet, that some families have a greater propensity to apomixis than most of the others. What we do know is that some of these apomictic species can be very successful and make a nuisance of themselves. The most infamous is the common dandelion (*Taraxacum officinale*), which is a global temperate weed and has been divided into hundreds of carefully named micro-species that are simply clones. Another apomictic species that is making a great nuisance of itself around the world in the bramble (*Rubus fruticosus*).

Is sexual reproduction important to plants?

This might seem like a strange question at the end of a chapter about flowers whose function is to facilitate sexual reproduction. However, many plants reproduce asexually, either through apomixis by a vegetative division such as runner. In fact, most of the world's most invasive species reproduce vegetatively. The root of the answer to this question lies in the fact that plants cannot move. This means that if a particular genome works for the parent, it will also work for the offspring if they live in the same place. In asexual reproduction, if the plant is the only one of its species, then it can create young plants to take its place.

Plants will often reproduce by a method that is available only to bisexual organisms, which is self-fertilization. While this stirs up the gene pool a bit, it does not introduce new versions of any of the genes, and in an animal population you might start to see inbreeding depression and a build-up of harmful mutations. However, because of the different way plants make their eggs and sperm, the most harmful mutations are purged from the population each generation, and so inbreeding depression is not as big a problem for plants as it is for animals. Indeed, it is fine for most plants to indulge in a bit of self-fertilization each year, and it may even be a good idea because it gives them what Darwin called 'reproductive assurance'. To produce some seed by self-fertilization is always better than producing no seeds at all.

⊙⊙ Bramble (*Rubus fruticosus*) is an apomictic success, although this can be problematical when it takes over as a weedy species.

Garden flowers

Our gardens are full of plants that, at some point of the year, produce flowers. These flowers bring colour and sometimes scent into our lives. They also feed pollinating animals, most of which are invertebrates and many of which are declining. And yet when you examine the flowers in gardens, they are often different from those that you see in fields, hedges and woods. It seems sometimes that our gardens are freak shows, where oddities and monsters are propagated.

Narcissus species have been distorted in many ways to produce different colour combinations and shapes to satisfy gardeners' appetites for curiosities and showy flowers.

A classic English herbaceous border is a good place to discover how gardeners have meddled with nature to produce bigger and brighter flowers.

Any colour so long as it is blue

The colours of flowers are selected by such considerations as pollinator availability and evolutionary history. However, some colours are 'missing' from the palette of major groups of garden plants. For example, for many years there has been a desire to breed a truly blue rose and a blue rhododendron, but despite the word 'blue' being included in the names of cultivars produced, the flowers remain stubbornly mauve. It has been shown that transferring the blue colour gene from delphiniums to roses is technically possible, but gardeners are traditionally opposed to genetic modification and such a blue rose has never become established in the nursery trade. However, new gene editing technology like CRISPR/Cas9 might make the breeding of new colours more acceptable.

Gardeners seek out 'black' flowers for their garden designs. *Tulipa* 'Queen of Night' is one of these successful modifications, fulfilling the desire for exotic flower colours.

THE CROSSED PATCH – HYBRIDS AND MULES

Another source of novelty is artificial hybrids between species that would not meet were it not for the gardener. Sometimes this is a case of reuniting sister species that share a recent common ancestor, in which case the resulting offspring might well be fully fertile. In other cases, while the seeds from the misalliances are fertile, the resulting plants are not able to produce viable sperm and eggs, and the hybrid has to be propagated vegetatively. However, the hybrid may benefit from a poorly understood phenomenon called hybrid vigour, which is the propensity of hybrids to be extra robust. This makes these creations very desirable garden plants.

Make mine a double

Not content with messing around with colours, gardeners also have a fascination with flowers that have an increased number of petals. Two classic examples of these 'double' flowers are carnations (*Dianthus* spp.) and dahlias (*Dahlia* spp.), although tulips (*Tulipa* spp.) and daffodils (*Narcissus* spp.) have been similarly altered. This is always achieved at the expense of other parts of the flowers – usually the sexual parts – and can be seen very clearly in hellebore flowers (*Helleborus* spp.), where the corolla normally comprises nectaries but in mutants these are replaced by petaloid structures. Other strange flowers are daffodils in which the corona is 'split'. The corona of a daffodil is actually fused stamen filaments, so for it to be split, the genes that fuse the filaments has to be broken.

The bigger the better

There is also a belief in some gardeners' minds that bigger is better. A common way to make super-sized specimens is to select a polyploid mutant in which the number of chromosomes has doubled. This is a relatively common event in plants and polyploid individuals are often able to thrive. However, they may not be perfect – the increase in flower size may not be accompanied by an increase in the strength of the flower stem, and the resulting plant will always need to be supported.

Plant breeders have successfully produced pom-pom dahlias, which lack the disc florets of the species dahlias and form near-spherical flowers.

Peony tulips have been bred to be 'double', producing a much more showy plant suitable for the garden border, but at the expense of its reproductive parts.

7

Seeds and fruits

A seed is a survival capsule, enabling a plant to endure through space and time. It consists of the plant embryo and a food supply enclosed in a tough coat, and may comprise more than three internal components to enable it to sense and respond to its environment. In addition, it may have external components to aid its movement away from the parent plant, although seed dispersal is also manipulated by the fruit.

A fruit is a structure that develops from the ovary, particularly its wall. To recap, the ovary is the region at the base of the carpel, which is the female part of the flower of an angiosperm. In this large group of species, it is inevitable that there will be a correspondingly large diversity of seeds and fruits in terms of size and structure.

The emergence of seeds, approximately 365 million years ago, was one of the most important events in the history of life on Earth. And yet for nearly ten times that length of time they did not exist, and for about 100 million years even land plants did perfectly well without them.

Before there were seeds

It is widely accepted that the first plants emerged about 2.1 billion years ago. This followed the 'endosymbiotic event' that resulted in a photosynthetic cyanobacterium being engulfed, but not digested, by another cell (see page 145). The two organisms began to cooperate, and the result was a photosynthetic organism powered by the chloroplasts that evolved from the cyanobacterium. These simple plants lived in oceans and freshwater lakes for 1.5 billion years, and so it is perfectly possible to be a plant and survive without seed. However, every plant needs a survival capsule.

The perils of moving a plant

Plants are immobile, and yet they are found all over the world, and from altitudes above 4,000 m (13,000 ft) to depths below 50 m (160 ft). This means that although they cannot move themselves, they can be moved. Moving a terrestrial plant that is rooted to the spot is a traumatic experience for the plant because it deprives it of its water supply. This water is required for a plethora of purposes, including cellular metabolism, photosynthesis and turgor, which keeps the plant upright. Aquatic plants have a different but equally stressful relationship with their water supply. It may be too salty, and thus able to draw water from the plant, or if it is fresh, excess water may enter the cells or nutrients may diffuse out. Clearly, isolating life from the medium in which it exists is a ubiquitous and unending activity. When a plant is moved, it may experience very inhospitable conditions.

Spores

Both aquatic plants and seedless land plants use spores as their survival capsules. (Seed plants also produce spores, but these do not leave the safety of the plant that produces them.) A spore is a unicellular stage in the life history of the plants. Generally, its production is preceded by cell division, when the number of chromosomes is halved and thus there is just one copy of the genome. Spores are capable of growing into the body that produces sperm and eggs in the plant's life history, but before they can do this the right conditions must be present. They are, however, amazingly resilient to the problems the environment throws at them thanks to a covering of vegetable Kevlar called sporopollenin (see box).

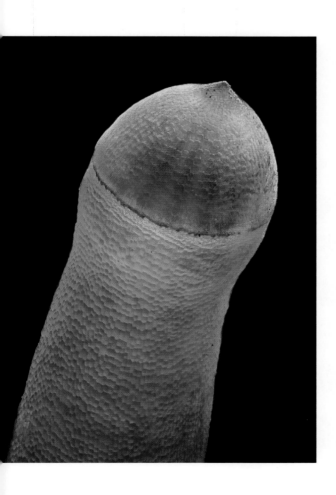

SPOROPOLLENIN

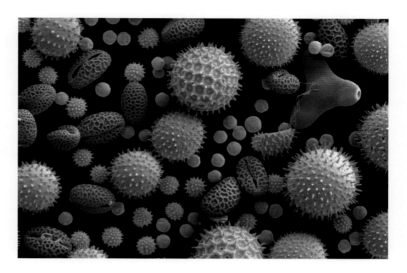

Sporopollenin is most commonly associated with pollen grains, where it is used on the outer wall (exine) to protect the sperm-producing male gametophytes of seed plants. However, the substance first evolved to protect the spores of algae. There is a lot yet to learn about sporopollenin, but we do

⌃ Pollen from a variety of common plants: sunflower (*Helianthus annuus*), common morning-glory (*Ipomoea purpurea*), Greek mallow (*Sidalcea malviflora*), lily (*Lilium auratum*), narrowleaf evening primrose (*Oenothera fruticosa*) and castor bean (*Ricinus communis*). The image is magnified ×500, so the bean-shaped grain in the bottom left corner is about 50 µm long.

know that it is a nitrogen-free polymer made up of monomers held together by ether linkage. It is chemically inert and chemically stable, and can resist attack from both chemical reagents and enzymes, meaning that it can fend off infection from fungi and bacteria. And it can tolerate not only freezing but also temperatures up to 300 °C (572 °F). Species can be distinguished by the quantity of sporopollenin they contain, and in the patterns created on the outer surfaces of their pollen grains and spores. Because some spores and pollen grains have remained unchanged for more than 500 million years, examination of fossilized finds has therefore been very useful in archaeological investigations.

The first seeds

⊳ Fossils of the Devonian plant *Chaleuria cirrosa* demonstrate heterospory, with both microsporangia and megasporangia being present on the fronds. Heterospory is a prerequisite for the evolution of seeds in the gymnosperms and angiosperms.

harles Darwin proposed that no organism has acquired a novel feature or adaptation for the benefit of any other than itself. He did, however, acknowledge that this does not preclude the possibility that the new feature is exploited by other organisms, and this is certainly true for seeds, which now provide the primary food supply for thousands of species. The seed evolved because it was beneficial for land plants to have a survival capsule, and initially it was not associated with modern structures like flowers and cones.

The lack of evidence

The evolution of seeds is not clearly understood. This is for several reasons, one of which is that it happened long ago. Perhaps as far back as 386 million years ago, the evolutionary ball started rolling with the production of pre-ovules, which are found in a range of fossils. However, not one of the fossils found to date is the missing plant that links the two extant groups of seed plants – the gymnosperms and angiosperms. The best we can say at the moment is that, a very long time ago, an extinct group of plants gave rise to two lineages, one of which is the present gymnosperms and the other the angiosperms. We have no clear idea what this missing link looked like, but we do know that several other groups of early seed plants have petered out and no longer exist.

⊽ A fossil specimen of *Archaeopteris halliana*, an extinct genus of tree-like plants with fern-like leaves dating from 383–323 million years ago.

⊲ Despite the similarity between the leaves of *Archaeopteris* and a fern frond, *Archaeopteris* is thought to be one of the earliest known trees and bears some resemblance to gymnosperms.

Where on the plant did the seeds evolve?

For seeds to be produced, a number of innovations are required. For a start, two types of spores are required (heterospory), to allow for the retention of the female egg-producing gametophyte within the sporophyte and the release of a smaller sperm-producing male gametophyte (the pollen grain) to seek out the female. Another essential innovation is the development of a structure to receive the pollen. You may have noticed that no mention has yet been made of flowers and cones. That is because it is generally considered inconceivable that the evolution of seeds occurred on the aerial parts of the plant, because that would have demanded the simultaneous development of too many novel features. Among the failed theories about the evolution of seeds is the idea that they evolved from the hard-coated sporocarps of aquatic ferns such as *Marsilea* species, which can survive for up to 100 years in mud. While these were not the first proto-seeds, they did evolve in the right place – at soil level.

Protection

All land plants produce embryos – hence their scientific name, Embryophyta or embryophytes. It is the character that clearly distinguished them from the ancestral algae. In all embryophytes, the embryo has to be fed, otherwise it could not survive. This means

SEED FERNS – AN OXYMORON

Ferns do not produce seeds, but the earliest seed plant fossils have leaves that resemble fern fronds. There are many examples of these, and they form an odd paraphyletic group of misfits and evolutionary adventurers, all but one of which (the missing ancestral link shared by angiosperms and gymnosperms) resulted in failure.

⌄ Although ferns do not bear seeds, the fossil record includes early plants with fertile fern-like leaves that bore seeds and pollen-producing organs at the base of those leaves.

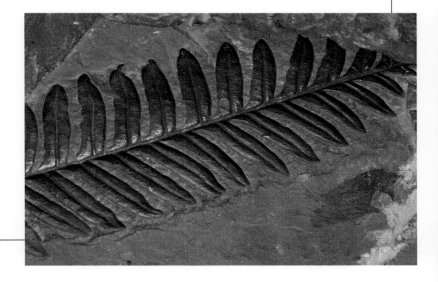

that the second of the three parts of a seed – the food supply – existed already. However, the embryo was not protected, at least not in a way that would prevent anything but desiccation. This required a new structure, the integument. In gymnosperms, the integument consists of one layer of tissue enclosing the female gametophyte and in angiosperms it consists of two layers. There is a small opening in the integument through which the pollen gains access to the female gametophyte and thus to its eggs. At its most simple and primitive, the seed is an integumented female gametophyte.

Gymnosperm seeds

Ŵhile the seeds of gymnosperms and those of angiosperms share the same name, that should not be taken as evidence of homology. And although they also share the same three basic parts (coat, embryo and food store), the origins of these components is not necessarily the same in both groups. Gymnosperm embryos are sustained by the female gametophyte, as in ferns, making them appear to be more primitive than the angiosperms. This led to the belief that evolution proceeded in a line from ferns to gymnosperms to angiosperms, but this is now known to be untrue.

⊳ A typical angiosperm flower, if there is such a thing, showing typical sexual parts of the flower and the development of the seed.

⌄ Gymnosperms such as *Cycas armstrongii*, an endemic of Australia's Northern Territory, have much larger seeds than most of the present-day angiosperms.

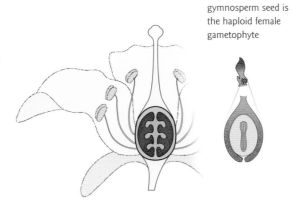

the food supply in a gymnosperm seed is the haploid female gametophyte

the food supply in an angiosperm seed is the triploid endosperm formed from the doubly haploid central cell of the female gametophyte and a single haploid sperm

Bigger on average

The differences between present-day gymnosperms and present-day angiosperms reflect differences in both their current and past biology. We can map these differences onto a phylogeny (an evolutionary tree) to try to decide the extent to which they are due to the past or to the present. One clear difference between the two groups is the size of their seeds: gymnosperm seeds are 59 times bigger on average and they do not vary as much as those of angiosperms. This difference goes back a long way, with the earliest angiosperm fossils having seeds with a volume of 1 mm^3, compared to a volume of 200 mm^3 for fossilized gymnosperm seeds of the same age. The smallest extant gymnosperm seeds weigh 0.63 mg, 10,000 times heavier than the lightest angiosperm seeds.

Determining seed size

It appears that there is something about the ways in which gymnosperms and angiosperms produce their seeds that imposes different upper and lower limits on their sizes. The ways in which the food supply for the embryo is produced varies greatly, and the fact that gymnosperms produce this food supply before the embryo is formed puts constraints on the size of the seeds. The size of the integument, embryo and food supply in a gymnosperm seed are linked, whereas in the angiosperms the three parts can develop independently. Surprisingly,

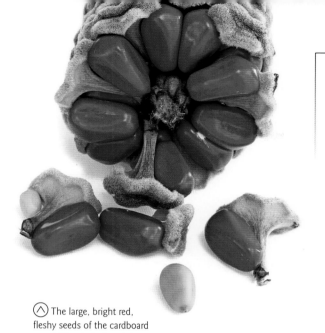

∧ The large, bright red, fleshy seeds of the cardboard cycad (*Zamia furfuracia*) are borne within a cone structure typical of the gymnosperms.

the gymnosperms are just as diverse when it comes to methods of dormancy. The size of seeds in relation to the size of the organism and the size of that plant's genome have also been investigated, and there is some correlation here although the significance of this is not yet clear.

DEDUCING THE PAST FROM THE PRESENT

Evolution is a tree, not a road. The organisms that live on Earth today are each on their own evolutionary trajectory, radiating out from events that have happened peripatetically over the past 3.8 billion years. Extant organisms all share ancestors, but there are parts of the tree that are missing – branches die and so do ancestors. It is believed that gymnosperms are all descended from a common ancestor, and likewise, it is believed that angiosperms also all descend from a common ancestor. These common ancestors themselves shared a common ancestor perhaps 325 million years ago, and here is the problem: we do not know what that common ancestor looked like. However, there is no reason to assume that the gymnosperms look more similar to the common ancestor than do the angiosperms. Despite our knowledge gaps, there are a couple of accepted notions about plant evolution: first, a group of plants that looked more like present-day gymnosperms than angiosperms dominated the world's ecosystems for 100 million years; and second, the angiosperms diverged from the ancestors of today's gymnosperms at least 140 million years ago, but perhaps much earlier. The structure of the plants that link these groups are the matter of much speculation and informed guesswork.

SMALL BEGINNINGS

Although a mature giant redwood tree (*Sequoiadendron giganteum*) weighs in at more than 2,500 tonnes, its seeds weigh just 6 mg. This means that each tiny capsule eventually gives rise to a plant 400 billion times bigger than itself.

⊳ The giant redwood (*Sequoiadendron giganteum*) is the sole living species in the genus *Sequoiadendron*, the world's largest trees and amongst some of the oldest living things on Earth.

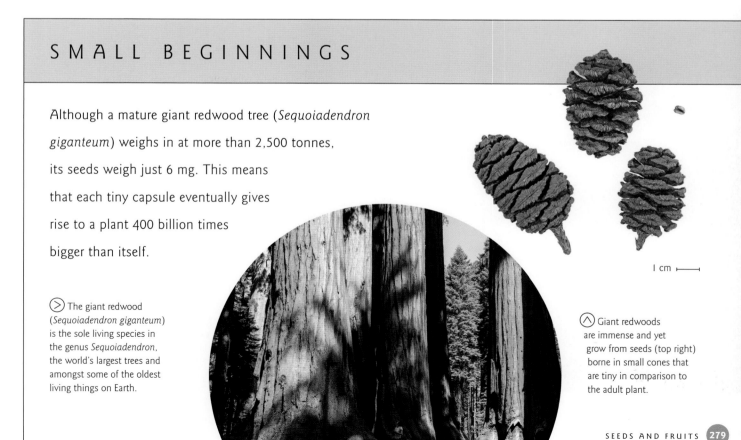

1 cm ⊢————⊣

∧ Giant redwoods are immense and yet grow from seeds (top right) borne in small cones that are tiny in comparison to the adult plant.

Fleshy conifer fruits

The term 'gymnosperm' is derived from the Greek words meaning 'naked seeds', because the seed is not enclosed in a structure. This means that the 1,000 or so species of gymnosperm do not produce fruits in the same way that the angiosperms do (even though the fleshy fruits of angiosperms are produced in a wide variety of ways). However, in many gymnosperm genera the species produce structures that, to the casual observer, look like a fruit. One of these, the Japanese plum yew (*Cephalotaxus harringtonia* var. *drupacea*), was even named by botanists after the drupe, a single-seeded fleshy fruit.

Different origin, same function

Many gymnosperms bear fleshy 'fruits', but these are produced in different ways. For example, the fleshy red aril on the seeds of European yews (*Taxus baccata*) is an outgrowth of the peduncle. The aril is the only part of the tree that is not toxic – even the seed coat contains some very nasty cyanogenic glycosides. In contrast to the yew tree, the fleshy parts of the 'fruits' of ginkgo (*Ginkgo biloba*) are derived from the outer of the two integuments that surround the seed. This is sometimes the way angiosperms make fleshy fruits, and when the genes controlling the development of ginkgo 'fruits' were identified, they turned out to be the same as those used by angiosperms.

⊲ Although superficially similar to angiosperm berries, the red, fleshy parts of yew tree (*Taxus* spp.) drupes are in fact formed from the peduncle and are the only edible parts of the plant.

⋀ The fleshy fruit of the Japanese plum yew (*Cephalotaxus harringtonia* var. *drupacea*) is remarkably similar to an angiosperm drupe, but is formed from the naked seeds typical of the gymnosperms.

An ancient strategy

Having fleshy fruits is thus an ancient strategy within land plants for the dispersal of seeds, and one that has evolved many times. Furthermore, the fruit is not a new structure belonging solely to the angiosperms, despite the fact that angiosperms are defined by having enclosed seeds. This is important from an evolutionary point of view, because it was once proposed that one of the reasons for the vast diversity of angiosperms is their novel use of animals to disperse their fleshy fruits. This is clearly not the case.

Fleshy fruits are today linked with seed dispersal by birds and mammals, particularly bats and primates, but we now know that the structures significantly pre-date these animals. Could any other animals have been co-opted into seed dispersal? The answer is yes, confirmed thanks to the discovery of a beetle with gymnosperm 'fruits' in its mouth preserved in amber, dated back to the Cretaceous period, 145–66 million years ago. Based on the fact that modern-day reptiles are often frugivores, it is also believed that pterosaurs and dinosaurs could have dispersed fleshy-fruited seed plants.

⟨⌐⟩⟨∧⟩ Gymnosperms with fleshy fruit such as ginkgo (*Ginkgo biloba*) adopt the same kinds of seed-dispersal strategies as angiosperms with similar fruit. The gymnosperm seed inside the ginkgo fruit is particularly prized in Asian cooking.

UNEXPLAINED ASSOCIATION

While most gymnosperms produce dry, woody cones, fleshy 'fruits' are found on many different species in this group. It turns out that very nearly all of these 'fleshy-fruited' gymnosperms are dioecious (with male and female reproductive organs on different plants), whereas the species that produce woody cones and dry seeds are monoecious. The reasons why this should be so have not yet been discovered.

Strange gymnosperms

The gymnosperms include 985 species in 79 genera and 12 families. By way of comparison, the beech family (Fagaceae), just one of the 416 angiosperm families, comprise 970 species in seven genera. Care should be taken not to read too much into these figures, but the gymnosperms are a very varied group that could be described as a bunch of oddballs. There is certainly more diversity in the gymnosperms than in the beeches, even though both contain almost the same number of species. A group like the gymnosperms is therefore beguiling, and consequently it has been studied in depth. Differences between the species may prove helpful when trying to explain the mystery of the origin of the angiosperms.

The gnetophytes

An unpronounceable name is a good start for a group of plants that seem to defy classification. The gnetophytes comprise three genera, each in its own family, and they are clearly variations on a theme. The genus *Welwitschia* includes one species, found in the Namib Desert. These plants can live for 1,500 years, and so can get away with a low fecundity. The ovule in the female cone has two integuments, the outer layer later developing into a wing that aids the dispersal of the seeds. There are about 40 species of

Ephedra, and despite some obvious differences between them, their ovules all have two integuments and several bracts, the outer layer of which may be fleshy or dry and winged. The third genus in the group, *Gnetum*, comprises 30 species. Their ovules have three integuments, the innermost extending into a tube to 'catch' the pollen, the middle layer becoming hard and the outer remaining fleshy. Both *Gnetum* and *Ephedra* exhibit double fertilization, a characteristic of the angiosperms, but the second fertilization results in either a second embryo or it aborts.

Welwitschia (*Welwitschia mirabilis*), a member of the gnetophyte group of gymnosperms, is a remarkably resilient plant that can live up to 1,500 years in the Namib Desert.

Ephedra species are unusual gymnosperms, not only because of their fleshy fruit, but also because they exhibit double fertilization.

LOOSE ARRANGEMENTS

When evolutionary biologists investigate the past, they look for relationships between species and try to explain features in terms of their selective advantage. This is sometimes referred to as adaptiveness. When it comes to the relationships between plants and animals, these can be much looser than some researchers hope, which makes biology look imprecise. This is not the case; rather, it is just another feature of the natural world that needs to be explained. There are several possible explanations for this so-called wriggle room in biological systems:

• History and phylogeny constrain distribution and morphology.

• Plants and animals rarely evolve at the same pace because they are subjected to different selective pressures.

• Plants species often have a longer half-life than animal species because of a higher level of phenotypic plasticity.

• Organisms with no previous interactions may exert an influence.

• Stochastic changes in selective pressures occur.

• There is not a perfect fit between the distributions of interacting species of plants and animals.

⌃ *Encephalartos ferox* has huge rugby ball-sized cones, but unlike many other gymnosperms the seeds contained inside are fleshy.

Ginkgo and the cycads

Ginkgo is a species out on a branch of its own, and appears elsewhere in this book. Like ginkgo, the cycads are often portrayed as living fossils, and yet recent work has shown that the 100 or so extant species in the order are actually less than 10 million years old. While the lineage itself is ancient, the species are clearly still evolving and speciating. This is a clear example of the ubiquitous fact that no matter how old the lineage, the species themselves are much younger. Of the five oddballs mentioned here, the cycads are the easiest to recognize as gymnosperms because their cones are very superficially coniferous. However, look closer and you will see that their seeds can often have a spongy, fleshy outer layer.

⌃ *Gnetum* species, such as *Gnetum scandens*, are one of the two gymnosperm genera that have double fertilisation, a feature more characteristic of angiosperms.

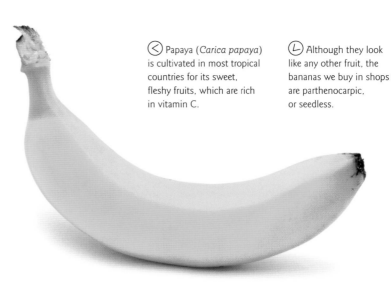

◁ Papaya (*Carica papaya*) is cultivated in most tropical countries for its sweet, fleshy fruits, which are rich in vitamin C.

↳ Although they look like any other fruit, the bananas we buy in shops are parthenocarpic, or seedless.

The origin of fruits

It is not known how fruits evolved in the angiosperms, because according to the fossil record the carpel – from which the fruit develops – turned up out of the blue about 140 million years ago. There appears to be an advantage attached to possessing a carpel, however, because the majority of plants on land (in terms of species numbers and biomass) have one. The definition of a fruit is a ripened ovary and any other structures that are attached to it and ripen with it; the inclusion here of 'any other structures' means that fruits can be very diverse and adaptable.

◁ Vegetable or fruit? Although sold in shops as a salad vegetable, tomatoes are in fact fleshy fruit containing numerous seeds.

Evolution and development

A rich seam of research that is capable of answering questions such as the origin of fruits is known as evo-devo, the snappy abbreviation of evolution and development. The basis of this research is that evolutionary relationships might be hidden in the genes controlling the development of organs that now look very different but share a common ancestor. This is being used to analyze and compare the development of the reproductive parts of the major groups of land plants, but

while it has provided some answers as to the evolution of fruits, sadly it has not come up with *the* answer. A current front runner is the mostly male theory, which proposes that the female reproductive parts were fused with the male cones of the ancestral seed plants to give the prototype for the bisexual flower with its carpel in the centre.

What is a fruit?

We may be in the dark with regards the evolutionary origins of fruits, but as we saw above, they can be defined. Fruits are ripened ovaries, but they are just the final stage and function of the carpel. Prior to ripening, the ovary protects the ovules, where the seeds develop. When the seeds have matured and are ready to be released into the wild, the ovary can do one of several things. The minimum it can do is to dry and open, so that the seeds simply fall out. Alternatively, the ripened ovary can develop into something that will benefit the seeds, and thus its genes, including

⊼ The tamarind tree (*Tamarindus indica*) produces seeds without fleshy outer coverings and therefore invest little energy in the promotion of seed dispersal by animal agents, relying instead on gravity.

⊃ Oak trees (*Quercus* spp.) are another group that primarily rely on gravity for dispersal of their acorns, but squirrels still find them irresistible and contribute to their dispersal far away from the parent plant.

offering protection and transport facilitation. As many as eight out of ten species choose the first option, releasing their seeds and allocating no resources to their dispersal. Examples of this are acorns falling from an oak tree (*Quercus* sp.), or seeds tumbling from the dried fruit of a foxglove (*Digitalis* sp.).

THE TROUBLE WITH WORDS

Fruits come in two basic types: dry and fleshy. Dry fruits are often overlooked, despite making up the majority of angiosperm fruits. One of the most familiar types of dry fruits are nuts, and herein lies a deeper problem in botany: words with precise botanical meetings have become colloquialisms. Tomatoes are sold as salad vegetables, but in botanical terms they are fruits. The fruit of members of the legume family (Fabaceae) are called legumes, but in French *légume* means 'vegetable' and greengrocers would definitely class French beans as vegetables! Does this matter? It can certainly cause confusion in the minds of students, and some botanists claim it has led to the decline of interest in botany in the education system.

Seed size

S eeds vary in weight from 0.3 μg to 18 kg, or 0.0000003 g to 18,000 g, or to put it another way, the biggest seed is 60 billion times heavier than the smallest. Seeds also vary in size, being anything from 0.18 mm to 500 mm long, nearly a 3,000-fold difference. It is a safe bet that these extremes are produced by very different plants living in very different places, with different selective pressures and conditions, and yet the function of all seeds is the same: to grow into another seed-producing plant.

Compromise, trade-offs and resource allocation

Seeds are energetically expensive to produce. To start with, the plant has to invest a lot of energy in the production of flowers, and it might even have to give energy away in the form of nectar to keep pollinators sweet. Once the egg(s) in the flowers have been fertilized, the embryo can develop and differentiate into the plumule and the radicle. Food for the embryo is stored in either its cotyledons (seed leaves) or in the endosperm that has resulted from the second fertilization event. Clearly there will be a link between the size of each seed and the number that are produced, which is generally an inverse relationship: the greater the number of seeds produced, the smaller they will be. The most seeds produced by one plant in a year is around 4 million, by the orchid *Cynoches ventricosum*, while the coco de mer (*Lodoicea maldivica*) produces at most a few

⟨<⟩⟨∧⟩ A single *Cynoches ventricosum* plant, a member of the orchid family, can produce a staggering 4 million tiny seeds in a single year.

⟨>⟩ A simplified seed embryo, showing the four main regions including the cotyledons (seed leaves) that nourish the developing root and shoot.

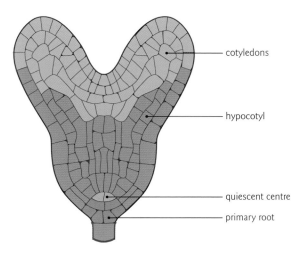

cotyledons

hypocotyl

quiescent centre

primary root

The coco de mer (*Lodoicea maldivica*) invests so much energy into producing its enormous seeds (the largest in the plant kingdom), that one plant will produce only a few a year and each can take up to seven years to develop to maturity.

increasing seed size is correlated with shade tolerance. Interestingly, the average weight of seeds increases by an order of magnitude (×10) for every 23 degrees in latitude you move closer to the Equator. Against this is the fact that within one Australian population of silver banksia trees (*Banksia marginata*) a fivefold variation in seed weight was discovered, and five very different floras from around the world have been found to have seed weights within very similar ranges. The prevailing conditions during seed development seem to affect seed size, with higher temperatures often leading to smaller seeds, and the position of the flower on the plant can also have an effect. Seed size therefore

seeds per annum. This is not dissimilar to the way birds vary in terms of the number of eggs they lay and how much parental care they offer to each chick. In plants, larger seeds represent greater parental investment.

Seed size, habitat and life history

Seeds vary greatly in their size, and understanding and explaining this has been one of the most vexatious questions in plant biology. We are still unable to detect universal rules governing seed size, although there are some tantalizingly robust relationships in specific circumstances. For example, in the United Kingdom,

is difficult to explain or predict, and perhaps this is the point. It may be that the ability of a plant to produce different-sized seeds in the same year and seeds of different sizes in different conditions is the best strategy, because small and large seeds have different advantages and constraints. A variety of seed sizes is best in an unpredictable world.

The seeds produced by a single plant are not always a uniform size and these size differences may confer different advantages and disadvantages to the germinating seed. The seeds of silver banksia trees (*Banksia marginata*) show a fivefold variation within a population.

ON BEING SUCCESSFUL

Seeds are thought to be one of the features of the gymnosperms and angiosperms that make them so successful, but this is based simply on the number of extant species. Evolutionary biologists argue long and hard about what constitutes biological success. At one level, just making it through the day without being eaten is success. A better long-term measure of success, also known as 'fitness', is lifetime reproductive success. This means that each individual should leave behind at least one descendent in order to be considered successful, and the more offspring you leave behind, the more successful in biological terms you have been.

Orchid seeds

The seeds of orchids are odd, which is something of a generalization bearing in mind that the orchid family (Orchidaceae) is the largest in the plant kingdom, with 25,000 species found in every habitat except deserts and submerged aquatic environments. This group of plants has been around for at least 75 million years, and their oddness is multifaceted. First, the seeds are very small, lack a carbohydrate store and have a coat that is just one cell thick. Second, the embryo grows into a protocorm rather than a seedling. And third, the orchids have formed an intimate and at times uneasy alliance with fungi.

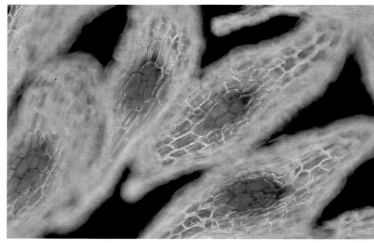

⋀ Walker's cattleya (*Cattleya walkeriana*) fruits contain thousands of tiny seeds, which are well suited to wind dispersal across long distances by virtue of their weight.

⊐ Viewed under the microscope, the tiny seed of this European terrestrial orchid is clearly visible through the one-cell-thick papery layer that surrounds it.

Why small is good for orchids

The seeds of orchids are 0.18–3.85 mm in size, but most are towards the lower end of this range and the largest is a hybrid and therefore the result of hybrid vigour (see page 271). Orchids began as terrestrial plants but many extant species are epiphytes growing on the branches of tropical trees. These plants take nothing from their host plant, so they are not parasitic, just benign lodgers. However, they share a problem with parasites like the mistletoes, namely how to get up into the crown of a tree. The majority of mistletoes use birds for transportation (see box on page 303), but orchids use wind. Dust-like orchid seeds, which also contain a lot of air, can easily be blown up onto tree branches. Among the very first plants to recolonize the Indonesian island of Krakatoa after the volcanic eruption in 1883 were four species of orchids, whose seeds must have been

carried 400 km (250 miles) to get there. Wind can be a very efficient dispersal vector, and the smaller a seed is, the further it will go.

Why small is bad for orchids

The seeds of orchids have made several sacrifices to become so small. First, their embryos are tiny and undifferentiated, so they have to do some catching up before they can put out their first photosynthetic leaf. Second, there are few supplies in the seeds to power the growth and development of the embryo. And third, they are not protected by a thick coat but by a papery layer, which is just one cell thick and about as waterproof as a piece of paper. On the plus side, the seeds are not an attractive meal for an animal, but these significant drawbacks still have to be addressed if the embryo is to grow into a mature flowering plant that produces more seeds.

The wrong way round – again

Orchid seeds look like dust and yet their pollen in normally stuck together in clumps the size of breadcrumbs. This is the reverse of the situation in most plants, where the pollen is like dust and the seeds are much bigger. Another example of orchids getting things the wrong way round is in the formation of the protocorm. A corm is a short, solid, vertical underground stem surrounded by thin, papery leaves, while a protocorm is the first stage in the development of a storage organ in orchids. Corms are generally produced months or years after the seed has germinated, when the plant has been actively photosynthesizing for some time – they act as storage organs. The protocorm of orchids, however, is the first structure produced by the embryo. It is typically 1 mm (1/32 in) in diameter and has no top or bottom to start with, although it will usually have some rhizoids to help with water uptake. Having made the protocorm, the new plant generally stops growing, because the pitifully small amount of energy stored in the cells of the embryo have been used up. It is possible that the protocorm becomes dormant at this point, waiting for a signal such as a period of cold to tell it that it is time to grow.

⌄ The tiny protocorms (*c.* 15 mm long) that develop when orchid seeds germinate rely on close mycorrhizal relationships with fungi. The growing protocorm very quickly exhausts the tiny reserves from the seed and must rely on its fungal partner for nutrients, a relationship called mycoheterotrophy.

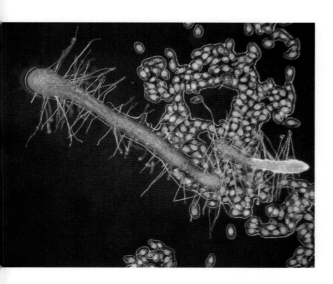

ORCHIDS AND FUNGI

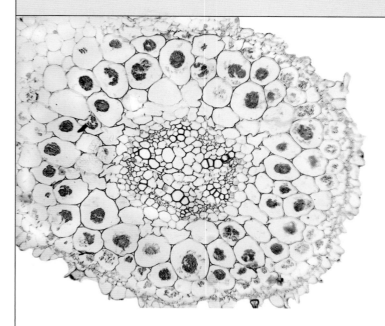

Most plants have a fungus growing around their roots. This symbiotic relationship involves the fungus facilitating nutrient uptake in return for some of the products of the plant's photosynthetic activities. When a fungus grows around the roots and between the cells in the roots, but its hyphae do not penetrate the cell membranes, it is known as an ectomycorrhiza. When the fungus in cahoots with the plant does penetrate its cell membranes, it is an endomycorrhiza. Orchids form endomycorrhizal relationships with fungi, whereby the fungus effectively infects the orchid and the orchid fights back, digesting the fungus. The outcome of the conflict can, and does, go either way. Whether infection of the embryo stimulates germination is unclear, and sometimes infection does not occur until there is a protocorm. As there are 25,000 species of orchid, there are many different variations in their relationships with fungi.

⌃ Mycotrophic relationships rely on the close symbiotic association between orchid roots and fungal hyphae. The degree of dependence varies greatly between orchid species.

Storage in seeds

(A)s mentioned earlier, seeds have three basic components that allow them to fulfil their role as a plant survival capsule: the embryo, which will grow into the seedling; the seed coat, which protects the embryo on its journey; and food for the journey and resources for the first few days after germination, before the roots and leaves of the seedling are fully functioning. The storage tissue is not the same in every seed in terms of location and composition.

AN ADVANTAGE OF BEING SMALL

Seed sizes vary hugely and there appears to be no perfect size; each size has its drawbacks and virtues. One of the virtues of being a small seed is that water can be imbibed quickly. The corollary of this is that large seeds, whose surface-area-to-volume ratio is significantly smaller, can have problems with becoming hydrated and thus germination. One of the largest seeds, the coconut, carries its own water supply.

The origin of the food store

In the thousand or so species of gymnosperm, the nutrient store in the seed consists of the tissue that was responsible for supporting the archegonium, or egg-producing organ. This means that the cells in the store have just one set of chromosomes. In the angiosperms, of which there are around 361,000 species, the food store is either in the cotyledons of the embryo or in the endosperm, which unusually has three sets of chromosomes. This is an innovation of the angiosperms and may be responsible in part for the fact that they are currently the dominant group of plants on Earth. Apart from the very different genetics of the two storage tissues, the gymnosperm seeds make their food before there is an embryo to feed, whereas the angiosperms wait until the egg has been successfully fertilized. The generally larger seeds of the gymnosperms, coupled with the presumptuous production of the nutrient store, makes this group of plants less efficient and more wasteful than the angiosperms.

(<) Due to the small surface area-to-volume ratio, the large seeds of the coconut (*Cocos nucifera*) would run the risk of not being able to hydrate themselves during germination if it were not for the water supply stored in each fruit.

Many seeds, such as those of peas (*Pisum sativum*), are rich in substances such as proteins, fats or carbohydrates. This is exploited by humans in agriculture where they are grown as food crops.

Baby food for plants

The nutrients in seed food stores come in many molecular forms and combinations. The major components are either carbohydrates (e.g. starch), proteins (in which case the seed, and thus the young plant, has a larger supply of nitrogen and sulfur) or fats (the most energy rich of the three). Another property of fats is that they are lighter for a given supply of calories, which makes them suitable for wind-dispersed seeds or seeds of species that need fast-growing seedlings. The legume family is known for its production of seeds rich in protein, which is often legumin. Unexpectedly, and inexplicably, the same molecule is used by ginkgo, an isolated, ancient lineage in the gymnosperms. Analysis of the mineral contents of seeds reveals that the proportions of these elements bear little or no relationship to the needs of the seedlings, which must become self-supporting as soon as they can. Nitrogen is most likely to run out first, followed by calcium.

THE HUNGRY SEEDLING

The largest seed in the world is the double coconut of the coco de mer (*Lodoicea maldivica*), endemic to the island of Praslin in the Seychelles, which unlike its single relative does not contain fluids. It does, however, contain as much as 10 kg (20 lb) of nutrients. This huge amount of food enables the developing seedling to grow up to 10 m (30 ft) horizontally away from the parent tree from which the seed fell. This unique form of dispersal is possible only because there are no animals on Praslin that are capable of eating the seed. This is just as well, because throughout its lifetime a female tree in the wild may produce an average of just one mature seed per year.

The seedling of coco de mer (*Lodoicea maliavica*) has such large reserves that the germinating embryo often grows sideways several metres before turning downwards in search of water.

Seeds and animals

S eeds did not evolve with the purpose of feeding animals, and yet many animals – including humans – depend on them for much of their calorific intake. The majority of seeds contain food for the embryo, the major exception to this rule being the orchids. This food, in the form of carbohydrates or fats, can provide energy for any organism – many gardeners will have seen birds such as finches feeding in the autumn on the seeds of plants in their gardens. Being eaten by an animal is, generally speaking, bad news, and yet plants have overcome this problem, and in some cases even turned it to their advantage.

The first seed-eaters

It is a reasonable proposition that the first seeds were produced at ground level. The fact that the majority of plants produce seeds shows that this is a very stable strategy from an evolutionary point of view. However, at some point since the emergence of the first seed plant, a passing animal must have taken an exploratory nibble at the seed and found that it was nutritious. Furthermore, the seed did not run away or fight back; it just accepted its fate. The plant clearly had to respond and adapt to this new threat. This response could, and did, take several forms.

A numbers game

The relationship between the density of a plant species and the density of the herbivores that feed off that plant is clearly a dynamic one. If the number of herbivores rises too high, for example, the number of plants may fall until it is no longer sufficient to support the animals and their population will crash. By producing large numbers of seeds, plant numbers can increase to acceptable levels as the likelihood is high that some seeds will avoid being eaten. One way to increase the odds further is to vary dramatically the number of seeds produced each year. This is known as masting, and is seen in tree species such as the European beech (*Fagus sylvatica*). In most years, this species will accept a high level of seed mortality. In a mast year, however, the production of seeds is so excessive that there is no way the herbivores that feed on them can consume all of them.

⌄ A male Brambling (*Fringilla montifringilla*) takes advantage of the copious quantities of seeds (right) produced in a beech mast year.

A PRECARIOUS RELATIONSHIP

One of the best-understood relationships between an animal and seeds is that between Clark's nutcracker (*Nucifraga columbiana*), a crow from North America, and the native pines, especially the eastern white pine (*Pinus strobus*) and its close relatives. The bird has a sublingual pouch in which it can store dozens of pine seeds, and it buries these in caches of just a few seeds at a time. Fortunately for the bird, it has a remarkable memory and navigational skills, so it can find enough of these stores to feed itself through winter. And fortunately for the tree, the bird rarely finds all of its caches. This is not surprising, because a single nutcracker may bury between 46,000 and 98,000 seeds in a year. Should either species in this close relationship go into decline, the other will suffer.

⌃ Clark's Nutcracker (*Nucifraga columbiana*), a member of the crow family, exploits the abundance of seeds produced by North American pine trees, pulling the seeds from the cone with its dagger-like bill, storing them temporarily in a pouch under its tongue, before hoarding them for the winter.

⌄ The Brazil nut tree (*Bertholletia excelsa*) relies on the sharp teeth and forgetful nature of the scatter-hoarding rodents called agoutis for its seed dispersal and survival.

The following year, there may be a crash in the herbivore population, when the reduced number of beech seeds produced cannot support the increased numbers of animals.

Seed sowing

Seed dispersal by animals can take many forms, but one of the best from the plant's point of view is where the seed is not only taken a safe distance from its parents, but it is also sown. Scatter hoarding is a well-documented phenomenon, and generally involves vertebrates as opposed to small creatures like ants that have permanent nests. In South America, agoutis gnaw out the seeds of the Brazil nut tree (*Bertholletia excelsa*) and then bury them to keep them safe for later. Fortunately for the trees, the agoutis do not recover every seed.

Protecting seeds

Seeds can fulfil their function as survival capsules only if they themselves are robust. The threats facing the embryo fall into two broad categories: there are those that are derived from the environment, such as fire, which are termed abiotic hazards; and there are those that come from other organisms, the major one being predation, which are termed biotic hazards.

⑦ Slow-growing agaves play the long game, producing flowers and seeds after many years.

⟩ Jequirity (*Abrus precatorius*), a member of the legume family, contains extremely toxic alkaloids, but potential herbivores associate the red-and-black colour of the seeds with toxicity and thus leave them alone.

Abiotic hazards

Weather can be a problem for any plant at any stage in its life. Drought, flood, frost, heat and fire can all retard growth and even kill a plant. Difficult conditions during the formation of fruits containing seeds can threaten the fecundity of a plant, and this is a particular problem for plants that fruit just once – monocarpic species. These species build up resources that can then be used for a once-in-a-lifetime final reproductive effort, and the time it takes them to do this varies with the habitat and size of the plant. For small herbs like hairy bittercress (*Cardamine hirsuta)*, it can take just a few weeks, but for some larger, slower-growing plants such as agaves (*Agave* spp.), it can take many years. Generally speaking, monocarpic species put a far higher proportion of their biomass into nurturing their embryos. For example, annual grasses may put 50 per cent of their biomass into raising their seeds, whereas in perennials the figure may be as low as 10 per cent.

⟨ Some plants, like hairy bittercress (*Cardamine hirsuta*), can squeeze many generations into the space of a year by completing their life cycle in a matter of weeks.

Deterring herbivores

In common with parents across the Earth, plants try to give their offspring the best start in life and prepare them for the threats that lie ahead, some of which come from other organisms. Seeds are a wonderful resource for animals, being full of carbohydrates, fats, proteins and minerals, so persuading animals not to eat them is an important survival strategy. One way plants do this is by making their seeds highly visible, advertising the fact that they are toxic in the hope that animals remember to associate the colour with toxicity. That said, if the animal does die, the seed will be inside a rotting carcass that may provide nutrients for the seedlings. Some of the world's most notorious toxins, including ricin and hemlock, are found in seeds. These molecules are expensive for the plants to synthesize, so it is a reasonable assumption that they have a survival benefit.

The benefit of a good coat

Chemical protection of seeds is one strategy for survival, but prevention is better than revenge. This is where the seed coat comes into the spotlight. The integument has several functions, one of which is undoubtedly to protect the embryo from enzymes in the intestines of birds and other herbivores. It is worth noting here that this is not an onerous task in the case of bird herbivory, because the gut environment of many birds is not as hostile as that of mammals. Another function of the seed coat is to endow physical dormancy,

which is broken by specific conditions and not by random damage or microbial attack, as is frequently stated. In addition, the seed coat physically protects the embryo from the environment, filtering out certain wavelengths of the light and keeping water in or out – parent plants grown in drought conditions will produce seeds with thicker seed coats. There is also clear evidence that in some seeds the integument includes chemicals with antifungal properties, and occasionally with antibiotics. One example is the wild parsnip (*Pastinaca sativa*), in which it has been shown that an increased levels of antibiotics significantly reduces the allocation of resources to seed production.

⌃ The seeds of wild parsnip (*Pastinaca sativa*) pack an antibiotic punch in their coats, but at the expense of the food resources produced in the seed.

⌐ ⌐ Castor-oil plants (*Ricinus communis*) are the source of the deadly toxin ricin, which was used to assassinate the Bulgarian dissident Georgi Markov in 1978. A minute quantity in the tip of a sharpened umbrella ferrule was sufficient to kill him within a few days.

The seeds of the Douglas fir (*Pseudotsuga menziesii*) have a wing that slows their descent to the ground, enabling the wind to blow them further from their parent plant.

Preventing germination

hen a seed is released by a plant, it will hopefully land in soil that is appropriate for its species, and that is moist and at a suitable temperature. The seed should germinate and be growing into a healthy seedling within a month. If it does not germinate under these conditions, there are two possible explanations. First, the embryo may be dead, or it never developed in the first place (the latter is surprisingly common in garden conifers). Or second, the seed may be dormant and will not germinate until that dormancy is broken.

Why dormancy is a good idea

Dormancy is found in many, although not all, plant species. It has evolved several times and so it is assumed that in some circumstances it has a survival value. It does, however, bring with it a level of risk, because the longer the seed lies around, the greater the chance an animal will

It is common practice in horticulture to dry seeds in paper bags to promote dormancy, or quiescence.

eat it. There are many potential reasons for the evolution of dormancy, which is a form of dispersal, but dispersal in time rather than space. Surviving in both unpredictable and predictable environments can be improved by optimizing the timing of seed germination, to coincide with the best conditions for seedling survival. This may involve avoiding cold, hot, wet or dry conditions, or avoiding predation. Dormancy can also be a way of preventing intergenerational breeding, a strategy that encourages genetic diversity. Habitat and life history both have an important influence on dormancy.

Types of dormancy

Dormancy types are commonly classed as either endogenous or exogenous. In endogenous dormancy, something in the embryo imposes the state, whereas in exogenous dormancy any other part of the seed and/or fruit can prevent germination. Exogenous dormancy may be physical, chemical or mechanical, whereas endogenous dormancy can be physiological, morphological or morphophysiological. Some researchers have also grouped together chemical and mechanical dormancies with the other physiological categories to create a new type of dormancy, which is a combination of physical and physiological components. Whichever system is used, the fact is that there is more than one way to achieve dormancy and no single system fits every situation. For example, combined physical and physiological dormancy may be the best strategy where fire is a regular component of the ecology of a habitat, as seen in the North American redstem ceanothus (*Ceanothus sanguineus*).

⌄ The dried seeds of okra (*Hibiscus esculentus*) will survive much longer than those left inside a moist fruit.

⌃ Redstem ceanothus (*Ceanothus sanguinea*) optimises its chances of survival in a habitat characterised by regular fire by adopting both physical and physiological seed dormancy.

NOT DORMANT, BUT WAITING

Seeds are often stored by farmers and gardeners for use the following year in a dark, dry, cold place where animals cannot find them. In these conditions, the seeds will not germinate. They might be dead or dormant, but they may also just be quiescent. A quiescent seed is one that will not germinate because it has not imbibed sufficient water and it is not at the right temperature for its biochemical processes to function. If a seed is dried to 5 per cent of its fresh weight and stored at −20 °C (−4 °F), it may be kept alive in this quiescent state for a very long time. This is the most important tool used in seed banks for plant conservation (see page 352).

Stimulating germination

F armers, gardeners and conservation biologists all need to know how to make seeds grow into vigorous plants. This may involve breaking the seed's dormancy, and in order to do that, the plant needs to be identified, at least to family level and preferably to genus or species. Important factors indicating the type of dormancy in a species new to you include how well developed or differentiated the embryo is, whether the embryo grows prior to germination, how impermeable the fruit and/or seed coat are to water, and finally whether the shoot and root appear at the same time or the root clearly precedes the shoot.

Breaking physical dormancy

This is perhaps the easiest type of dormancy to crack, because it may simply involve breaking the hard, woody layer of integument cells that are impregnated with substances such as lignin and cutin, or waxes and other lipids. A typical family that uses this type of dormancy is the legume family, and wise vegetable growers know that soaking their bean or pea seeds in warm water for 24 hours before sowing can help to soften the seed coat. Some tropical woody members of the family have large seeds that can be carefully cracked with a hammer.

Breaking physiological dormancy

In some seeds there is no physical barrier to germination, but a physiological inhibitor instead. This might be a chemical that prevents the cells of the embryo from dividing and thus stops the root from emerging, an example being the plant hormone abscisic acid (ABA). A common way in which physiological dormancy is broken is through a minimum period of warm or cold weather, or periods of both, known as stratification. Common ash (*Fraxinus excelsior*) and sycamore (*Acer pseudoplantanus*) seeds contain inhibitors

⋀ Many members of the legume family, such as beans (*Phaseolus* spp.), have thick, hard seed coats that must be ruptured before germination can begin.

⋗ Seeds of the common ash (*Fraxinus excelsior*) contain inhibitors that must be washed or leached out before germination can begin.

Morphophysiological dormancy, which involves a combination of morphological and physiological components, is found in a number of members of the broomrape family (Orobanchaceae), all of which are parasitic plants. For dormancy to be broken in the parasite's seed, there must be a period of good conditions during which the embryo finishes developing, plus a chemical stimulation from the host plant to trigger germination.

⌄ Parasitic broomrapes (*Orobanche* spp.) demonstrate a combination of two types of dormancy: physiological and morphological.

⌃ Broomrape seeds will geminate only when both types of dormancy are broken and the environmental conditions are amenable.

that simply have to be washed, or leached, out. Sometimes, a molecular change in a layer of the fruit can be brought about by a period of cold, and this facilitates the emergence of the root. This is found particularly in members of the rose family (Rosaceae).

Breaking morphological dormancy

Morphological dormancy is when the embryo is either undifferentiated or differentiated but undeveloped at the point when the plant releases its seeds. For such dormancy to be broken, the correct conditions must be present for the embryo to continue its differentiation and development. This means that there must be a moist substrate at the right temperature and the right light conditions, which might be dark (buried) or light (not underneath other plants).

Breaking combinational dormancy

To overcome combinational dormancy, an impenetrable barrier must be broken for water to enter the seed and then a physiological barrier such as an inhibitor must be removed by cold temperatures, or there must be after-ripening of the embryo at high temperature. This being biology, and therefore nothing being simple, the order in which the two types of dormancy are broken differs between species.

Seeds and fire

⌃ Forest fires are not uncommon in certain habitats, for example in Mediterranean-type climates, but many plants can survive and reproduce even in these extreme conditions by adopting various survival mechanisms.

(W)hen plants emerged from their watery environment about 470 million years ago, they remained wet enough to resist burning for millions of years. However, once they became lignified and grew into trees, they contained sufficient combustible material to fuel a decent conflagration. It is reasonable to suppose that any innovation that enables a plant to get through a fire has been naturally selected. Such innovations include protecting seeds.

▷ Giant redwoods (*Sequoiadendron giganteum*) are just one type of gymnosperm that are able to survive fire by protecting their seeds in woody cones. Once the fire has passed, the cones open to release their seeds into the freshly prepared and nutrient-rich soil.

Where do you keep yours?

There are two places where seeds survive fire. First, there is the obvious place – in the soil. How deep is deep enough for seed survival depends on the temperature of the fire, but a good rule of thumb is a depth of 50 mm (2 in). Second, plants can 'hide' their seeds in cones that are held in the canopy. Such woody cones have evolved several times in both gymnosperms and angiosperms. Pine trees (*Pinus* spp.) are the best-known examples of serotiny (the release of seeds from a cone prompted by an environmental stimulus, e.g. fire) in the gymnosperms, but Australians will be just as familiar with the cones of banksias (*Banksia* spp.) in the angiosperms.

No germination without smoke

Seeds are commonly buried by simply falling into a fissure in the soil, but many are also buried by ants in their nests. Such seeds often have a fat body known as an elaiosome attached to them, which is thought to contain a chemical that is irresistible to ants. Irrespective of the method by which seeds come to be buried, researchers are exploring why they germinate following a fire. While heat was the first and obvious suspect, it has been shown that in many cases it is smoke that triggers germination, and specifically the chemical butenolide in that smoke. Following a bush fire, it is common for the ground to be covered with seedlings.

The right sort of fire

The features of wildfires are numerous and vary with every event. The time of year and the size of the area burnt are two, but the major variable is intensity, which itself is determined by factors such as the type of plants (i.e. the calorific value of the fuel) and the length of time since the last fire (i.e. the amount of fuel lying on the soil surface). The intensity, which is a combination of temperature and the length of time the fire stays in one place before burning out, is critically important to plants. If the fire is too intense, the seeds in the ground will get too hot to survive. Seeds that are held in cones in the canopy are released when heat from the fire melts the wax that is sealing the scales of the cone, but if the fire is too intense, the flames will reach into the canopy and the cones and their seeds will be incinerated.

⬅ In Australia, the heat of a bush fire causes *Banksia* cones to open and release their seeds after the fire itself has passed.

➤ After a bush fire, seeds stored in the seed bank in the soil germinate en masse in response to the chemicals in the smoke.

Fruits and animals

(A) long with flowers, fruits are a defining feature of the angiosperms. The ovary at the base of the carpel encloses and isolates the ovules, protecting them from the environment through a delicate phase. In common with any part of a plant, a fruit can be eaten, but it appears that some plants – particularly tropical woody species – go out of their way to attract herbivores, while others do the exact opposite by being toxic. There must be some advantage to the plant in developing fruits in these ways, because they often contain energy-rich carbohydrates, demanding an investment of resources.

(∧) Bohemian waxwings (*Bombycilla garrulus*) rely on the fleshy fruit of many plants to survive the winter, and in doing so disperse the seeds away from the parent plant.

Dispersal syndromes

The most common explanation for the evolution of fleshy fruit is that fruit characteristics are associated with the dispersal of seeds, in much the same way that flower characteristics are associated with pollination. In pollination, the correlation between pollinator and flower is often so predictable that there are well-defined pollination syndromes (see page 240). Despite various efforts over that past 100 years, however, no one has come up with a comparable set of seed-dispersal syndromes. For example, tubular red flowers with lots of sweet nectar are often pollinated by birds, but sweet-tasting, bright red fruits are not

the preserve of birds. It seems that many animals eat many different fruits, and that many fruits are eaten by many different animals. Pollination in particular, and herbivory in general, are very well studied, but frugivory and seed dispersal are lagging behind.

It is not all about dispersal

Some pollinator–plant relationships involve the fruits as well as the other components of the flowers. Fig wasps are able to lay their eggs in some of the ovaries of female fig (*Ficus* spp.) flowers because the styles attached to these ovaries are short enough for the ovipositor to reach them (see page 259). Flowers with long styles are not parasitized but are pollinated. *Yucca* flowers are similarly parasitized by the egg-laying females of yucca moths (see page 259), although there appears to be some restraint on the part of the moths because they lay eggs in only a few of the ovules. This means that some viable seeds will be produced to continue the host plant lineage.

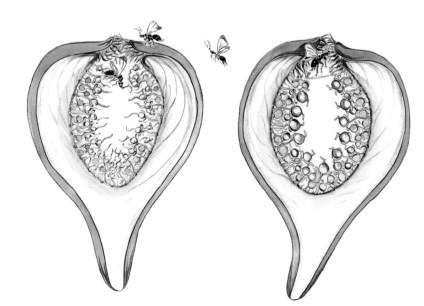

(⟨) The variation in length of the ovaries in the female flowers of the fig (*Ficus* sp.), mean that some are parasitized and some are pollinated by visiting fig wasps.

A STICKY ISSUE

There are approximately 700 species of mistletoe around the world, belonging to three closely related families, and they are all parasites of woody plants. The majority of species produce fleshy fruits that are eaten by birds, but neither these fruits nor their seeds are normal. The seeds have no seed coat and instead are enclosed in a sticky, gloopy tissue known as viscin. The viscin can be translucent, and in some cases the mistletoe seed is photosynthetic before it starts to grow on a new host. The fruit of common mistletoe (*Viscum album*) is traditionally eaten by the mistle thrush (*Turdus viscivorus*). Although the viscin can prevent some birds from eating the fruit, this is not the case with the thrush. However, the viscin contains a powerful laxative, ensuring that the uncoated seed does not remain in the gut long enough to be digested. The viscin also survives passage through the gut, and helps the excreted seed – now dispersed from the parent plant – stick to a new branch.

In recent years, warmer winters in the United Kingdom have attracted migrating blackcaps (*Sylvia atricapilla*) from Germany. Being smaller than a mistle thrush, a blackcap cannot swallow common mistletoe fruits whole, and the viscin sticks to its beak. In the process of rubbing the fruit off, the blackcap sows the seed in a bark crevice more effectively than the thrush. As a result, climate change is indirectly leading to an increase in the common mistletoe in the United Kingdom, thus reversing decades of decline. Indeed, the blackcap is so efficient at sowing, that the common mistletoe is now parasitizing another mistletoe, *Loranthus europaeus*, as the bird moves between plants in its quest for food.

⊘ Mistle Thrushes (*Turdus viscivorus*) are the most common dispersal agent for the seeds of common mistletoe (*Viscum album*), whose sticky seed coat provides protection and an anchor for the seed to its new host.

⋀ Common mistletoe (*Viscum album*) is the subject of many folk tales and superstitions that involve all kinds of properties from enhancing fertility to warding off witches.

⋖ Blackcaps (*Sylvia atricapilla*) are now known to be an effective alternative dispersal agent for mistletoe.

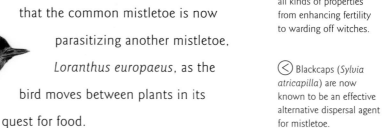

Dry fruits

The majority of fruits produced by plants are dry fruits, including all the grains harvested from cereal crops. These are a particular type of dry fruit called a caryopsis, in which the seed and ovary wall are fused together. Dry fruits are often associated with abiotic methods of dispersal, because they lack a nutritious fleshy tissue to act as an inducement to an animal. There are, however, many well-documented examples of animals scatter hoarding dry fruits. This is clearly similar to the dispersal of seeds inside fleshy, edible fruits, but it differs in one very important way: the reward is the seed itself, and the parent plant 'sacrifices' some of its offspring so that others may survive.

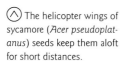

Ⓐ The helicopter wings of sycamore (*Acer pseudoplatanus*) seeds keep them aloft for short distances.

Ⓛ The seeds of goat's-beard (*Tragopogon pratensis*) take advantage of their parachutes, which let them float through the air away from the parent plant.

Ⓥ Ash (*Fraxinus* spp.) seeds, or keys, have dry wing-like appendages that aid wind dispersal.

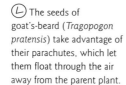

CONFUSING CONVERGENCE

A familiar sight in the autumn in hedgerows around the world is old man's beard or traveller's-joy (*Clematis vitalba*). This species is a member of buttercup family (Ranunculaceae) and has flowers with many unfused carpels. The fruits that develop from the ovary are small, hard, single-seeded achenes. However, the style persists and develops into a hairy appendage that aids wind dispersal by acting as a simple parachute. A very similar fruit – an achene with a feathery style – is found in the very distantly related *Dryas* genus in the rose family, low-growing plants that are a familiar sight in the northern hemisphere. Despite the extraordinary similarity of their fruit, these plants have nothing much else in common.

On a wing and a prayer

The wind can transport seeds long distances – 400 km (250 miles) in the case of orchid seeds to Krakatoa after its eruption (see page 288), and 100 km (60 miles) for a conifer seed in North America. These are probably exceptions, and most wind dispersal is over far shorter distances because adaptations to flight are more to do with slowing the rate of vertical fall than movement in a horizontal direction – as seen in the hairy, parachute-like structures on the fruits of many members of the daisy family (Asteraceae). The samaras (winged fruits) of maples like the sycamore

(↑) Mountain avens (*Dryas octopetala*) has small seeds whose dispersal is aided by a hairy style that acts as a parachute.

(∧) The silky appendages of the seeds of traveller's-joy (*Clematis vitalba*) demonstrate why the plant is also called old man's beard. The appendages increase the surface area of the seed to help with wind dispersal.

look like the rotary blades of a helicopter but they do not keep the seeds in the air permanently. Pine (*Pinus* spp.) and ash (*Fraxinus* spp.) trees have fruits that are a variation on the theme of helicopter blades, while birch (*Betula* spp.) fruits have two wings and look like a small butterfly. An important consideration here is the lack of control over the direction of travel. Prevailing winds will result in a seed shadow, with distribution being asymmetrical. This may sound trivial, but the colonization of the oceanic Canary Islands off Africa has been significantly influenced by the fact that the prevailing winds come from the northeast, blowing seeds from southwest Europe and northwest Africa.

Hanging on

The fruits of many families of flowering plants have small hooks so that they can attach themselves to passing mammals and birds. So successful is this strategy that the ability of the tiny hooks on the fruits of burdock (*Arctium* spp.) to hang onto clothing was the inspiration for the invention of Velcro by Swiss engineer George de Mestral. The fruits of plantains (*Plantago* spp.) do not have hooks but become adhesive when wet, sticking to mammal fur and bird feathers. Plants in this genus are pollinated by wind but dispersed by animals. There appears to be very little relationship between choice of pollinator and dispersal vector in plants.

WATER DISPERSAL

The use of waterways by plants as agents of dispersal is perhaps by mistake, because this is the ultimate unidirectional transport. Transport across oceans is similar, in that the plant is at the mercy of the currents. A particular hazard associated with water transport is the fact that the water may stimulate premature germination, before the seed has been delivered to dry land. Because of this, seeds that are transported successfully by sea have thick, waterproof fruit and seed coats.

(>) Water dispersal may not be common, but it is the method used by mangroves, whose seeds float just beneath the water surface, and have thick, waterproof seed coats.

(>) Seeds of burdock (*Arctium lappa*) adopt a 'Velcro' method of dispersal, using hooks to attach themselves to passing animals. The burs are fruit heads covered in involucral bracts that become woody and hooked after fertilization.

Fleshy fruits

F leshy fruits are familiar to us because we eat them, as do many other animals (see also pages 342–43). In fact, this may be the main function of fleshy fruits – to be eaten. However, it is not out of an altruistic desire to feed animals that the plant puts resources into fruit, but to have its seeds dispersed. Many parts of the plant associated with flowers can become fleshy, and to an animal it matters not one jot where its food comes from. To a botanist, however, a fleshy fruit is one in which the fleshy tissue is derived from the wall of the ovary.

ANALOGY, HOMOLOGY AND FLESHY FRUITS

There are clearly many ways in which plants can make fleshy fruits. To the plant, function is everything, but to the botanist, function is of secondary importance to position and origin. Structures can look very different and have different functions but occupy the same position on the plant. These structures are homologous. Structures can have similar functions, such as being fleshy to attract and reward animals, and be derived from completely different parts of the plant. These structures are analogous. There is a lot of analogy in fruits, an extreme example being a fleshy seed such as that of the pomegranate (*Punica granatum*). Its fruit is a berry that contains seeds with a fleshy outer layer, making them doubly fleshy.

Drupes

Fleshy fruits with just one seed, called drupes, are found in many families of plants across the evolutionary tree. Within the monocots, drupes are common in the palm family (Arecaceae) and asparagus family (Asparagaceae). The bright orange fruits of cuckoo-pint (*Arum maculatum*), in the

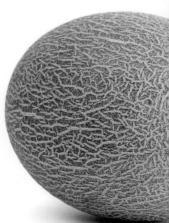

 Confusingly, the blackberry (*Rubus fruticosus*) is not a berry but is an aggregate of drupes, individual fleshy fruits each containing a single seed. On the other hand, the melon (*Cucumis* sp.) is a berry, because it is a fleshy fruit containing many seeds.

monocotyledonous arum family (Araceae), are a familiar sight in European woodlands, and look remarkably similar to the seeds of another shade-dwelling plant, the stinking iris (*Iris foetidissima*). The arum fruits are seriously toxic and it appears that the iris may be mimicking them to avoid predation. Other familiar drupes are cherries, plums and peaches. When lots of little drupes, or drupelets, are arranged on a receptacle, it is known as an aggregate fruit – blackberries and raspberries are good examples.

Berries

This being botany, raspberries are not berries, while tomatoes, bananas, grapes and aubergines are. The definition of a berry is a many-seeded fleshy fruit, and they include countless variations that look very different from one another. Citrus fruits like oranges and lemons are a type of berry, in which the fleshy tissue is made up of fluid-filled hairs. Melons and cucumbers are berries with thick rinds and are known as pepos.

 Purple aubergines, red tomatoes, yellow lemons and green melons. Fleshy fruit comes in a wide range of different colours that may or may not have a bearing on their function in biology.

The colour of fruit

Fleshy fruits tend to be colourful, whereas dry fruits tend to be brown. While some biologists like everything that has been adapted by natural selection to have a function, naturalist Alfred Russel Wallace cautioned in 1879 that, when it comes to fruit characters such as colour, looking for a function could be a waste of time. However, that has not stopped people speculating and researching the area. Colour in fruits may be a signal, encouraging animals to eat them, or discouraging animals from doing so because they are toxic. The signal could change with time from the latter to the former, to ensure that the seeds are removed only when ripe. This might be a change to red from green to make the fruit more obvious when ripe, because green and red are contrasting colours that make each other more saturated. As Wallace warned, however, colour transformation might just be a consequence of chemical changes during ripening or varying temperatures at night. How plants work is still a mystery in many ways.

 Naturalist Alfred Russel Wallace (1823–1913) warned against relying on function as a cause of fruit colour.

true fruit: nutlet, e.g. borage

true fruit: schizocarp, e.g. mallow

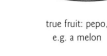

true fruit: pepo, e.g. a melon

true fruit: nut, e.g. a walnut

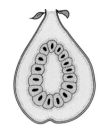

accessory fruit: syconium, e.g. a fig

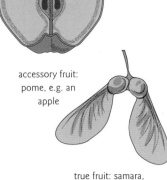

accessory fruit: pome, e.g. an apple

true fruit: samara, e.g. a maple

true fruit: legume e.g. a peapod

False fruits

Ⓐlthough a fruit is defined as a ripened ovary and any other structures attached to it that ripen with it, there are times when a line has to be drawn and structures that appear to be attached to the ovary must be declared ineligible. These 'fruits' were previously referred to as false fruits, which would not have been a problem except that many of them are found in any greengrocer's shop, including strawberries, figs, apples and pears. The term false fruit has now largely been replaced with accessory fruit, which is more appropriate because it implies correctly that the structure is a fruit with some additional structures.

⋀ Some examples of true and accessory fruits, demonstrating that accessory fruits are very often the ones we search for in the fruit section of the supermarket.

◡ Medlars (*Mespilus germanica*; top) and hawthorns (*Crataegus monogyna*; bottom) are accessory fruits, with the flesh derived from the hypanthium, a structure characteristic of Rosaceae species.

Accessorizing fruits

Being imaginative with fruits is a trait that members of the rose family have taken to the highest level. Some species bear fruits that are unambiguously, botanically, fruits – for example, cherries, nectarines, and peaches are drupes. Another example of a conformist is the genus *Dryas* (see box on page 304), whose members produce achenes. However, within the rose family we also find perhaps the easiest accessory fruit to understand, the strawberry. Rosaceae species have many things in common, including the hypanthium and receptacle. The hypanthium is a cup in which the carpels sit, and around the edge of which are attached the perianth and the

stamens. The receptacle is a cone in the centre of the hypanthium on which the carpels sit. In the strawberry (*Fragaria* spp.) flower, the carpels develop into achenes – true fruit. These achenes are embedded in the receptacle, which turns bright red when the achenes are ripe.

Apples and pears

The fruits of apples (*Malus* spp.) and pears (*Pyrus* spp.) are examples of a type of accessory fruit found only in the rose family – the pome. It appears in several other genera and consists of a core, which is the fruit, and an expanded receptacle, which makes up the flesh of the fruit. In cotoneasters (*Cotoneaster* spp.), hawthorns (*Crataegus* spp.) and medlars (*Mespilus* spp.), the flesh of the fruit is derived from the hypanthium rather than the receptacle.

◡ Strawberries may look a 'normal' fruit, but the sweet, juicy flesh is derived from the receptacle. To be a true fruit requires that the flesh is derived from any part of the ovary.

FIGS – AGAIN

As we saw earlier (see page 302), figs have an intimate and unique relationship with wasps in their pollination, and the theatre on which this is acted out is the hollow interior of a concave receptacle that has just one small entrance. The crunchy parts of the fig 'fruit' (called a syconium) are the ripened ovaries of the female flowers. If you are thinking that this sounds like the arrangement of the parts in a rose hip, there are a couple of very significant botanical differences to note. The rose hip is a hypanthium that contains many carpels of just one flower, whereas the syconium of the fig is a receptacle.

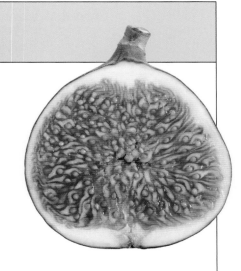

⊼ Figs (*Ficus* ssp.) are formed from the receptacle of the flower and contain the ovaries from many flowers.

⊲ In rose hips the hypanthium, a structure typical of members of the Rosaceae family, forms the fleshy part of the fruit and contains seeds formed from the carpels derived from one flower.

Pineapples and mulberries

Another group of fruits that are not quite what they seem are those formed from many flowers, which then merge together or are contained in one fruit-like structure. One subset of these are two multiple fruits that are in very distantly related groups: pineapples in the bromeliad family (Bromeliaceae) and mulberries in the mulberry family (Moraceae). In both of these, many small flowers on the receptacle each contribute a carpel and thus fruit to the 'fruit'.

ⓛ ⓥ Pineapples (left) and mulberries (right) are examples of aggregations of many smaller fruits.

Seed dispersal

Seed dispersal occurs through both time and space. Dispersal in time enables plants to ride out unfavourable conditions – the most extreme example being 30,000 years (see opposite). In contrast, dispersal in space enables plants to get their offspring away from their roots. The record here is 18,000 km (11,000 miles), verified for a species of *Acacia* that appears to have made the journey from Hawai'i in the Pacific Ocean to Réunion in the Indian Ocean. It is assumed that the seeds were in mud on the body of a bird, which travelled between the islands around 1.4 million years ago.

⌃ Dispersal of seeds through time relies on the longevity of the seed. 2,000-year-old seeds of the date palm (*Phoenix dactylifera*) have been shown to be still viable.

Why leave home?

Dispersal is something into which plants invest a good deal of resources, and so it is a reasonable assumption that it is an evolutionarily stable strategy. But why is this? A plant living in a habitat that has everything it needs to survive and reproduce is a successful, fit plant. Would it not want the same for its offspring? The answer here is yes and no. If it is a monocarpic plant that dies after fruiting, there is no need for it to disperse its seeds because it will leave a gap on its death. However, if it is a long-lived perennial, it needs to send its offspring away because the most severe competition any organism faces is from members of the same species. Dispersal also has the advantage of enabling the species to try out new habitats that might suit it better. In addition, it helps to reduce breeding with close relatives, which is never a good idea. Finally, it can enable plants to get away from predators and pathogens.

The dispersal options

The options for seed dispersal fall into two categories: abiotic dispersal, which uses wind or water, sometimes aided by modifications to the seed such as wings or buoyancy tanks, respectively; and biotic dispersal, which relies on an animal as the vector. When animals are used, there are three subcategories: endozoochory, where the seed is transported on the inside of the animal following ingestion but (hopefully) not digestion; epizoochory, where the seed is attached to the outside of the animal, such as the adhesive fruits of goosegrass (*Galium aparine*); and dyszoochory, or scatter hoarding of seeds, where the hoarder fails to recover all the seeds. All of these options may be aided by ballistic dispersal, in which the seeds are ejected explosively, as in the caper spurge (*Euphorbia lathyris*).

⌄ Seeds of goosegrass (*Galium aparine*) are covered in tiny hooks and are dispersed in space by sticking to the coats of passing animals.

SURVIVAL RECORD HOLDER

In 1994, a 1,300-year-old seed of the sacred lotus (*Nelumbo nucifera*) that had been sitting in mud was germinated and grew into a plant, making it the oldest viable seed recorded at that time. In 2008, this plant lost the record to 2,000-year-old seeds of the date palm (*Phoenix dactylifera*). Then, in 2011, the bar was raised to 30,000 years when seeds of a species of *Silene* recovered from frozen ground in central Asia were germinated and grown into flowering plants. These examples of extraordinary longevity show clearly how seeds enable plants to be dispersed through time. It also shows how the food supply in the seed is able to keep the embryo alive for very long periods. Finally, it shows that faith in artificial seed banks for the conservation of plants is very well founded.

(>) The record for the oldest viable seeds lies with a species of *Silene*, at 30,000 years. Shown here is white campion (*Silene latifolia*).

(<) Seeds of the sacred lotus (*Nelumbo nucifera*) can survive for 1,300 years and still germinate.

THE ECOSYSTEM ENGINEERS

There is a branch of seed dispersal by animals that is so important it has its own name – myrmecochory, which is dispersal by ants. It is found in more than 80 botanical families and in many different habitats. Seeds of members of the large genus *Euphorbia* have an attachment known as an elaiosome, which is an energy-rich fat body that acts as a reward for the ant. Ants pick up *Euphorbia* seeds and carry them back to their nests, where they remove and consume the elaiosome. The seed is then discarded on the nest rubbish heap, which provides the correct conditions for germination.

(>) Energy-rich fat bodies called elaiosomes attached to seeds of plants such as euphorbias provide a nutritious reward for the animals that disperse them.

(∨) Ants are one of the animals that find elaiosomes irresistible. They will haul *Euphorbia* seeds back to their nests where the elaiosome is removed and stored and the seed discarded.

Fruit dispersal

Fruit dispersal is influenced by fruit structures in the same way that flower pollination is influenced by flower structures. Of those fruits and seeds that are modified to enhance dispersal, dry fruits and seeds have become smaller and lighter to be moved passively by wind and water, or have adopted hooked or sticky surfaces to cling to the outsides of animals. Fleshy fruits and seeds, irrespective of the origin of the flesh, are modified to be eaten by animals to promote dispersal. Pollination and fruit dispersal by animals are considered to be examples of mutualism relationships, whereby both parties benefit. Darwin proposed that the flower–pollinator relationship is very close, and there has been a great deal of research carried out both to support and, recently, refute this idea. Frugivory and seed dispersal are lagging behind in this respect, but research in these areas is catching up.

∧ Fruit dispersal is not the exclusive domain of mammals and birds – reptiles such as tortoises are not averse to eating fleshy fruits.

⊓ Reptiles such as iguanas are particularly attracted to colourful, smelly, oily fruits that are easily accessible at ground level.

The start of the affair

As soon as fleshy fruits (here meaning any seed(s) with a squidgy coat) appeared, animals tried to eat them. Initially, those animals might have been invertebrates, choosing rotting ovary tissue as an appealing place to lay their eggs. The plants with the tissue most likely to rot may have been preferentially selected, and from this fleshy fruits may have evolved. Once there were fleshy fruits, more animals would eat them, but these animals were very different from the herbivores we see today. Reptiles and dinosaurs did and, in the case of the former, still do eat fruits. The role of reptiles in fruit dispersal today is often dismissed, yet it exists and it has a name – saurochory. This group seems to be particularly attracted to oily fruits and those that are smelly and colourful. Furthermore, the fruits eaten by reptiles are often attached to the plant near the ground. The significance of this is that the first seeds and fruits were attached close to the ground. This might be serendipity, but it might also account for the importance of ancestral reptiles as fruit dispersers before mammals and birds diversified.

HOW SAFE IS ENDOZOOCHORY?

Plants do not seem to have a problem with passage through the gut of a
vertebrate, particularly the guts of birds. But why is this, and how have the
plants adapted to the relationship? Although mammalian guts are long and
harsh, those of birds are not, so passage through them is relatively quick.
The pulp of a fruit may include a laxative to increase the speed of passage, or it
may include an emetic. The role of endozoochory in dormancy and early seedling
growth appear to be greatly overrated. Some fleshy fruits do contain germination-
inhibiting chemicals that enable the seed to register when it has been through
a gut and thus removed from the location of its parents. However, the idea
that the gut environment helps to soften up the seed coat, thereby enabling
it to imbibe water when released, and that being released in faeces helps
the initial growth of the seedling, both lack experimental support.

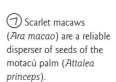

⑦ Scarlet macaws
(*Ara macao*) are a reliable
disperser of seeds of the
motacú palm (*Attalea
princeps*).

Mammals and birds

Saurochory predicts which fruits will appeal
to reptiles. Birds, which tend to have a poorly
developed sense of smell, generally disperse
fruits that are odourless but are brightly
coloured red, black or blue. Frugivorous
mammals tend to be nocturnal, and the fruits
they disperse are therefore generally dull in
colour but highly fragrant. Studying dispersal
by animals is difficult, because proving a
connection between a specific animal and
a specific seedling requires meticulous
observation. The reliability of vertebrates
as seed dispersers has also been questioned,
because where an animal deposits the seed is
determined by the needs of the animal and
not the needs of the plant – roosts, nests and
breeding or display territories may all be
unsuited to the plant. However, one research
project has clearly and unambiguously
revealed that macaws (three species in the
genus *Ara*) are the perfect dispersers of seeds
of the motacú palm (*Attalea princeps*).

⋀ Frugivorous animals
such as this crowned lemur
(*Eulemur coronatus*) have yet
to be proved as reliable seed
dispersers, and plants are very
often at the mercy of their
behaviour.

▽ This Galloway cow
has become the unwitting
dispersal agent for the
Velcro-like fruit of a burdock
(*Arctium* sp.) plant.

People and plants

Our dependence on plants is absolute. They have shaped and defined the world we live in, and will always be essential for our very survival.

By capturing the energy of the sun through photosynthesis, plants are at the heart of the major cycles that created and sustain the biosphere. Not only are they the base of almost all food chains, they also provide essential ecosystem services: mediating the water cycle and the balance of gases in the atmosphere, making this a world in which we and other animals can live.

From the beginning, our species has selectively exploited the vast diversity of plants to meet our daily needs and as a source of inspiration in art and gardens. In our relationship with plants we have often changed them, shaping their morphology and biochemistry to our ends. At first this was a slow process, reflecting the harvesting and selection of seeds yielding the largest and most nutritious crops. Today, we have powerful techniques for selecting and introducing valuable traits into plants, to meet growing requirements for food. However, the demands on nature made by our burgeoning global population threaten the diversity of plants, and many cultivated plants and wild species are threatened with extinction. The race is now on to protect plant diversity, our life-support system.

Cultural connections

(T)he ancestors of modern humans relied heavily on plants in their diet, but our species has taken the relationship with them even further. Since the earliest times, people have shown remarkable inventiveness, finding uses for a multitude of plants. All parts of a plant, from above or below ground, have been evaluated and appropriated to meet our needs.

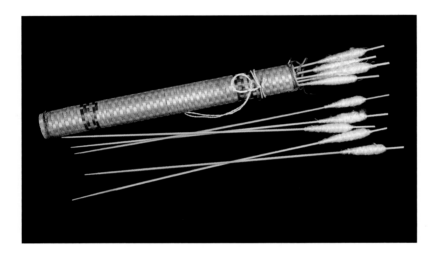

(∧) Plants for hunting: a woven quiver and cane blowpipe darts for use with curare, an alkaloid plant toxin. Made by Pemon indigenous people in Venezuela.

Survival skills

The ability to find and recognize many kinds of plants was once an essential skill for survival. Humans rely mainly on visual cues to distinguish between plants, identifying them by their morphology, although our senses of smell and taste also play important roles. As the properties and qualities of different plants were recognized, they became central to human life and powerful cultural connections were established. Knowledge of the growth of plants through the seasons enabled people to migrate and disperse around the globe, where they encountered different floras as they occupied new lands. The highest priority would always have been to forage for edible plants, usually to fill out a diet of meat, fish and seafood obtained by hunting.

Through time, plants came to be used in every aspect of daily life. Wood was burnt for warmth and fires lit for cooking and defensive purposes. Later, we learnt how to make charcoal, still an important fuel in many countries. Branches and timber provided everything from temporary shelters to permanent dwellings of increasing sophistication. Leaves, from palms to rushes, provided thatch for roofs and gave shelter from the elements. The properties of distinct types of timber were exploited in the manufacture of tools, furniture, and household and decorative items.

Plants for everything

A wide variety of plant fibres proved useful for making threads and ropes. Some, such as cotton (from *Gossypium* spp.) and linen (from the common flax, *Linum usitatissimum*), could be spun and woven into fabrics. Clothing could also be fashioned from large leaves such as those of the Chusan palm (*Trachycarpus fortunei*) or from bark cloth, usually made from the beaten bark of figs

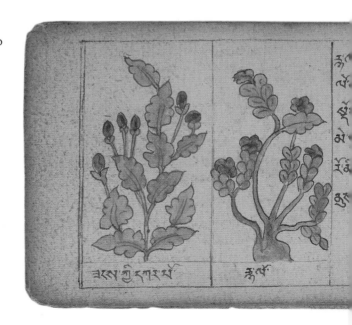

⊃ Deep blue indigo dyes are valuable and much sought-after plant products traditionally extracted from a wide variety of species.

In China, the leaves of several species of *Strobilanthes* in the acanthus family (Acanthaceae) are used to dye fabrics.

(*Ficus* spp.) and their relatives. Since Neolithic times, pigments extracted from plants have been used to colour and decorate fabrics, while those from henna (*Lawsonia inermis*) and woad (*Isatis tinctoria*) were used to dye hair and skin.

Other plant extracts have been used a wide variety of beverages, medicines and drugs. The refreshing and mildly stimulating infusions made from leaves of the tea plant (*Camellia sinensis*) leaves and beans from coffee (*Coffea* spp.) are consumed by billions around the world. We use herbs and spices to preserve our food and to enhance its flavour. Fermentation of grains and grapes produce a great diversity of intoxicating beers and wines. Stronger drugs, such opiates from the opium poppy (*Papaver somniferum*) and psychoactive alkaloids from peyote (*Lophophora williamsii*), were traditionally used in ceremony, magic and divination. The power of plants as a source of poisons has also been known, and applied, since antiquity. Thousands of plant species are listed in the materia medica of cultures around the world, with at least 7,000 being used in Chinese traditional medicine alone.

Today, we usually buy the plant products we use and many would struggle to recognize the species from which they come. Although foraging for wild food is fashionable, few of us could survive if forced to rely on nature rather than retail. In an era of globalization, plants used for everything from dietary staples to traditional medicines are traded around the world, and demand often outstrips supply, especially for those harvested directly from the wild.

⌄ Illustrated pages from an ancient Tibetan text, with instructions on the preparation and uses of traditional medicines extracted from plants of various kinds.

Ethnobotany

(T)he study of traditional knowledge of plants is called ethnobotany. It is important for many reasons, not least as an important part of our cultural heritage and because our exploration of the plant kingdom is far from complete. Recognizing that most biodiversity-rich areas are inhabited by indigenous or local communities, Article 8 of the United Nations Convention on Biological Diversity emphasizes traditional knowledge, innovations and practices.

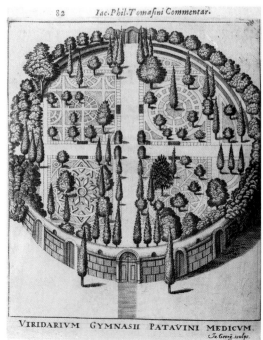

their ecology, to understand when and where they could be found. The most expert practitioners possess knowledge extending into healing, divination, and spiritual and religious affairs, and are variously referred to as medicine men, shamans and witch doctors.

Recording plant wisdom

The invention of written languages allowed plant knowledge to be recorded and shared in ways that now provide powerful glimpses into the past. In ancient China, medicinal plant knowledge is traced back to 2700 BCE and the legendary Emperor Shen Nung, whose discoveries were documented in later

A Socotran man, with an encyclopaedic knowledge of the diversity and uses of the plants of the remarkable island he calls home, carefully harvests roots from a tree traditionally used as a medicine.

The botanical garden in Padua, Italy, established in 1545 for the cultivation and study of medicinal plants, provided a model that inspired the establishment of many other European physic gardens.

An oral tradition

As civilizations around the world acquired knowledge of plants, it became increasingly important to codify and communicate information, passing it from one generation to the next. Initially, this was an oral tradition, taught by elders to younger generations alongside the other skills they needed for survival. In many indigenous or aboriginal cultures, men and women maintained different traditions of plant use, passed on separately to sons and daughters. Common themes in such traditions were the need to know the morphology of plants, to recognize and differentiate between them, and to know writings. Clay tablets dating from seventh-century BCE Assyria recorded the use of more than 250 plant drugs, many of which are familiar today. Western traditions of plant medicine have their roots in ancient Greece and such works as *Historia Plantarum* and *De Causis Plantarum* by Theophrastus of Eresos, and *De Materia Medica* by Dioscorides. The latter was to influence many of the later herbals that followed the invention of the printing press in the fifteenth century, perhaps the best known of which is John Gerarde's *Herball* of 1597.

Although western and Chinese medicine may seem worlds apart today, the early literature shows very similar emphasis on, for example, the 'hot' or 'cold' properties of specific plants and on correcting imbalances in the body. Both traditions have versions of the doctrine of signatures, in which similarities between plants and parts of the body are interpreted as a sign of special curative properties. The roots of mandrake (*Mandragora officinarum*), for example, have hallucinogenic properties and their resemblance to human bodies has been associated with powerful magical properties.

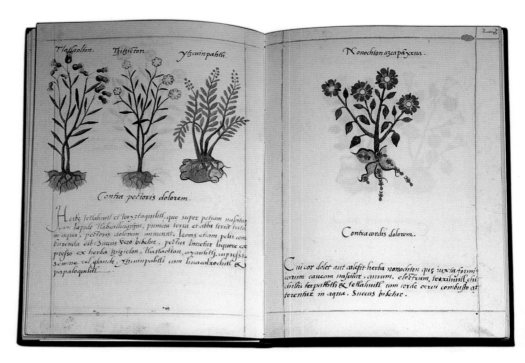

From physic garden to modern botany

It was the use of plants in physic, or medicine, that gave rise to the establishment of scientific botanic gardens in Europe. In 1447, Pope Nicholas V established a physic garden in the Vatican, and others soon followed in Pisa (1544), Padua (1545) and elsewhere in Europe. What began as gardens for the cultivation and study of medicinal plants gradually became centres of botanical science, examining the diversity and classification of plants in addition to their uses. In the modern era, the search for new medicines follows several different paths, from synthetic chemistry to natural product discovery, often guided by ethnobotany.

∧ The Badianus Codex, the earliest book on the medicinal plants of the New World, was written by the Aztec physician Martin de la Cruz and translated into Latin by Juan Badiano, another Aztec Indian, in 1552.

GINSENG

The English name 'ginseng' (*Panax ginseng*) comes from the Chinese *rén* (meaning 'person') and *shēn* (meaning 'plant root'). Swedish botanist Carl Linnaeus understood the importance of the plant in Chinese medicine, and when classifying it he named the genus *Panax*, meaning 'all-healing', after Panacea, the goddess of universal remedy. The global market for ginseng exceeds US$2 billion annually, with China, South Korea, the United States and Canada producing almost all of the traded product.

Purple maize (*Zea mays* ssp. *indurata*), rich in anthocyanin pigments, is an ancient Meso-American variety of what is now the world's most productive crop.

Domestication of plants

etween 15,000 and 10,000 years ago, and with a remarkable synchronicity around the world, agriculture began to develop in widely separated civilizations. There would have been strong parallels in each case, with people who gathered food from the wild storing some of the seed and using it to sow a new crop close to home. Keeping seeds of the best-performing, most edible and nutritious crops introduced selection, shaping plants to meet human needs. An agricultural way of life enabled significant population increases, leading to larger settlements and, ultimately, the first cities.

Early crops – cereals

Our understanding of the origins of domestication was originally dependent on finding the archaeological remains of ancient crops, or graphical representations of their cultivation. More recently, however, greater precision has come from the ability to analyze the genome for DNA markers. This has revealed, for example, that the cultivation of rice (*Oryza sativa*) in Asia began about 13,500

An ancient Egyptian stone carving from the Temple of Horus at Edfu depicts a priest carrying stalks of wheat (*Triticum* sp.), demonstrating the long-standing importance of this crop.

years ago with the domestication of wild rice (*O. rufipogon*) in the Pearl River valley region of China. Soon afterwards, around 10,000 years ago, wheat (*Triticum* spp.) and barley (*Hordeum* spp.) cultivation began in the Fertile Crescent of the Middle East (see page 338). Meanwhile, around the same time in southern Mexico, selection of the most edible forms of a wild grass called teosinte (*Zea mays* ssp. *parviglumis*) gave rise to maize (*Z. mays* ssp. *mays*), the most important grain to have originated in the new world (see box on page 339). The common feature of these critical steps in human history is that the crops were cereals, annual grasses with nutritious grains, capable of being dried and stored. Cereals are named after Ceres, the Roman goddess of agriculture, and many distinct species are consumed. Those of African origin include sorghum (*Sorghum bicolor*), fonio (*Digitaria* spp.) and teff (*Eragrostis tef*) from the Ethiopian highlands. More familiar, and more widely traded, are barley (*H. vulgare*), oats (*Avena sativa*), sugar cane (*Saccharum* spp.) and rye (*Secale cereale*). However, it is maize, rice and wheat that are the staple foods for much of humanity.

Rank	Crop	World production 2012 (tonnes)	Average world yield 2010 (tonnes per hectare)	World's most productive countries 2012 (tonnes per hectare)	World's largest producing countries 2013 (tonnes)
1	Maize (corn)	873 million	5.1	United States (25.9)	United States (354 million)
2	Rice	738 million	4.3	Egypt (9.5)	China (204 million)
3	Wheat	671 million	3.1	New Zealand (8.9)	China (122 million)
4	Potatoes	365 million	17.2	Netherlands (45.4)	China (96 million)
5	Cassava	269 million	12.5	India (34.8)	Nigeria (47 million)
6	Soya beans	241 million	2.4	Egypt (4.4)	United States (91 million)
7	Sweet potatoes	108 million	13.5	Senegal (33.3)	China (71 million)
8	Yams	59.5 million	10.5	Colombia (28.3)	Nigeria (36 million)
9	Sorghum	57.0 million	1.5	United States (86.7)	United States (10 million)
10	Plantain	37.2 million	6.3	El Salvador (31.1)	Uganda (9 million)

∧ Ten staples that feed the world (by annual production).

Root crops

Cereal crops are highly nutritious because their dry, one-seeded fruits contain the endosperm, provided to nourish the developing embryo, but the characteristics of other plant organs also make them highly suitable for cultivation. The underground parts of many plants, whether roots, rhizomes or tubers, are adapted as storage organs for carbohydrates, and as such they are potentially important as highly calorific crops. In the tropics, root vegetables such as taro (*Colocasia esculenta*) corms, and yam (*Dioscorea* spp.) and cassava (*Manihot esculenta*) tubers, are often more important than cereals. Because they contain toxins intended to reduce predation by animals, all three root crops – and many others – require careful preparation before they are edible. As with cereals, selective breeding has led to the development of thousands of selected lines, known as landraces, maintained and handed on within communities.

THE ORIGINS OF PLANT BREEDING

Out of the ancient practice of selecting and keeping the best varieties of cultivated plants emerged the science of plant breeding. At its simplest, this involves propagating plants and retaining only those with the most desirable characteristics. A more targeted approach involves cross-breeding with other species or varieties to introduce new genes and select for desirable traits, such as increased yield, improved flavour, resistance to pests, improved capacity for storage, or tolerance of drought or salinity.

< The starchy tubers of cassava (*Manihot esculenta*), native to South America, are widely grown in the tropics. The plant thrives on low rainfall and in poor soils unsuitable for growing more nutritious crops.

⊲ Peanuts (*Arachis hypogea*) ready for harvesting. The yellow flowers, like those of other legumes, are borne above ground, but after fertilization the ovary stalk becomes greatly elongated, forming a 'peg' that pushes the developing pods underground.

An underground bounty

Most of the underground plant organs we use derive from roots and underground shoots, an exception being the peanut or groundnut (*Arachis hypogaea*), a legume that develops underground seedpods. In the life of a plant, most underground organs store energy, in the form of carbohydrates, or water. If edible, they are useful sources of dietary calories, but are usually low in proteins and other nutrients. In addition to true roots, a wide variety of bulbs, corms, rhizomes and tubers are cultivated.

⊲ Red onions (*Allium cepa* cultivars) have been selected for their mild, sweet flavour. When the bulb is sliced in half, its overlapping fleshy leaf bases are clearly visible.

Bulbs and corms

Bulbs are almost exclusively found in monocotyledons, and consist of fleshy leaf bases growing from a compressed and very reduced stem. Being underground leaves, they do not carry out photosynthesis and, lacking chloroplasts, are never green. Many familiar ornamental garden flowers are grown from

bulbs, including daffodils (*Narcissus* spp.) and tulips (*Tulipa* spp.). The most widely used edible bulbs belong to the mainly northern hemisphere genus *Allium*, which includes onions (*A. cepa*), garlic (*A. sativum*) and chives (*A. schoenoprasum*). These have been cultivated for more than 5,000 years and are valued in cooking for their pungent taste – from chemical compounds intended to deter herbivores – and, to a lesser extent, in medicine.

Corms are short, swollen stems, usually surrounded by protective dried petioles, or leaf bases. When cut open, a solid mass of tissue comprising parenchyma cells rich in stored starch is exposed. Two species in the arum family (Araceae) that are widely cultivated for their nutritional corms are taro (*Colocasia esculenta*), originally from India and Southeast Asia, and eddoe (*C. antiquorum*), from China. Sometimes considered a single species, these important tropical crops were carried by the original settlers of Hawai'i and Polynesia. The cells of their corms contain needle-shaped raphides, crystals of calcium oxalate, which if eaten without careful preparation damage the tissues of the mouth and throat, making swallowing difficult. Soaking in cold water partially removes the crystals, and cooking turns the corms into a starchy mass, making them less harmful.

Rhizomes and tubers

Rhizomes are also underground stems. They have extended horizontal growth, with shoots and roots emerging at the nodes. Most can readily be cut into pieces and used for vegetative propagation. The edible rhizome of the sacred lotus (*Nelumbo nucifera*) grows in the soil of shallow ponds and contains numerous air-filled cylindrical channels. Rhizomes of members of the ginger family (Zingiberaceae), including turmeric (*Curcuma longa*) and ginger (*Zingiber officinale*), are widely used as spices in cooking and in traditional medicine.

A tuber is a specialized kind of rhizome or stolon. Much thickened in places, it acts both as underground storage and vegetative propagule, with the capacity to sprout new shoots and roots. The potato (*Solanum tuberosum*) is the most important tuberous crop, ranked fourth in its contribution to the global food supply. Its wild ancestor (a member of the *S. brevicaule* complex) was domesticated by the Inca people of South America between 5,000 and 8,000 years ago. There are now more than 5,000 varieties in cultivation, and breeding programmes draw upon genes from related species. Cassava, originally from Brazil, is ranked fifth in the global food supply, and is of immense importance in the tropics.

Lotus (*Nelumbo nucifera*) rhizomes have long, hollow channels that allow air to reach the submerged root system of the plant. A fast-growing and nutritious crop with many distinct varieties, the rhizomes are harvested by hand two to three months after planting.

A colourful selection of traditional Andean potato (*Solanum tuberosum*) varieties cultivated by the Quechua people of Chuquis district in Huánuco province, Peru.

Useful roots

(R)oots, the primary underground organs of plants, provide a surprising diversity of useful products, from food and beverages to medicines and rubber. As is the case for underground stems, those we exploit are mainly adapted for the storage of energy and water. This enables the plant to survive in desert and high alpine habitats, where liquid water is scarce or it needs to regrow rapidly after being grazed by animals.

What connects latex and coffee?

Making rubber from the latex of the Kazakh dandelion (*Taraxacum kok-saghyz*) root was initially a wartime response to shortages of supplies from rubber trees (*Hevea brasilinesis*). The latex is produced in specialized elongated cells called laticifers, which form a secretory system extending throughout the plant as a defence against herbivory. Aside from being characteristic of the Cichorieae tribe of the daisy family (Asteraceae), laticifers are found in several other families of flowering plants, including the spurges (Euphorbiaceae). Cichorieae also includes several species with useful tap roots. Chicory (*Cichorium intybus*) roots are roasted as an additive to, or substitute for, coffee, and dandelion (*Taraxacum* spp.) roots are less commonly used in the same way. Those of purple salsify (*Tragopogon porrifolius*) and black salsify (*Scorzonera hispanica*) can grow up to a metre (3 ft) long and are minor vegetable crops in Europe. All four of these tap roots contain the polysaccharide inulin, rather than starch, as a storage product, and this is extracted for use as soluble dietary fibre. Inulin gives the roots diuretic properties, widely exploited in traditional medicine around the world and reflected in common names such as *pissenlit* in France and piss-a-bed in England.

Tasty tap roots

The carrot family (Apiaceae) is a large group of aromatic dicotyledons with flowers arranged in umbels. Most have large tap roots and although some, like the water dropworts (*Oenanthe* spp.), are extremely poisonous, others are among the most familiar vegetables. The carrot (*Daucus carota* ssp. *sativus*), whose tap root is not always orange but can be purple, red or yellow, was domesticated in central Asia and selectively bred for reduced woodiness and improved flavour. Different

⌃ A pneumatic truck tyre manufactured by Continental from Taraxagum, a brand of latex produced from dandelion (*Taraxacum* sp.) roots and developed as a potentially sustainable alternative to conventional rubber (*Hevea brasiliensis*) by the Fraunhofer Institute for Molecular Biology and Applied Ecology in Germany.

⌄ Although the most widely cultivated carrots (*Daucus carota* ssp. *sativus*) are the familiar orange colour, new varieties with many different pigments are becoming ever more popular.

shapes and sizes have also been selected – shorter carrots can grow in heavy or rocky soils and be harvested early, while longer-rooted varieties are better suited to sandier soils and take more time to mature. The closely related parsnip (*Pastinaca sativa*) has a similar geographical origin; its root was historically used as a sweetener. Parsnips that have stayed in the ground over winter are sweetest, because more of the starch in the root has been converted to sugar.

Neat beets

Sea beet (*Beta vulgaris* spp. *maritima*), a widespread old world coastal member of the amaranth family (Amaranthaceae), is the wild ancestor from which five cultivar groups of crops have been developed. Most of the present-day diversity is the work of crop breeders from the eighteenth century onwards, a key element of which was the selection of a rare allele reducing sensitivity to day length and resulting in a biennial life cycle. The tap roots of sugar beet (*B. vulgaris*, Altissima group of cultivars) contain up to 20 per cent sugar by weight, making it a competitor, in cold temperate regions, to the tropical sugar cane. The dark red beetroot (*B. vulgaris*, Conditiva group of cultivars) is the most familiar culinary vegetable, while the field beet or mangelwurzel (*B. vulgaris*, Crassa group of cultivars) is an important fodder crop for livestock in temperate regions.

THE SCREAMING MANDRAKE

According to folklore, the human-shaped narcotic root of the mandrake screams and kills those who pull it from the ground. Yet this tropane alkaloid-containing member of the nightshade family (Solanaceae) was variously prized for bringing fertility, acting as a talisman for protection and having anaesthetic properties. Author J. K. Rowling kept the tradition alive in the Harry Potter books, which faithfully recount the hazards of harvesting mandrakes.

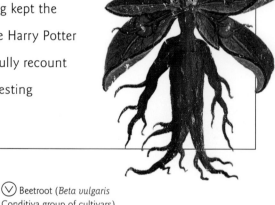

⋁ Sea beet (*Beta vulgaris* ssp. *maritima*), seen here growing wild on a sandstone cliff in the United Kingdom, is the wild species from which many different beet crops have been developed.

⋁ Beetroot (*Beta vulgaris* Conditiva group of cultivars) is primarily grown for its deep red taproots, although the leafy parts of the plant are also edible, resembling spinach.

Stems for bark and fibres

The strength and flexibility of stems aids the growth of plants towards the light, and humans have found uses for every part of them, from the outer epidermis to the wood within. The straight, flexible stems of many species lend themselves to specific applications. Archaeological evidence shows that at least 12,000 years ago, around the time people adopted an agricultural way of life, they wove baskets from the most suitable plants to hand and used species with thorny branches as living fences, to protect and enclose livestock.

Barking up the right tree

Bark forms a protective layer around trunks and branches that can enable trees to survive physical damage and fire. The thick bark of cork oak (*Quercus suber*), a tree of fire-prone Mediterranean habitats, can be harvested repeatedly every 12 to 15 years to make corks for bottling wine. The inner bark of cinnamon (*Cinnamomum cassia* and related species), obtained from coppiced branches, is harvested as a fragrant spice. The woody outer bark is discarded and the inner bark, which contains cinnamaldehyde, is dried and rolls into quills. One of the most familiar of all medicines, aspirin, was developed as a synthetic version of the naturally occurring salicylic acid, traditionally extracted from the bark of willow trees (*Salix* spp.). The medicinal use of willow bark for treating fevers was recorded by Dioscorides and earlier writers. Quinine, isolated from the bark of *Cinchona* species (especially *C. officinalis*) and used traditionally in South America for treating fevers, became highly coveted around the world as a cure

⊲ A cork oak (*Quercus suber*) growing in the Alentejo region of Portugal, soon after the outer bark has been cut and peeled away. A new layer of bark regrows and can be harvested in about ten years.

⌃ Basket weaving, using plants ranging from bamboo, to cane, palm and wood, is one of the oldest human crafts with different traditions developed on every continent.

EFFICACIOUS EXUDATES

The dragon's blood tree (*Dracaena cinnabari*), native to Socotra Island in the Arabian Sea, and *Aquilaria* trees from Southeast Asia are the source of precious exudates. Dragon's blood is a bright red sap used as a pigment and in medicines, while agarwood or oud is the fragrant resin produced by *Aquilaria* trees in response to wounding followed by fungal infection. Top-quality agarwood perfumes are worth more than their weight in gold, and consequently many species of *Aquilaria* are now endangered. Other useful exudates are produced by plants to deter herbivores or to heal mechanical wounds. Some, like rubber and chicle (used to make chewing gum and obtained from sapodilla trees, *Manilkara zapota*), are obtained by tapping: cutting the bark of the tree and collecting the latex that runs from the wounds. To obtain turpentine, a resinous exudate from coniferous trees, especially pines (*Pinus* spp.), sheets of bark are cut away. The wound response material, oleoresin, is then distilled, producing a liquid solvent and solid rosin.

for malaria. It also gives the bitter flavour to tonic water, which gained popularity in British India, where it was taken with gin to prevent malaria. Other chemicals found in bark have deadlier uses. The bark of the southern African ordeal tree (*Erythrophleum suaveolens*) contains a poison used on arrows for hunting and to determine guilt or innocence. If the accused died after drinking the bark extract, they were guilty; if they survived, they were innocent.

Functional fibres

Bast fibres, elongated cells that form a protective layer of inner bark outside the phloem, can be spun into yarn for rope making or weaving into fabric. The main sources are hemp (*Cannabis sativa*), jute (*Corchorus capsularis* and *Corchorus olitorius*), common flax and kenaf (*Hibiscus cannabinus*). The fibres are obtained through the process of retting, which involves keeping the harvested stems under water for several weeks. During this time, naturally occurring fungi break down the materials that bind cell walls together.

(⋀) Dragon's blood trees (*Dracaena cinnabari*) are long-lived and often show numerous scars from the repeated harvesting of their resin. A sharp knife is used to make a shallow circular hollow in the trunk, to which the tree responds by exuding a wound-sealing resin.

(∟) Hemp (*Cannabis sativa*) thread, rope and canvas spun from stem fibres are prized for their strength, durability and resistance to rotting.

Wood

(W)ood is one of the earliest materials to have been used by mankind. Today, we distinguish between softwoods, from conifers, and hardwoods, from angiosperms. However, the actual density of timber – a major factor in the uses to which it can be put – does not divide into these two convenient categories. Among the conifers, yews (*Taxus* spp.) yield hard timber, while in the flowering plants, the balsa tree (*Ochroma pyramidale*) produces timber lighter than cork.

(∧) The wood produced by different trees varies greatly in physical properties such as density and strength, and in wood grain, the pattern of alternating light and dark areas revealed in sawn and polished timber. When sawn longitudinally, many kinds of timber are figured rather than uniform, with different patterns being prized for such uses as musical instruments.

(∨) Woodchips, often from plantations of coniferous trees, are an important raw material used in a wide variety of applications, from biomass fuel, to organic mulch in gardening and, as seen here at a paper mill, in paper making.

Softwoods

Softwoods make up more than 75 per cent of commercially traded timber and derive from a diversity of coniferous trees, which reflects their wide distribution, especially in boreal forests. Much of the timber from conifers is used in construction, providing key structural elements, especially in timber-framed buildings. Softwoods, especially those grown in plantations, are also widely used to make wood pulp and woodchips as raw material for the manufacture of paper, fibreboard and chipboard. The bark is discarded, but both the heartwood and sapwood can be used. Woodchips are also utilized as a renewable biofuel in heating or combined heat and power applications. In recent years, wood pulp has been used to make a new material, nanocrystalline cellulose (NCC), which has a strength-to-weight ratio eight times that of stainless steel. NCC can be used to make paper and board, composite materials for packing, water-absorbing products and even food, where it can provide a low-calorie alternative to carbohydrate thickening additives.

Pine and spruce (*Picea* spp.) are two of the most familiar softwoods. In addition to construction applications and furniture making, these timbers are used for the soundboards of many acoustic stringed instruments, where their light weight performs well.

of timbers lend themselves to different uses in sports equipment and musical instruments. Cricket bats are traditionally made from white willow (*Salix alba*), and baseball bats from hickory (*Carya* spp.) or ash (*Fraxinus* spp.). Hardwoods such as maple (*Acer* spp.), rosewood (*Dalbergia* spp.) and mahogany (*Swietenia* spp.) have long been preferred for the necks and bodies of acoustic instruments, including guitars and violins. The densest and most finely grained hardwoods, such as ebony (*Diospyros* spp.), which will not float in water, are used for the black keys of pianos and for fingerboards on the necks of stringed instruments.

Although straight planks of sawn timber are preferred for construction, wood with a distinctive figured grain is often prized by woodworkers and furniture makers. It is obtained from burls (also known as burrs), rounded outgrowths on the trunks of trees that develop in response to stress from injury or infection. The anatomy of the burl comprises densely packed lateral branches, each of which defines a knot within the wood.

Hardwoods

The presence of xylem vessels, in addition to tracheids, distinguishes angiosperm hardwoods from coniferous softwoods. They are also generally slower growing and produce finer-quality timber, which is used to make more complex and subtly crafted items, from furniture, barrels and ships to decorative instruments. The specific physical properties

⊲ The burled bark of a Peruvian pepper tree (*Schinus molle*) indicates the presence of complex figured timber within, suitable for making veneer and turned objects. Walnut (*Juglans* spp.) is one of the most prized burlwoods, seen here in a vase.

⊽ Rosewood, from various species of *Dalbergia* and other genera with richly coloured, fragrant timber, is highly valued in furniture making, as in this table with mother-of-pearl inlay. Demand for this slow-growing hardwood greatly exceeds supply.

TIMBER TRAFFICKING

There is now a global shortage of the finest hardwoods, and in recent years a growing market in illegally felled timber has emerged. Rosewood is now the world's most trafficked wild product. Demand in China for *hongmu*, rosewood used in classical furniture, now threatens important old-growth tropical forests in countries from Southeast Asia to Africa and Latin America, and has resulted in all *Dalbergia* species being listed in Appendix II of the Convention on Trade in Endangered Species.

HORSERADISH TREE

The horseradish tree (*Moringa oleifera*), a fast-growing native of northwestern India long used in Ayurvedic medicine, has been widely planted in semi-arid regions, where it is regarded as a 'miracle tree' in the fight against malnutrition. The trees are easily propagated from cuttings, become rapidly established and are often grown as living fences. The leaves, which are rich in vitamins and minerals, are eaten and, thanks to their antiseptic properties, are also powdered and used as soap.

Edible leaves

Around the world, leaves of many hundreds of plant species have been eaten since the beginning of human history. The young leaves of many conifers can be eaten, as a survival food, although those of yews are poisonous. In contrast, the leaves of many flowering plants are less toxic, and one of the earliest uses of plants by humans was for foliage gathered as fodder for livestock. Most of the major edible leaf crops come from herbaceous flowering plants, including domesticated members of the mustard family (Brassicaceae).

Versatile vegetables

Brassicas are one of the most important groups of vegetables, rich in vitamins, minerals and antioxidants. They provide an excellent source of dietary fibre, and contain sulforaphane and isothiocyanates, which can protect against some forms of cancer. It is these chemical defences that give the flavour of mustard we value in brassicas. No other plant genus has been bred to provide so many useful products, and the tolerance of these vegetables to cold – even freezing – conditions means that they can be cultivated in temperate as well as warmer regions of the world.

⌄ White leadtree (*Leucaena leucocephala*) is a small, fast-growing tree from Central America that is widely grown in tropical regions for firewood and livestock forage. Sometimes called a miracle tree because of its nitrogen-fixing root nodules, it produces high-protein fodder while also improving soil fertility.

FEASTING ON FERNS

The young 'fiddlehead' fronds of some ferns are edible, with ostrich fern (*Matteuccia struthiopteris*) being one of the most palatable. All ferns should, however, be consumed in moderation, because some of the chemicals they contain to defend them from herbivores are carcinogenic. Because of their toxicity, ferns have traditionally been used in many cultures as anthelmintics for treating parasitic worms. Ferns are prominent in Maori culture – several species are eaten as *pikopiko*, and the *koru* symbol of the unfurling frond symbolizes new life, growth and peace. The silver fern (*Cyathea dealbata*) is a widely used emblem of New Zealand.

⊘ Ostrich fern (*Matteuccia struthiopteris*) fronds must be boiled in water for at least five minutes to be edible.

A bounty of brassicas

One of several mustard family species that has been domesticated, the wild cabbage (*Brassica oleracea*) is a coastal plant of southern Europe and the Mediterranean region. Its early cultivation in ancient times provided animal fodder as well as human food and medicines. Sea kale (*Crambe maritima*) and sea rocket (*Cakile* spp.), other strandline members of the same family, are still consumed in their original wild form as minor vegetables. Collard greens (*B. oleracea* var. *viridis*) and kale (*B. oleracea* var. *acephala*) are open, leafy crops that are close in form to their wild ancestor, the variety name *acephala* indicating that these plants have not been selected to develop a distinct head. Instead, selection in cultivation favoured the plants with the largest leaves. The densely overlapping leaves of the cabbage (*B. oleracea* var. *capitata*) head were developed by selecting for shorter internodes and an enlarged terminal bud. Brussels sprouts (*B. oleracea* var. *gemmifera*), in contrast, were bred for their enlarged axillary buds, which grow in tight spirals around the stem.

The versatile brassicas even merit a place on display. Ornamental horticulture has selected varieties of kale with colourful white, green and red foliage and interesting variations in leaf shape.

⊘ Just some of the great variety seen in brassica leaf crops. Left to right: Brussels sprouts (*Brassica oleracea* var. *gemmifera*), red cabbage (*Brassica oleracea* var. *capitata*), kale (*Brassica oleracea* var. *acephala*) and green cabbage (*Brassica oleracea* var. *capitata*). Red cabbages have been selected for their higher levels of anthocyanins.

Leaves that satisfy the senses

I n our long relationship with plants, humans have found many uses for leaves that tap into the diverse chemistry of the secondary metabolites they contain – organic compounds that are not used in the growth and development of the plant, but play important roles in ecology. Although formerly considered mere waste products, secondary metabolites are now known to be part of a complex system, mediating the relationship between the plant, its physical environment and its competitors or predators.

Aromatic leaves

In members of the mint family (Lamiaceae), volatile oils are involved in temperature regulation, reducing water loss and mechanical damage, and defending the plant against herbivores and pathogens. The glandular hairs that secrete these chemicals have even been shown to act as detection sensors, responding to walking insects and triggering the activation of defence genes. The effects of the secondary metabolites they produce can be quite remarkable. Leaves of purple sage (*Salvia leucophylla*), for example, release monoterpenes into the air, or from dead leaves into the soil, which inhibit the germination of seeds of

nearby competitors. Leaves of thyme (*Thymus* spp.), oregano (*Origanum vulgare*) and basil (*Ocimum basilicum*) have all been shown to have antimicrobial properties, disrupting the cell walls and metabolism of bacteria and fungi. Little wonder then that Lamiaceae species and other plants rich in secondary metabolites have been used for thousands of years as culinary herbs and in traditional medicine. In addition to flavouring food, mint plants (*Mentha* spp.) have a long tradition of use in many cultures for their anti-inflammatory, analgesic and diuretic properties. Similarly, lavender (*Lavandula angustifolia*) oil is widely used as a fragrance, is noted for its

The Valensole Plateau in the Alpes-de-Haute-Provence of France, known as the 'Lavender Country', is renowned for the beauty of its landscapes, which are dominated by extensive fields of lavender (*Lavandula angustifolia*) and lavandin (*Lavandula × intermedia*). The Mediterranean climate and sunny slopes lead to abundant yields of essential oils.

calming effect and, more recently, has been found to be effective against protozoal *Giardia* and *Trichomonas* pathogens. The presence of essential oils in the leaves of culinary herbs adds not only flavour to food, but also acts as a natural preservative agent.

Stimulating leaves

Alkaloids (nitrogen-containing heterocyclic compounds) are secondary metabolites produced in different plant organs as a defence against herbivory. Some alkaloid-containing leaves have a long history of human use thanks to their psychoactive properties. Caffeine is present in the mildly stimulating beverages prepared from the leaves of tea plants, where it occurs with theobromine, and in yerba mate (*Ilex paraguariensis*), in which a second alkaloid is xanthine. The consumption of cocaine, a much more potent tropane alkaloid extracted from the leaves of coca (*Erythroxylum coca*), dates back more than 8,000 years in Peru, and today generates billions of dollars in illegal trade.

Ornamental leaves

The shapes and colours of leaves are a principal element in ornamental horticulture and garden design. In addition to drawing directly on the diversity of foliage found in nature, gardeners have selected numerous forms with unusual shapes and variegation. Often these are discovered as sports, or chance mutations, and can be maintained only by

vegetative propagation because the distinctive feature will not be inherited. Horticulturists exploit the ability of different plant organs to regenerate or to be grafted onto the stems of close relatives. In variegated hydrangeas (*Hydrangea* spp.), for example, root cuttings are taken and new plants generated from these, while variegated shell gingers (*Alpinia zerumbet* var. *variegata*) are grown from divided rhizomes. In the case of trees such as the fern leaf beech (*Fagus sylvatica* var. *asplenifolia*), cuttings are taken and grafted onto rootstocks from the parent species.

⊤ Bunches of culinary herbs, representing just part of the diversity found in the mint family (Lamiaceae), being preserved for later use by air drying.

∧ Tea (*Camellia sinensis*) pickers in the hills of Sri Lanka harvest the youngest leaves from new growth. Originally cultivated in China, tea is now grown in many countries around the world.

Floral extracts

Compared to roots, stems and leaves, or seeds and fruits, few, if any, flowers can be considered significant edible crops. Many are, in fact, edible, including nasturtium (*Tropaeolum* spp.) and pumpkin (*Cucurbita pepo*) flowers, day lilies (*Hemerocallis* spp.) and the young inflorescences of banana plants (*Musa* × *paradisiaca*). However, none of these species has been bred or developed to enhance their flowers for the purpose of consumption, mainly because these structures have relatively limited nutritional value.

∧ Nasturtiums (*Tropaeolum* spp.) are popular garden plants grown for their showy, colourful flowers and attractive foliage. The flowers are edible and, like the leaves, have a peppery rocket-like flavour.

Floral teas

The pleasing aromas of flowers, used by the plant to attract pollinators (see page 242), have led to many kinds being used to make herbal teas, consumed both as beverages and as traditional medicines. Chamomile (*Matricaria chamomilla*) tea, which promotes relaxation and sound sleep, is one of the most familiar. Lime (*Tilia* spp.) flower teas are a rich source of flavonoids, and are used to treat coughs, colds and high blood pressure.

Chrysanthemum (*Chrysanthemum morifolium* and *C. indicum*) tea, used in China and elsewhere in Asia, is both a refreshing drink and a treatment for ailments such as sore throats and circulatory problems. Hibiscus tea from roselle (*Hibiscus sabdariffa*) is popular throughout Egypt, Latin America, the Caribbean, Africa and Asia. The bright red flowers contain anthocyanin pigments that are said to help reduce blood pressure.

Hops and brewing

One significant crop that is a flower (or, strictly, an inflorescence) is the hop (*Humulus lupulus*), used in brewing to impart a bitter flavour to beer. It has been the subject of extensive breeding programmes, which rather than influencing the morphology of the plant, a dioecious vine, have been targeted towards developing a distinctive taste. More than 80 commercial hop varieties have been bred, each giving a particular regional flavour to the resulting beer. Only the cone-like female inflorescences are used in brewing. Situated in the axils of leafy bracteoles, these each

< Dried rose (*Rosa* spp.) flowers for sale in a street market in Kunming, China, are used for their fragrance to flavour hot drinks, either on their own or together with tea leaves.

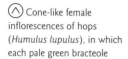 Cone-like female inflorescences of hops (*Humulus lupulus*), in which each pale green bracteole enacloses the ovules, which provide the bittering agent used in brewing.

contain a single ovule and the outer surface of the ovary is covered with numerous multicellular, bright yellow glandular hairs known as lupulin glands. These produce the terpenoid essential oils and terpenophenolic resins that are used to balance the sweet flavour of malt and, thanks to their antibiotic properties, to prevent beer from spoiling during storage. Hops are also used in traditional medicine to promote sleep, and some of the many compounds produced by the lupulin glands are now being tested as potential anti-cancer drugs.

Precious perfumes

The most precious products obtained from flowers are the perfumes prepared by concentrating essential oils from the petals, which in the plant serve to attract pollinators. The oil from petals of Damask rose (*Rosa × damascena*), lily of the valley (*Convallaria majalis*) and ylang-ylang (*Cananga odorata*) are three classic examples, blended with animal musks to make premium perfumes. In ancient times, fragrances were obtained by expression, compressing flowers and other parts of plants

to squeeze out essential oils. By medieval times, distillation was used to extract floral fragrances and perfumes became increasingly important as luxury products. As with floral teas, most of the essential oils used as perfumes also have traditional medicinal applications. Lavender (*Lavandula* spp.), for example, has been used since antiquity as an antiseptic agent, and today is a common ingredient in soaps and cleaning products. Fragrances themselves were long thought to protect against disease, and in medieval times plague doctors carried flowers and aromatic leaves in protective face masks.

Top: The flowers of the cananga tree (*Cananga odorata*) from Indonesia are the source of the perfume ylang-ylang. Below: Antique copper perfume still from the museum of Maison Fragonard, Grasse, France, used to extract essential oils by passing steam through a layer of flowers and condensing the watery distillate. The essential oils float to the surface and are collected.

The beauty of the flower

(T)he astonishing diversity of forms and colours by which flowers advertise themselves to pollinators has long been prized for aesthetic reasons in ornamental horticulture. Flowers have been selected and bred to increase their visual impact, with the result that the choicest examples are often quite different from their wild ancestors. During the golden age of exploration, wealthy European landowners sponsored plant-hunting expeditions around the world to discover and bring back new plants to be admired in their gardens.

The more the merrier

Simply bringing the beauty of nature into cultivation at home has never been enough to satisfy some gardeners. When it comes to petals, the more the merrier is the rule for many, and there has been a long fascination with double-flowered forms. Roman naturalist Pliny the Elder wrote of the hundred-petalled rose (*Rosa* sp.), double peonies (*Paeonia* spp.) were prized in China and written about in the ninth

(<) The double-flowered form of Maltese cross (*Lychnis chalcedonica* 'flore pleno') produces dense heads of scarlet flowers.

(v) Himalayan blue poppies like this *Meconopsis* × *sheldonii* 'Slieve Donard' are short-lived perennial plants that require rich, moist soil; they are prized for their dramatic flowers.

OUT OF THE ORIENT

The early wave of garden introductions from the new world was followed in the mid-nineteenth century by the discovery of the hugely diverse Chinese flora, with its hundreds of species of rhododendrons (*Rhododendron* spp.), magnolias (*Magnolia* spp.), primulas (*Primula* spp.) and gentians (*Gentiana* spp.). The arrival of these unknown blooms in Europe heralded a new era of garden romance and excitement. Among the most exotic and celebrated of these discoveries were the striking Himalayan blue poppies (*Meconopsis* spp., right), described by English botanist Frank Kingdon-Ward in the memoires of his two-dozen expeditions to China, Tibet and Southeast Asia.

TULIP MANIA

The desire to seek out and grow novel and dazzling garden flowers has long been a pursuit of the wealthy, with the most unusual blooms regarded as desirable status symbols. In early seventeenth-century Holland, 'tulip mania', a craze for tulips, swept through that country, with the rarest varieties commanding exorbitant prices. Speculation in the market skyrocketed in November 1636, but in February the following year the bubble burst, leaving many dealers bankrupt.

The depiction of tulips in still-life paintings from the period is a legacy of this costly fascination.

⊙ In *The Tulip Folly* by Jean-Léon Gérôme (1882), a nobleman stands guard over his most precious tulip while soldiers trample flowerbeds to reduce the supply and stabilise the market.

century, and Gerarde's *Herball* illustrated several double forms of flowers. Double-flowered varieties are often given the Latin name *flore pleno* and result from a homoeotic mutation, which causes the stamens to be replaced by additional whorls of petals. Commercial breeding has developed many double-flowered forms, especially of roses, camellias (*Camellia* spp.) and carnations (*Dianthus caryophyllous*).

Improving on nature?

Hybridization – one of the main ways of introducing new colour schemes into flowers – is achieved by manually transferring pollen from one parent to the stigma of the other. Amateur and professional breeders have created more than 3,000 named varieties of day lilies in this way, which have been selected for their colour, pattern and flower size, which is now twice as large as in the original wild plants. Once the desired outcome has been created, new plants can be produced sexually from seed or, to reduce the potential for variation, asexually by tissue culture. The latter technique involves growing plants in test tubes on an agar growth medium, and has also been used to great effect in making moth orchids (*Phalaenopsis* spp.), once the preserve of specialist growers, readily available in an enormous variety of colours.

Since the late 1980s, researchers in Australia have been using genetic engineering to transfer the genes that produce the bright blue pigment delphinidin in delphiniums (*Delphinium* spp.) to roses, carnations and chrysanthemums. These three targets are the most valuable in the cut flower industry but none has naturally occurring blue pigments. Whether success will result in significant sales remains to be seen, but much energy has been put into what some consider the holy grail of horticulture.

⊙ Thanks to breeding programmes and mass production using tissue culture, moth orchids (*Phalaenopsis* spp.) have gone from exotic rarities to readily available house plants.

Our daily bread

Cereals are staple foods and the foundations of civilizations around the world. In morphological terms, how have they been transformed from wild grasses? Most obvious is the increase in kernel size, improving the yield of the crop. But in the newest varieties, grain size is only part of the story, as each organ of the plant has been fine-tuned and enhanced. Ultimately, increased yield depends on overall performance and ability to grow rapidly with minimal input of fertilizers and pesticides, even in conditions of drought and salinity.

⊘ Bread has been one of the world's most popular staple foods since the origins of agriculture. Yeast spores, naturally occurring on the surface of cereal grains, act as a leavening agent.

The growth of wheat

The Neolithic people of the Fertile Crescent exploited several kinds of wild annual grasses bearing nutritious seeds that could be ground into flour, usually after toasting them to remove the husks. By keeping and sowing the best seed, they began the process of domestication. One of the earliest kinds of wheat was first cultivated from wild einkorn (*Triticum boeoticum*) around 8000 BCE in southeastern Turkey. Selection by early farmers gave rise to the cultivated einkorn (*T. monococcum*), a diploid with slightly larger grains and a valuable mutation. In the wild type, the delicate rachis shatters to disperse the spikelets during ripening, whereas in the mutant form the rachis is toughened and remains attached to the ear. This made harvesting and processing much more efficient.

The same mutation was selected for in a second of several other early wheats, emmer (*Triticum dicoccon*), a naturally occurring tetraploid hybrid of two diploid species, *T. urartu* and a goatgrass (*Aegilops* sp.). An important later development was the hybridization of emmer and Tausch's goatgrass (*A. tauschii*) to produce the cold-tolerant

hexaploid bread wheat (*T. aestivum*), whose seeds have a higher gluten content, making the flour suitable for leavened bread. Bread wheat and durum wheat (*T. durum*) have also been selected to produce naked or forms with thinner, more delicate glumes, removed as chaff during threshing.

Ⓛ Cultivated emmer wheat (*Triticum dicoccon*) was originally selected for because of the ease with which its grain can be harvested. Unlike those of its wild relatives, the spikes do not 'shatter' into separate grains when ripe.

⌄ Wild einkorn wheat (*Triticum boeoticum*) from the Fertile Crescent was one of the first cultivated crops. It produces low yields of grain but can be grown on poor, dry soils.

⌄ A golden field of ripening wheat (*Triticum aestivum*), the world's third-most important crop in terms of calories consumed. This reflects the adaptability of wheat, which can be grown from sea-level to altitudes above 3,000 m (10,000 ft) and on a wide variety of soils, with both spring and winter varieties being available.

Reaping the harvest

As wheat cultivation spread, other grasses travelled with the seeds as weeds, including some that turned out to be better suited to growing in the new conditions they found themselves in. Wild barley (*Hordeum spontaneum*), for example, which occurs from North Africa through the Middle East to Tibet, grew even better than bread wheat in cold climates and was the progenitor of cultivated barley. Similarly, oats, millet (various species in the tribe Eragrostidae) and sorghum were first dispersed as weeds before thriving in cultivation in the warmer climate of Africa.

Successful selections

Other early varieties of wheat, including spelt (*Triticum spelta*), durum wheat and their wild relatives, serve as a valuable gene pool for modern plant breeders. Important traits introduced during the Green Revolution (see box on page 347) include short stems that do not collapse, as early varieties would, when grown with high levels of chemical fertilizer. Reduced height (Rht) dwarfing genes from the Norin 10 wheat cultivar developed in Japan in the 1930s have reduced sensitivity

to the plant hormone gibberellin, which regulates cell extension in the stem. In addition, winter and spring wheat varieties have been developed for cultivation in different regions, and drought and salinity tolerance in wheat are being increased through marker-assisted selection (MAS). Complete DNA sequencing of the wheat genome has identified more than 96,000 genes, originating from five distinct genomes of ancestral species. A longer-term grand challenge for plant scientists is to introduce nitrogen-fixation systems from legumes into wheat, the world's most important crop, and other cereals.

∨ In ancient times, querns – small hand-turned grindstones – were used to mill cereal grains into flour. Water- and wind-powered mills, serving entire communities, were among the first applications of machinery.

Non-cereal seed crops

$\left(\text{T}\right)$he seeds that contribute most calories are undoubtedly cereals, but other important seed crops with equally long histories provide different nutritional benefits. Some, such as rapeseed (*Brassica napus*), have been grown for oil since antiquity. Oil from the original form of this plant is unsuitable for consumption, but selective breeding to increase yields and reduce harmful chemicals led to the development of canola oil, now the third most important plant oil in commercial trade. Oils from rapeseed and the seeds of other plants are also used in the manufacture of biodiesel.

Amaranths

The ancient Aztecs of Central America cultivated amaranths (*Amaranthus* spp.) for their starchy seeds. Like maize and sorghum, amaranths are C4 plants (see page 154), photosynthesizing efficiently at elevated temperatures and in a wide range of soils and climates. The grain they produce is higher in protein and the amino acid lysine, which is scarce in most other grains. In South America, the Inca people domesticated quinoa (*Chenopodium quinoa*), a close relative of the amaranths and with similar qualities. The seeds of both are naturally toxic until boiled to remove the bitter-tasting saponins. Neither has been subject to much improvement, except to reduce bitterness. Numerous landraces, locally selected and maintained in cultivation, represent a rich source of genetic variability.

$\left(\wedge\right)$ Although rapeseed (*Brassica napus*) is an ancient crop, its cultivation has increased enormously in recent years as its uses have diversified. It is now a familiar sight in the landscape.

$\left(>\right)$ A local farmer alongside her colourful crop of amaranth (*Amaranthus* spp.), with its red and yellow flower spikes, growing together with the root crop taro (*Colocasia esculenta*) in the mountainous Ghandruk region of Nepal.

A finger on the pulse

The legume family (Fabaceae) is one of the most versatile and important to mankind, because the root nodules of the plants capture and fix atmospheric nitrogen, which increases the fertility of the soil. The characteristic pods contain seeds high in protein, and those harvested as dry grains are known as pulses. Like cereals, the fact that the seeds are dry enables them to be stored for extended periods after harvesting. Despite the importance of pulses as a source of protein for humans and animals, global consumption has been slowly declining. To promote their role in achieving food security, 2016 was declared the International Year of Pulses by the Food and Agriculture Organization, which recognizes 11 major types of pulses.

The distinctive biconvex shape of lentil seeds gave rise to the species' scientific name, *Lens culinaris*. This ancient crop, grown by Neolithic farmers in the Fertile Crescent alongside einkorn and emmer, contains around 25 per cent protein – significantly more than cereals – and is drought tolerant and capable of growing on marginal land.

⌃ A field of soya bean (*Glycine max*), a nutritious legume used to make tofu, soy sauce and other human foodstuffs, as well as animal feeds. Its cultivation in Amazonia has spread at the expense of rainforest.

SPILLING THE BEANS

Controversially, soya beans are grown on a vast scale, often in areas cleared of biodiversity-rich tropical rainforest and *cerrado*. Much of the global crop has been genetically engineered, for example to introduce tolerance to glyphosate, used as a proprietary herbicide. Offsetting the often negative perception of herbicide-resistant genetically resistant organisms (GMOs) is the fact that they can be grown in low tillage systems, which helps to conserve topsoil.

(See also box on page 349.)

Soya bean (*Glycine max*), another long-cultivated pulse, originating in east Asia and grown in China since at least 7000 BCE, contains up to 45 per cent protein. Despite their nutritional value, many pulses have indigestible integuments, used to protect the seed from insect predation. They therefore often require boiling and a long cooking period to degrade these compounds before they can be eaten, and much of the effort put into breeding pulses is aimed at reducing the levels of these substances.

⌄ Numerous varieties of lentils (*Lens culinaris*) have been developed. Left to right: petite estoria, split red lentils (the most familiar form in cooking) and green lentils.

Edible fruits

(W)hereas seeds are generally useful because, as plant propagules, they store energy for germination, fruits are adapted to facilitate and enhance seed dispersal. They come in a wide variety of forms. Some, like hazelnuts, with a toughened shell formed from the ovary wall, are used only for the nutritious seeds within. Others, such as green beans and snow peas, have fleshy pods that are harvested before they are fully ripe. However, most edible fruits offer a sweetened fleshy reward to animal dispersal agents.

⋀ The diversity of form nd flavour in fruits of the rose family (Rosaceae) is remarkable. The apple (*Malus domestica*) fruit is a pome, while both blackberry (*Rubus fruticosus*) and strawberry (*Fragaria* × *ananassa*) have aggregate fruits. The blackberry is an aggregate of many, whereas the strawberry in an aggregate accessory fruit in which the fleshy part develops from the receptacle of the flower. The cherry (*Prunus avium*) fruit is a drupe.

Fruits of the rose family

The rose family (Rosaceae) is one of the most economically important in the plant kingdom, with a long history of domestication. The fruits of apples and pears, known as pomes, are formed from the swollen, fleshy calyx tube and receptacle within which the carpels are embedded. The apple (*Malus domestica*) originated in central Asia and was probably the first tree to be domesticated; it is now the most widely cultivated fruit tree of temperate regions. To ensure the desired characteristics of a cultivar are maintained, propagation involves grafting onto rootstocks, the choice of rootstock determining the size of the tree. Commercially, compact trees on dwarfing rootstocks are preferred. This technique has a long history – around 300 BCE, Alexander the Great of Macedonia sent dwarf apples to Athens, and grafting was also widely practised in ancient Rome.

Cross-breeding between apple varieties has resulted in an enormous number of cultivars suited for three main uses. Dessert apples have been selected for colour, size, texture and flavour. Many early varieties had russeting, or areas of rough brownish skin; while this character is preserved in some modern cultivars, it has been carefully bred out of most. Cooking apples are generally larger, with a tart flavour, while cider apples have been bred for the highest levels of sugar, required for fermentation, combined with tannins for taste.

Pear trees (*Pyrus* spp.) have similar origins and diversity to apples, with three main species in commercial cultivation and fruits that can be 'pear-shaped' or resemble apples. Other pome-bearing species with long histories of cultivation include quince (*Cydonia oblonga*) and medlar (*Mespilus germanica*) from southwest Asia, and loquat (*Eriobotrya japonica*) from China. Medlar fruits are unusual in that they are not eaten until they have been bletted, a process of softening that goes beyond ripening and involves the breakdown of cell walls after the fruits have been left on the trees and exposed to frost.

FRUIT FORMS

Fleshy fruits – many of which have been domesticated – can be divided into three main categories. Simple fruits ripen from flowers with a single simple or compound ovary, and include berries like gooseberries and blackcurrants, and drupes like plums and peaches. Aggregate fruits, such as strawberries and raspberries, develop from a single flower with many unfused carpels. Multiple fruits, like the pineapple, mulberry and breadfruit, develop from an entire inflorescence of multiple flowers.

Going bananas

Bananas (*Musa* spp.), tropical and subtropical species with a starchy fruit, are now the fourth-largest cultivated crop in the least developed countries. They were first domesticated in Southeast Asia and Papua New Guinea, perhaps as early as 8000 BCE, from two ancestral species, *M. acuminata* and *M. balbisiana*. Both of these have numerous large seeds, so selection by early farmers focused on propagating parthenogenetic plants, which develop sterile fruits without fertilization taking place. A complex history of hybridization and dispersal of banana cultivars saw different lines migrating into India and Africa, and onwards around the world. Current annual production totals more than 100 million tonnes, but is based on a narrow genetic base, with widely grown cultivars such as 'Cavendish' being highly susceptible to pests and diseases.

(⋁) *Musa balbisiana*, one of the wild ancestors of the modern cultivated banana. Although the flesh contains so many large seeds that it is now considered inedible, it is assumed that the fruits were cooked and eaten as a starchy food before the development of modern varieties.

An African oil palm (*Elaeis guineensis*) plantation on recently cleared forest land in Thailand. Since its introduction to Java in 1848, the palm has become widely grown in Asia, often at the expense of rainforest.

Palm oil comes from the reddish mesocarp of the African oil palm fruit, while the white inner kernel produces palm kernel oil.

Oils and spices

Some important plant oils, used as lubricants and for foods, are produced by crushing ripe fruits. Olive (*Olea europaea*) was first domesticated in the Fertile Crescent and has been a traditional tree crop in the Mediterranean since at least 8000 BCE, while oil from the African oil palm (*Elaeis guineensis*) has been used since at least 5000 BCE to produce a highly saturated fat. Spices – dried aromatic plant products, often highly valued for their culinary uses and medicinal properties – also have a long history, being some of the earliest commodities traded around the world.

Mixed pink (*Schinus* sp.) and black (*Piper nigrum*) peppercorns, the latter having been harvested at different stages of ripeness.

Black peppers

The world's most traded spice today is black pepper, the fruit of the black pepper plant (*Piper nigrum*). This perennial woody vine, native to south and Southeast Asia, has been used in Indian cuisine since at least 2000 BCE. The fruit is a drupe, with an outer fleshy part surrounding a hard endocarp and a single seed. Domestication led to increased yields and enhanced quality, without significantly changing the morphology of the plant. Black pepper was carried from India across the Indian Ocean to Egypt and Rome, along a route that persisted for thousands of years. The fall of the Roman Empire reduced this trade, but it expanded again during the Crusades, with Venice controlling and monopolizing operations in Europe. The seagoing nations of Europe explored new routes to break the monopoly. When Vasco da Gama reached India via the Cape of Good Hope in 1498, Portugal gained control of imports from the East Indies. The Dutch and British soon followed, establishing their own East India companies to import many commodities, including spices.

THE HEAT IS ON

Chilli peppers (*Capsicum* spp.) belong to the nightshade family (Solanaceae) and were domesticated in Mexico by at least 7500 BCE, before being spread widely through Central and South America. The Italian explorer Christopher Columbus, who in 1492 was trying to find a westerly route to the Spice Islands in the East Indies, called them peppers because they tasted like the black peppers he knew. Chillies soon spread throughout the old world, reaching India in late fifteenth century through the Portuguese port of Goa. Five species of *Capsicum* are widely cultivated today, with bell peppers (*C. annuum*) being the most widely consumed – and usually the least spicy. The hottest chillies, measured at more than a million heat units on the Scoville scale (developed to compare capsaicin levels), include varieties such as *naga jolokia* and *naga morich*.

⌃ Chilli peppers (*Capsicum* spp.) drying in the sun at a farmstead in Nepal. They are also consumed fresh, often directly from the plant.

⌄ Spices in a marketplace in Anjuna in the North Goa district of India, formerly a Portuguese possession and a vital link in the spice trade to Europe.

A plethora of peppers

Fruits of long pepper (*Piper longum*) have a stronger taste than the closely related black pepper due to a higher content of the alkaloid piperine, and are used in Ayurvedic medicine. The long, dry catkins of minute fruits were also used as a medicine in ancient Greece. Other native species of *Piper* are consumed in west Africa, Mexico and Southeast Asia. The new world has its own native, pink peppercorns, from the Peruvian and Brazilian pepper trees (*Schinus molle* and *S. terebinthifolius*, respectively). These are often included, for their bright colour, in mixtures of whole peppercorns. The Chinese Sichuan peppercorn (*Zanthoxylum bungeanum*) is a member of the citrus family (Rutaceae), and is harvested for its intensely spicy, lemony peppercorns, which produce a tingling numbness in the mouth.

Advances in plant breeding

(A)s we saw earlier, plant breeding began with the straightforward selection for reproduction of plants considered to have desirable attributes. This process begins with the subjective choice of, for example, the plant specimens with the heaviest crop of fruits, and is based upon identifying and keeping these out of a wider pool of genetic variation. Over successive generations, this selective cultivation changes the genetic composition of the plants. Another key technique involves overcoming natural barriers to cross-breeding, to create hybrids, usually between closely related species, but sometimes between genera.

Mendel's peas

It was the Augustinian friar Gregor Mendel who, in 1856, began the experimental hybridization of different varieties of pea plants (*Pisum sativum*) and discovered the rules of what came to be known as genetic, or Mendelian, inheritance. He crossed peas that exhibited seven independent variable traits and found that in the resulting progeny one trait was always dominant. Tall pea plants crossed with short pea plants always resulted in tall plants, a condition Mendel described as dominant. When the resulting plants were crossed with each other, one in four of the next generation would be a short plant.

(∧) Pods, like those of the pea (*Pisum sativum*), are the typical fruits of the legume family (Fabaceae). The seeds within, each developed from a single ovary, are highly nutritious and can be dried for long-term storage.

(>) Flower colour in peas follows scientist Gregor Mendel's rules of heredity. Purple and white flowers each breed true when crossed with their own kind, but when crossed together, purple colour dominates in the first generation of offspring, with white reappearing in the second.

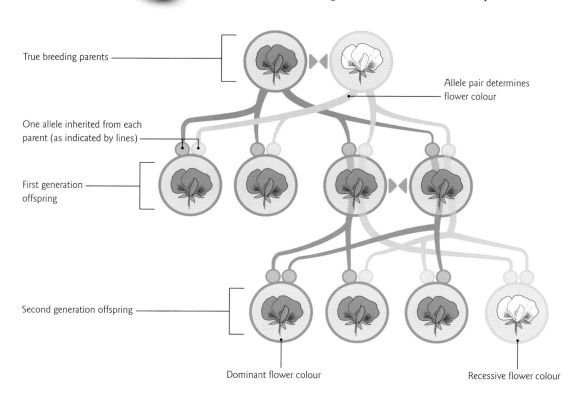

True breeding parents

Allele pair determines flower colour

One allele inherited from each parent (as indicated by lines)

First generation offspring

Second generation offspring

Dominant flower colour

Recessive flower colour

Mendel called the factor that caused short plants to reappear recessive. Recessive factors, or genes as we now know them, do not exert their influence when the dominant gene is also present, but only when both parents pass on the recessive gene. The importance of Mendel's discoveries was only fully appreciated in the early twentieth century, when his work provided a new understanding of the processes of natural selection and ushered in the science of genetics.

Applied genetics

Equipped with a growing understanding of genetics, plant breeders began to develop much more targeted approaches to bringing traits of interest into cultivated plants. These include drawing upon the genetic diversity present in locally adapted landraces, in cultivars or in the wild relatives of crops. There are, however, many challenges to this approach, not least that traits frequently involve complex interactions between genes expressed in different plant organs at distinct stages in the life cycle.

Crosses between selected parents are usually performed manually, by transferring pollen from the male parent to the stigma of an emasculated female flower (the stamens are removed to prevent self-pollination). After this artificial pollination is complete, the flower is

covered with a bag to prevent contamination by other unwanted pollen grains. Hybridization often results in infertile progeny containing an uneven number of chromosomes. In many cases, fertility can be restored by doubling the chromosome number, which is achieved with the use of colchicine to inhibit cell division. Further advances have since followed, making it increasingly possible to understand and interact with the entire genome of the plant.

(∧) To select for specific characteristics, pollen can be transferred by hand, on a cotton bud, from the anthers of the male parent and transferred directly onto the stigma of the chosen female plant.

THE GREEN REVOLUTION

In the late 1960s, the development of high-yielding dwarf varieties of wheat and rice, combined with new agricultural technologies using chemical fertilizers and irrigation, led to the so-called Green Revolution. This has been credited with increasing food production, especially in developing countries, and saving many from famine. However, there have been negative long-term consequences of pesticide and fertilizer use associated with the intensive agriculture that developed, and in some cases it led to narrower genetic diversity in crops.

(>) Triticale, a high-protein crop grown mainly for animal fodder, is a laboratory-bred hybrid between wheat (*Triticum* spp.) and rye (*Secale* spp.).

The era of genomics

O ur ability to manipulate plants in the laboratory has advanced rapidly in recent years, thanks especially to the ever-falling cost of sequencing entire genomes. The first plant genome to be sequenced, in 2000, was that of thale cress (*Arabidopsis thaliana*), chosen a decade earlier as a self-pollinating diploid with a short life cycle and comparatively small genome (135 megabase pairs). We can now assign functions to the 27,000 genes and the 35,000 proteins they encode in the species, and the complete genome sequences are available for numerous breeding lines.

⑦ With its well-understood genome and short life cycle, thale cress (*Arabidopsis thaliana*) is the model used for understanding how specific genes function in plants.

Genetic engineering

Genetic engineering involves transferring the genes responsible for a specific trait directly into plants that lack those genes. One of the first applications involved transfer of the genes responsible for producing insecticidal proteins in the soil bacterium *Bacillus thuringiensis* (referred to as Bt). In 1995, Bt potato, containing transgenes producing the CRY 3A Bt toxin, became the first pesticide-producing crop

to be approved for use in the United States, followed by Bt cotton and Bt maize. The area of land cultivated worldwide with genetically modified (GM), or transgenic, crops has increased dramatically, reaching more than 185 million hectares (700,000 square miles) in 2016. The most commonly planted biotech crops are maize, soya bean, canola and cotton, with 54 per cent grown in 19 developing countries and the balance in seven developed, or industrial, countries. More recent transgenic crops include proprietary

⋀ During spore production, the bacterium *Bacillus thuringiensis* (Bt) naturally produces insecticidal proteins.

⧽ Harvesting a genetically modified crop of insect-resistant Bt upland cotton (*Gossypium hirsutum*) in Vidarbha, Maharashtra, India.

Roundup Ready varieties, resistant to the broad-spectrum herbicide glyphosate.

Initial applications of GM technology were primarily directed towards more profitable production, while second-generation GM foods aim to offer benefits to consumers. 'Golden Rice', for example, is a variety of *Oryza sativa* that is being genetically modified with two transgenes to synthesize beta-carotene as a dietary precursor to vitamin A. Despite these benefits, genetic engineering continues to be contentious (see box).

New and emerging technologies

Plant breeding is an important and fast-moving area of research, in which new ways of understanding and manipulating genes and their regulation are emerging all the time. Marker assisted-selection, for example, accelerates plant breeding by using DNA-based markers for the genes of interest, or marker sequences situated close to them on the chromosome. It can be applied to complex traits such as yield, drought tolerance and disease resistance involving multiple genes, and uses high-throughput DNA screening to determine rapidly whether the desired traits are present in the progeny produced by conventional crossing methods rather than through the transgenic approach.

CONTROVERSIAL CROPS

Despite the benefits they bring, GMOs remain controversial. Concerns have been expressed over the safety of consumers and the environment, and over the application of intellectual property law to the modified organisms. While investigations into the consumption of GM foods have found no additional risks compared to conventional crops, there have already been instances of, for example, weeds developing resistance to the herbicide glyphosate because of intense selection pressure. There are also instances of transgenes transferring from nearby GM crops to non-GM crops or wild relatives.

Genome editing makes it possible to remove or replace DNA sequences using engineered nucleases, which cause double-strand breaks at a specific point in the DNA and then repair the break with the insertion of new DNA. This enables trait stacking, in which several different genes can be brought together on the chromosome so that they will co-segregate during cell division, remain together during breeding and be inherited in successive generations. The gene-editing technology CRISPR, for example, has been adapted from a bacterial immune system that protects against viral attack.

⌄ Gene-editing techniques use a DNA-cutting enzyme to remove a precisely targeted region of DNA from a specific gene and replace it with a short strand carrying a different DNA sequence, thereby altering or repairing the function of the gene.

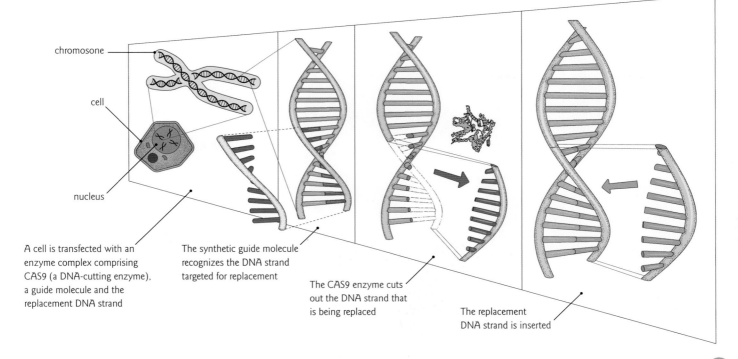

chromosone

cell

nucleus

A cell is transfected with an enzyme complex comprising CAS9 (a DNA-cutting enzyme), a guide molecule and the replacement DNA strand

The synthetic guide molecule recognizes the DNA strand targeted for replacement

The CAS9 enzyme cuts out the DNA strand that is being replaced

The replacement DNA strand is inserted

Conservation of plant diversity

International agreements such as the United Nations Convention on Biological Diversity recognize that, against the background of a growing global population and accelerating environmental change, we need to preserve as much biodiversity as possible for the future. The highest priority of the global community has been to protect the plants we rely on most – those that feed us. However, it is important to conserve diversity at the ecosystem, species and genetic level, regardless of whether we find individual plants useful for specific purposes or not.

Conserving crops

Since the 1960s, preserving the diversity of crops and developing new varieties has been the remit of a global network of 15 Consultative Group on International Agricultural Research (CGIAR) centres, dedicated to specific crops, ecosystems or regions. The public domain collections in their care are held in trust to guard against the dangers of agricultural dependence on limited genetic diversity. When we have relied on too few varieties, the results have sometimes been catastrophic. At least a million people died between 1845 and 1852 in Ireland's Great Famine, when potato blight (caused by the fungus *Phytophthora infestans*) swept through Europe. Overdependence on a few cultivars, especially 'Irish Lumper', which was not resistant to potato blight, was a major factor. In the 1970s, an epidemic of southern corn leaf blight fungus (*Bipolaris maydis*) destroyed a sixth of the maize harvest in the United States as a result of overdependence on highly susceptible Texas cytoplasm (*cms-T*) maize. It is now clear that we must preserve genetic diversity in crops as a source of future resilience and adaptability.

The International Rice Research Institute in Manila contributed to the Green Revolution (see box on page 347) through the development of semi-dwarf varieties, and is now pursuing the development of 'Golden Rice' (see page 349). The Centro Internacional de la Papa, in Lima, specializes in potatoes, sweet potatoes and other Andean roots and tubers. In 2004, the Crop Trust was established as an independent organization under international law to ensure the conservation and availability of crop diversity for food security worldwide. Its Svalbard Global Seed Vault on Spitsbergen holds duplicate collections from thousands of seed banks around the world as insurance for the future (see opposite).

Wider plant diversity

A compelling case for preserving all plant diversity has yet to secure such support. Although we know that plants created, and maintain, the conditions necessary for animal life in the biosphere, it is sometimes argued that there is a level of redundancy, making it unnecessary to conserve all species. This rationalization is usually based on assigning

plant species to functional groups, as though they were interchangeable pieces in a machine. However, a stronger moral case is that it would be indefensible to suggest that a species needs to prove its utility to us for it to be worth saving from extinction. It is often said we should preserve biodiversity out of enlightened self-interest. In truth, little enlightenment is needed to understand that ecosystem services – our life-support systems – depend on plant diversity. Fortunately, the global community of 3,000-plus botanic gardens and other organizations are rallying around the internationally agreed Global Strategy for Plant Conservation. The world's botanic gardens already conserve and manage a greater proportion of the known plant species than any other sector.

⌃ The Famine statues by Rowan Gillespie stand in Ireland's Dublin Docklands in memory of almost a million people who died in the Great Famine of 1845–1849.

Ⓛ Harvesting experimental beans (*Phaseolus* spp.) near the Colombian town of Darién for the International Center for Tropical Agriculture (CIAT).

Ⓥ Samples of tropical forages conserved *in vitro* at the CIAT gene bank in Colombia. Seeds of these plants have been sent to the Global Seed Vault in Svalbard for conservation.

Preserving plant diversity

(P)lants have many strategies for survival, including multiplying through sexual and asexual means and dispersing themselves widely. Efforts to preserve plant diversity tap into, and exploit, the fundamental biology of plants. Seed banks play a prominent role in this, given that many seeds can remain dormant for centuries. But not all seeds share this facility, so other forms of living collection are also required. An important consideration in each of these methods of *ex situ* conservation is to make sure that as much genetic diversity as possible is captured and preserved.

⋁ Samples of DNA from wild plant species are stored in carefully labelled and databased packets, kept dry with silica gel, at China's Germplasm Bank of Wild Species in Kunming.

Seed banks

⋁ A visually stunning display of seeds welcomes visitors to China's national seed bank, the Germplasm Bank of Wild Species at the Kunming Institute of Botany in Yunnan province. With seeds provided by Kew's Millennium Seed Bank, these fibre-optic rods originally appeared on the UK Pavilion at Expo 2010 Shanghai.

Seed banks are a very cost-effective way of conserving plant diversity for the 75 per cent of plants with orthodox seeds – those that can tolerate desiccation and freezing. Such seeds are generally small, have hard seed coats and have a dormant state that is broken only when the right environmental cues trigger germination. The standard protocol is to dry seeds to a moisture content of about 7 per cent before storing them at –20 °C (–4 °F). The world's most diverse seed bank, the Millennium Seed Bank at the Royal Botanic Gardens, Kew, in London, developed viability tests and germination protocols suitable for wild plant species, many of which have never previously been brought into cultivation.

A key role of any seed bank is to supply plant material. The Svalbard Global Seed Vault has already done this, providing duplicate collections of seeds which were destroyed by the conflict in Syria to the International Center for Agricultural Research in Dry Areas. Recalcitrant seeds, which are often large, and have a thin skin and short dormant phase, cannot be banked in the conventional way, and must be conserved by cryopreservation or tissue culture (see box), or as living plants. Unfortunately, recalcitrant seeds occur in many important groups of plants, including oaks and their relatives in the oak family (Fagaceae), and in many groups of tropical trees.

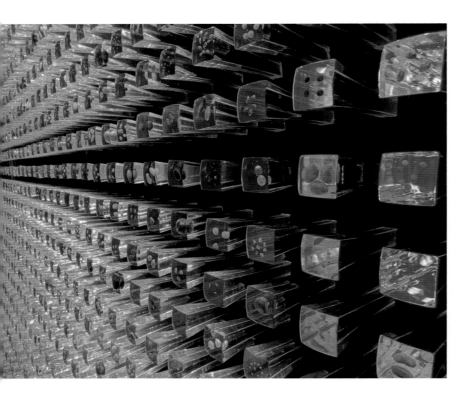

The nursery at Kadoorie Farm and Botanic Garden, Hong Kong, raises thousands of endangered plants for ecological restoration, including a rare hornbeam, *Carpinus insularis*, discovered in 2013.

Living collections

Botanic gardens hold internationally important living collections of plant diversity. The PlantSearch database, maintained by Botanic Gardens Conservation International, has records of 1.3 million accessions held in around 1,100 institutions, and estimates that at least one-third of all plant species are kept in botanic gardens and arboreta. While that represents a significant resource, it is not yet sufficient to safeguard the future of plant diversity. Well-coordinated strategic actions are required to make the world's living collections of plants fully representative of species and genetic diversity, combined with major programmes of collecting to acquire seeds, cuttings, bulbs and other propagules. A convergence of missions between the crop and wild plant sectors is highly desirable to secure plant diversity for the future.

CRYOPRESERVATION AND TISSUE CULTURE

Cryopreservation is a more expensive conservation option for seed banks, requiring the isolation of embryo axes from seeds and their treatment with cryoprotectant chemicals to reduce the impact of physical damage from freezing in liquid nitrogen. An alternative strategy is to cryopreserve dormant buds or small pieces of shoot axis, which can be recovered by grafting or direct rooting, or by *in vitro* culture using the same methods as tissue culture. Tissue culture and micropropagation have been used to maintain large numbers of small living plants such as orchids.

Discovered in southern Mexico as recently as 1973, *Deppea splendens* was extinct in the wild by 1986 owing to habitat destruction. Fortunately, botanic gardens raised plants from seed and today the shrub survives in cultivation.

Plants and the future

The future of plants is our future; our fates are inextricably intertwined. What, then, does the future hold? Many indicators show we are putting an unprecedented burden on the capacity of our planet to sustain life. The rate of increase in the global population has slowed from a peak in the 1960s but continues to rise, and the United Nations predicts that there are likely to be 11.2 billion people by 2100. The best solution for achieving sustainable development will be to create plant-rich environments everywhere on the planet.

⌃ Seedlings provide hope for the future, enabling us to restore degraded landscapes, capture carbon in forests, keep rivers flowing and feed a growing global population.

Sustainable Development Goals

In 2015, United Nations member countries approved an agenda on sustainable development that subsequently led to the adoption of 17 Sustainable Development Goals, or 'Global Goals', with associated targets to be reached by 2030. Governments, businesses and civil society have started to mobilize efforts to achieve these goals, which are intended to improve the lives of people everywhere. Many of the goals engage with issues that depend fundamentally on plant diversity, especially those addressing food security, biodiversity loss, sustainable cities and climate change.

Goal 2: Zero Hunger

Since the 1990s, an estimated 75 per cent of crop diversity has been lost from farmers' fields. There is enormous scope to increase the diversity of existing crops, drawing on the genes they and their wild relatives contain. This will require further development of seed and plant banks, and more sustainable agricultural practices. Agricultural intensification has tended to be at the expensive of biodiversity, soil quality and ecosystem services. The key will be to reverse the decline by integrating biodiversity into purposefully redesigned agricultural landscapes. Advances in plant breeding hold many promises for the future, providing issues of equity and access can be surmounted.

Goal 11: Sustainable Cities and Communities

Given that more than half of humanity lives in urban areas and that rural–urban migration is increasing, Goal 11 calls for safe and affordable housing, improved public transport, more public green spaces, and improved urban planning and management. This is an opportunity to rethink and redesign the relationship between plants and the city through initiatives such as urban farming and creating high-quality green spaces. What is to prevent the green roofs, green walls and urban parks of the future from holding collections of rare and threatened crop and wild plants?

United Nations Sustainable Development Goals	
1 No Poverty	10 Reduced Inequality
2 Zero Hunger	11 Sustainable Cities and Communities
3 Good Health and Well-being	12 Responsible Consumption and Production
4 Quality Education	13 Climate Action
5 Gender Equality	14 Life Below Water
6 Clean Water and Sanitation	15 Life on Land
7 Affordable and Clean Energy	16 Peace, Justice and Strong Institutions
8 Decent Work and Economic Growth	17 Partnerships to Achieve the Goals
9 Industry, Innovation and Infrastructure	

Goal 13: Climate Action

Climate change poses an existential threat to our species. Technological solutions, such as migrating from fossil fuels to renewable energy, must play their part. However, it was photosynthesis on land and in the oceans that first created the breathable atmosphere. Forests, savannahs, grasslands and carefully managed agricultural landscapes are therefore fundamental in sequestering carbon. Keeping this a green world is one of the best ways of minimizing the impact of climate change.

Goal 15: Life on Land

Goal 15 aims to 'protect, restore and promote sustainable use of terrestrial ecosystems, sustainably manage forest, combat desertification, and halt and reverse land degradation and halt biodiversity loss'. These are enormous challenges, all of which hinge around our understanding of, and care for, plant diversity. There is a growing, but under-resourced, movement towards ecological restoration. At present the most significant, rate-limiting obstacle is the lack of availability of seeds of native plant species. Much could be done to overcome this constraint, such as establishing high-diversity seed orchards for native species.

The global challenges we face are daunting. Although mankind has relied on plants for millennia, there is so much more to be done to make the most of our relationship with them. There has never been a more exciting time to be a botanist.

∧ The quest for sustainable cities, combined with an understanding of the health benefits of contact with nature, is driving the development of new and innovative landscapes like this urban green space in Hong Kong.

∨ Workers water the Widu tree nursery in Senegal's Louga region, which is part of the Great Green Wall (GGW) initiative. The intention of the project is to create a lush 15-km (10-mile) wide strip of different plant species to span the 7,600 km (4,700 miles) across Africa to the south of the Sahara. It will stretch from Senegal in the west to Djibouti in the east to halt desertification.

Glossary

achene Small, dry, indehiscent fruit containing a single seed.

adventitious root Root that develops from non-root tissue such as a stem.

allelopathy Production of biochemicals to inhibit the development and growth of neighbouring plants.

alternation of generations Life cycle comprising alternating haploid gametophyte and diploid sporophyte generations.

anther Pollen-producing part of the stamen of a flower.

antheridium (pl. antheridia) Organ that produces and holds male sperm cells in bryophytes, fern allies and ferns.

antheridiophore Upright structure in some bryophytes that bears antheridia.

apomixis (adj. apomictic) Asexual reproduction involving the formation of an embryo without the formation of eggs or sperm. The resultant offspring is a clone. Also known as agamospermy.

archegonium (pl. archegonia) Organ producing and holding female egg cells in bryophytes, fern allies, ferns and some gymnosperms.

aril Outer seed appendage, usually fleshy and brightly coloured.

bract Modified leaf, often growing below a flower or inflorescence.

cambium Thin layer of tissue between the xylem and phloem whose cells actively divide to produce secondary growth.

capitulum Composite flowerhead in members of the daisy family (Asteraceae).

carpel Female reproductive organ of a flower that encloses the ovules.

caryopsis Achene whose seed coat is joined to the ovary wall, as in grasses.

caudicle In orchids, the thread attaching pollinia to the flower column.

chloroplast Specialized chlorophyll-containing plant organelle responsible for photosynthesis.

cladode Modified stem that resembles and functions like a leaf, carrying out photosynthesis.

collenchyma cells Specialized parenchyma cells with selectively thickened cellulose walls, providing flexible support in stems (cf. sclerenchyma cells).

corm Swollen underground stem base that acts as a storage organ.

cotyledon Embryonic leaf in angiosperms, usually the first leaf present on germination. Monocotyledons have one cotyledon, while dicotyledons have two.

cyme (adj. cymose) Inflorescence whose main axis terminates in a single flower that blooms before those on the lateral stems below.

dioecious Having male and female reproductive organs on separate plants (cf. monoecious).

diploid Having two complete sets of chromosomes, one inherited from the sperm cell and the other from the egg cell (cf. haploid).

drupe Fleshy fruit with a thin skin and single hard stone enclosing a seed, e.g. a plum. Drupelets are small drupes, as seen in aggregate fruits, e.g. raspberries.

early wood Wood produced at the start of the growing season, characterized by large cells with thin walls (cf. latewood).

elaiosome Fat- or protein-rich fleshy seed appendage, often used to attract ants for seed dispersal.

endodermis Single cell layer surrounding the inner vascular tissue in roots.

endosperm Food reserve inside a seed that nourishes the embryo.

epiphyte Non-parasitic plant that grows on another for support.

gamete Mature haploid sex cell of an organism, either the female egg cell or male sperm cell. Gametes of the opposite sex unite during sexual fertilization to form a diploid offspring.

gametophyte Haploid, gamete-producing generation in the life cycle of a land plant (cf. sporophyte).

gemma (pl. gemmae) Cluster of cells that can separate from a parent plant and produce new individuals in a form of asexual reproduction.

genome Complete set of an organism's genetic information, including its DNA and genes.

haploid Having a single set of chromosomes, half the diploid number (cf. diploid).

haustorium (pl. haustoria) Specialized modified stem or root of a parasitic plant that penetrates the host plant to extract nutrients.

hypanthium Usually cup-like flower structure that bears the sepals, petals and stamens, e.g. in rose family (Rosaceae) members.

hypocotyl Part of the plant embryo above the radicle and below the cotyledons.

indusium Tissue that covers and protects the sorus of a fern.

inflorescence Flowering stem of a plant with more than one flower, classified by the arrangement of flowers.

integument Tough jacket-like coat of a seed.

krummholz Vegetation comprising trees that are stunted through exposure to high winds in extreme subalpine and subarctic environments.

lamina (pl. laminae) Leaf blade, where the majority of photosynthesis takes place.

latewood Wood produced later in the growing season, characterized by small cells and thicker walls (cf. early wood).

lignin Decay-resistant cell wall polymer that forms wood and provides support.

lignotuber Swelling at the stem base containing buds and food reserves from which new growth arises when the plant is damaged, e.g. by fire.

meiosis Process by which diploid cells divide to produce four haploid cells. It involves two cell divisions, during the first of which there is an exchange between chromosomes of maternal and paternal genetic material.

meristem Area of undifferentiated tissue where active cell division causes growth to occur. Apical meristem is located at root and shoot tips.

mesophyll Leaf tissue, comprising palisade mesophyll and spongy mesophyll cells.

micropyle Narrow channel in the ovule of a seed-bearing plant though which the pollen tube gains entry.

monoecious Having male and female reproductive organs on the same plant (cf. dioecious).

mycorrhiza (pl. mycorrhizae) Fungus that forms a symbiotic relationship with a plant through its roots. In ectomycorrhizae the fungal hyphae do not penetrate the plant's roots, whereas in endomycorrhizae the hyphae enter the root cells.

nectary Organ that produces nectar as a reward for pollinators.

nucellus Central part of the ovule, containing the embryo sac.

palisade mesophyll
Upper layer of the mesophyll, comprising elongated cells whose chloroplasts absorb the majority of sunlight for photosynthesis.

panicle Many-branched inflorescence in which the lower flowers are oldest.

paraphyletic Group of organisms that contains a common ancestor and some but not all of its descendants.

parasite Organism that lives in or on another and gains a benefit at a cost to the host.

parenchyma cells
Thin-walled cells that form the ground tissue of a plant.

petal Flat, coloured, non-fertile structure outside the reproductive parts of a flower.

petiole the stalk of the leaf, which connects the leaf blade or lamina to the stem.

phloem Living vascular tissue that transports sugars around a plant (cf. xylem).

photosynthesis
Process by which green plants use energy from sunlight to convert carbon dioxide and water into sugars and oxygen.

phyllode Modified flattened petiole that functions as a leaf.

pistil Female part of a flower, comprising the stigma, style and ovary.

pit membrane Relatively thin membrane across the centre of a pit in the cell wall of a tracheid, through which water passes.

plumule First shoot produced on germination, arising above the hypocotyl and bearing the cotyledon(s).

pneumatophore
Aerial root of some waterlogged plants, e.g. mangroves, through which gas exchange occurs.

pollen Microscopic male gametophytes of seed plants produced by male cones and flowers, containing the male sex cells.

pollinium (pl. pollinia)
Cluster of numerous coherent pollen grains dispersed as a unit in some flowers, e.g. orchids.

pome Fruit comprising a fused hypanthium surrounding a central seed-containing core, e.g. an apple.

raceme (adj. racemose)
Simple inflorescence with individual flowers borne on short stems off a central main axis.

rachis (pl. rachides)
The main axis of a compound structure, e.g. central axis of a fern frond or elongated axis of an inflorescence.

radicle Part of the plant embryo that develops into the first root on germination.

receptacle Enlarged tip of the stem that bears the flower organs.

rhizobium (pl. rhizobia)
Nitrogen-fixing bacterium that colonizes cells within the root nodules of some plants, e.g. legumes (Fabaceae).

rhizoid Thread-like unicellular protuberance on the underside of a bryophyte gametophyte that acts like a root.

rhizome Root-like stem that grows horizontally underground.

samara Simple dry, hard fruit with one or more wings to aid dispersal.

sclerenchyma cells
Specialized parenchyma cells with lignified secondary walls, providing rigid support in stems (cf. collenchyma cells).

seed bank Seed-storage facility established as a means of preserving plant diversity.

self-fertilization
In which the male and female sex cells of a single individual fuse. No new genetic material is introduced but many offspring can be produced.

sepal Usually green, leaf-like structure below the petals of a flower.

serotiny (adj. serotinous) Release of seeds in response to an environmental trigger, e.g. fire.

sorus (pl. sori) Cluster of sporangia in a fern.

spathe Large bract protecting the flower spike in some plants, e.g. arum family (Araceae).

sporangium (pl. sporangia) Structure in bryophytes, fern allies and ferns in which spores are formed.

sporophyll Leaf in spore-bearing plants that carries sporangia.

sporophyte Diploid, asexual generation in the life cycle of a land plant (cf. gametophyte).

stamen Male reproductive structure of a flowering plant, comprising the anther and stalk-like filament.

stolon Stem growing horizontally across the ground surface that produces roots and branches – and hence new plants – at specialized nodes. Also called a runner.

stoma (pl. stomata)
Microscopic opening, usually on the underside of a leaf, through which gas exchange occurs.

strobilus (pl. strobili)
Cone-like structure in lycophytes and horsetails that bears sporophylls; also refers to cones in gymnosperms.

style Part of the pistil between the ovary and stigma.

suberin Waterproof wax-like substance that impregnates cork cell walls.

syconium (pl. syconia)
Hollow receptacle whose interior contains many tiny flowers, e.g. a fig.

syncarp (adj. syncarpous) Fruit that develops from a female organ with fused carpels.

tepal Outermost organ of a flower in which the sepals and petals are undifferentiated.

thallose Having a flattened structure with no differentiation into leaves or stems.

thermogenesis Heat production within an organism, e.g. in flowers.

thigmonasty
Movement of a plant organ in response to touch. Also called seismonasty.

torus-margo pit
Specialized cell wall pit in a conifer tracheid that acts like a valve to regulate water flow.

trichome Hair-like structure in some plants, whose structure and function varies widely.

tracheid Rigid, tubular xylem cell through which water is transported.

xylem Vascular tissue with lignified cell walls comprising tracheids or vessels that transports water from the roots upwards to the rest of the plant.

zoospore Haploid dispersal cell of an alga, which uses one or more flagellae to move.

zygomorphic
Of a flower, bilaterally symmetrical, able to be cut into two equal halves in only one way.

zygote Diploid cell produced following the fertilization of a haploid egg cell by a haploid sperm cell.

Further reading

Arber, A. 1950. *The Natural Philosophy of Plant Form.* Cambridge University Press.

Ambrose, B. A. and Purugganan, M. D. 2012. *The Evolution of Plant Form.* Wiley-Blackwell.

Balick, M. J. and Cox, P. A. 1996. *Plants, People and Culture: The science of ethnobotany.* Scientific American Library.

Beck, C. B. 2005. *An Introduction to Plant Structure and Development: Plant anatomy for the twenty-first century.* Cambridge University Press.

Bell, A. D. 1991. *Plant Form: An illustrated guide to flowering plant morphology.* Oxford University Press.

Cardon, Z. G. and Whitbeck, J. L. (eds). 2007. *The Rhizosphere: An ecological perspective.* Elsevier.

Dacey, J. W. 1980. *Internal winds in water lilies: an adaptation for life in anaerobic sediments.* Science, 210(4473): 1017–1019.

de Kroon, H. and Visser, E. J. W. (eds). 2003. *Root Ecology.* Ecological Studies 168. Springer.

Essig, F. B. 2015. *Plant Life: A brief history.* Oxford University Press.

Evert, R. and Eichhorn, S. 2013. *Raven Biology of Plants.* 8th edn. W. H. Freeman/Palgrave Macmillan.

Fenner, M. and Thompson, K. 2005. *The Ecology of Seeds.* Cambridge University Press.

Glover, B. 2014. *Understanding Flowers and Flowering: An integrated approach.* Oxford University Press.

Goodman, R. M. 2004. *Encyclopedia of Plant and Crop Science.* M. Dekker.

Gregory, P. J. 2006. Plant Roots: Growth, activity and interaction with soils. Wiley-Blackwell.

Hacke, U. (ed.). 2015. *Functional and Ecological Xylem Anatomy.* Springer.

Hickey, M. and King, K. 2001. *The Cambridge Illustrated Glossary of Botanical Terms.* Cambridge University Press.

Isnard, S. and Silk, W. K. 2009. *Moving with climbing plants from Charles Darwin's time into the 21st century.* American Journal of Botany, 96(7): 1205–1221.

Kingsbury, N. 2009. *Hybrid: The history and science of plant breeding.* University of Chicago Press.

Lack, A. and Evans, D. E. 2005. *Plant Biology.* Taylor & Francis.

Langenheim, J. H. 2003. *Plant Resins: Chemistry, evolution, ecology, and ethnobotany.* Timber Press.

Lewington, A. 2003. *Plants for People.* Eden Project Books.

Mabberley, D. 2017. *Mabberley's Plant Book: A portable dictionary of plants, their classification, and uses.* 4th edn. Cambridge University Press.

MacAdam, J. W. 2009. *Structure and Function of Plants.* Wiley-Blackwell.

Murphy, D. J. 2007. *People, Plants and Genes: The story of crops and humanity.* Oxford University Press.

Nabhan, G. 2016. *Ethnobiology for the Future. Linking cultural and ecological diversity.* University of Arizona Press.

Niklas, K. J. 2010. *Plant Biomechanics: An engineering approach to plant form and function.* University of Chicago Press.

Niklas, K. J. and Spatz, H. C. 2012. *Plant Physics.* University of Chicago Press.

Proctor, M., Yeo, P. and Lack, A. 2003. *The Natural History of Pollination.* Timber Press.

Raven, P. H., Evert, R. F. and Eichhorn, S. E. 2017. *Biology of Plants.* 8th edition. Macmillan.

Rosell, J. A., Gleason, S., Méndez-Alonzo, R., Chang, Y. and Westoby, M. 2014. *Bark functional ecology: evidence for tradeoffs, functional coordination, and environment producing bark diversity.* New Phytologist, 201(2): 486–497.

Russell, G. 2003. *Plant Canopies: Their growth, form and function.* Cambridge University Press.

Sperry, J. S. 2003. *Evolution of water transport and xylem structure.* International Journal of Plant Sciences, 164(S3): S115–S127.

Spicer, R. and Groover, A. 2010. *Evolution of development of vascular cambia and secondary growth.* New Phytologist, 186(3): 577–592.

Taylor, E. L., Taylor, T. N. and Krings, M. (2009). *Paleobotany: The biology and evolution of fossil plants.* Academic Press.

Tortora, G. J., Cicero, D. R. and Parish, H. I. 1972. *Plant Form and Function: An introduction to plant science.* Macmillan.

Vogel, S. 2012. *The Life of a Leaf.* University of Chicago Press.

Index

Picture credits

(t = top, m = middle, b = bottom, l = left, r = right)

Illustrations on pages 15t, 18, 25, 30t, 36l, 37l, 38l, 39t, 46br, 48b, 52t, 53tm, 57t&b, 59b, 62, 65b, 94t, 95r, 97t, 98t&b, 100b, 103t&br, 107b, 108b, 112t, 113t, 122b, 123l, 127t, 128b, 141t, 142t, 145t, 146, 152t, 155t, 157b, 165t, 171tr, 182b, 183m, 184, 185t, 195t, 198, 199t, 207t&m, 208t, 209, 211m, 229, 278t, 286br, 308t, 346b, 349 by Robert Brandt.

All images copyright the following:

Alamy Stock Photo: 27t Heritage Image Partnership Ltd; 27b Science Photo Library; 50 blickwinkel; 51b Yon Marsh Natural History; 59t Martina Simonazzi; 66b Denis Crawford; 73t Nigel Cattlin; 75 Krystyna Szulecka; 76t Zoonar GmbH; 77b Adrian Weston; 80l Natural Visions; 84b M I (Spike) Walker; 88t Science History Images; 123r blickwinkel; 136tr buccaneer; 164br Zoonar GmbH; 170b Steffen Hauser; 173b imageBROKER; 177t Garden World Images Ltd; 180t Natural Visions; 186t Sabena Jane Blackbird; 189t FloralImages; 197b Bob Gibbons; 202l Science Photo Library; 203b studiomode; 210r Custom Life Science Images; 236tl The Natural History Museum; 246b The Natural History Museum; 258t Premaphotos; 259t Minden Pictures; 262bl Bob Gibbons; 277r Scott Camazine; 288r buccaneer; 293b Andy Catlin; 298t Witold Krasowski; 303t Duncan Usher; 310b Arterra Picture Library; 313t robertharding; 313l Zoonar GmbH; 323b imageBROKER; 329t Richard Mittleman; 330b Stephanie Jackson; 336l Andrew Kearton; 348bl Mediscan; 348br Joerg Boethling; 350b dpa picture alliance.

AL Baker: 231t.

Josef Bergstein (MPI-MP: Research Group Kraemer): 77t.

Stephen Blackmore: 281r, 316t, 317t, 318t, 327t&m, 334b, 336t, 339t, 340b, 344b, 345t, 352t&b, 353t, 355t.

BlueRidgeKitties: 74t.

Craig Boase: 267t.

Joel Brehm & Daniel Schachtman (University of Nebraska-Lincoln. ©2017 The Board of Regents of the University of Nebraska): 88b.

Brendan Choat: 99mr.

Continental: 324t.

Cornell University Plant Anatomy Collection: 43b, 44b, 101tl&ml.

PG Davison: 186bl&br.

Sylvain Delzon: 108m.

Diego Demarco: 213m.

JC Domec: 116br&bl.

LA Donaldson, J Grace & GM Downes: 127ml&bl.

Doranakandawatta: 291b.

Dreamstime: 23t Sociologas; 32b Nancy Kennedy.

Andrew Drinnan: 14b, 24t&b, 34b, 36t, 38r, 40r, 41b, 43tr, 46t, 48tl,bl&br, 53tl,tr&b.

AR Ennos, H-Ch Spatz & T Speck: 121b.

Stefan Eberhard: 11.

Robert Eplee: 81t.

Fred Essig: 211t, 212m&b, 214r.

GAP Photos: 237b JS Sira; 264t Tim Gainey; 270t Marcus Harpur; 296b Jonathan Buckley.

Getty Images: 32t Ed Reschke; 35m Nastasic; 44t Garry DeLong; 46m Photos Lamontagne; 70t Ed Reschke; 112b Dr Richard Kessel & Dr Gene Shih; 113b Nigel Cattlin/ Visuals Unlimited, Inc.; 216t Susumu Nishinaga/Science Photo Library; 216b Ed Reschke; 355b SEYLLOU DIALLO/Stringer.

Lorna J Gibson: 105t&m.

Kari Greer and the USDA Forest Service: 132t.

Uwe Hacke & Steven Jansen: 109l,m&r.

Michael Hough (from Master thesis, 2008, State University of New York): 85t.

Ian_MC99: 237t.

Sandrine Isnard & Wendy K Silk: 118t.

Anna Jacobsen: 110tr, 111b.

Steven Jansen: 99br.

Agata Jedrzejuk et al. (The Scientific World Journal, Vol. 12, Article ID 749281): 121tl.

Jon E Keeley (from Israel Journal of Ecology & Evolution, 2012, pp.123–135): 133t.

John Kinross: 188.

C Leitinger: 111t.

Jennifer Mahley: 102b.

Rui Malho: 217b.

Ciera Martinez: 101bl.

Stefan Mayr: 110tl.

MC McCann, B Wells & K Roberts: 106b.

Joel McNeal: 219br.

Nature Photographers Ltd: 303b Paul Sterry.

Olivia Messinger: 252b.

Oak Ridge National Laboratory: 69.

Dr S Orang: 50t.

Neil Palmer/CIAT: 297b, 351bl&br.

Alann J Pedersen: 189b.

G Pilate, et al.: 127mr&br.

Jarmila Pittermann: 95l, 99t, 99bl, 100t, 104b, 110bl,bm&br, 114tm&tr, 116t, 118b, 119t&b, 125tl.

Libor Pitterman: 115tr&b.

George Poinar, JR Finn & N Rasmussen: 263b.

D. Price-Goodfellow: 7bl, 120r, 170t, 171b, 219bl, 272.

Peter Richardson: 199b.

Chris Rico: 105b.

Julietta Rosell: 114tl&b, 115tl.

Catarina Rydin & Kristina Bolinder: 204bl&br.

Science Photo Library: 20b Dennis Kunkel Microscopy; 22r Steve Gschmeissner; 42 Dr Keith Wheeler; 60l Dr Jeremy Burgess; 60r Dennis Kunkel Microscopy; 61t Dennis Kunkel Microscopy; 63t Omikron; 64 Nigel Cattlin; 66t Biodisc, Visuals Unlimited; 67b Ted Kinsman; 82 Eye of Science; 83t USDA/Science Source; 83b Wim Van Egmond; 90l Astrid & Hanns-Frieder Michler; 90r Eye of Science; 91t Nigel Cattlin; 91b Dennis Kunkel Microscopy; 97b Dr Keith Wheeler; 142b Power and Syred; 145b Biology Pics; 154l Ramon Andrade 3Dciencia; 158t&b Dr Jeremy Burgess; 159tl Power and Syred; 168b Steve Gschmeissner; 194b Noble Proctor; 196b Dr Keith Wheeler; 197t Claude Nuridsany & Marie Perennou; 217t Eye of Science; 265b Photo Insolite Realite; 289t Dr Keith Wheeler.

John D Shaw: 132bl.

Clive Shirley: 190t.

Shutterstock: 3 Africa Studio; 4 Palokha Tetiana; 6tl Dr Morley Read; 6tr Pongwisa Dechapun; 6br LutsenkoLarissa; 7tl Brian Maudsley; 7tr Blue Rose photos; 7br Jakob Fischer; 9 Nagib; 10b Tropper2000; 12 Dr Morley Read; 14t Manfred Ruckszio; 15b Ghing; 16l Pablo Rodriguez Merkel; 17 THPStock; 19t primola; 20t Grimplet; 22l D Kucharski K Kucharska; 25b Lucky-photographer; 29 Catmando; 33t Orest Iyzhechka; 33m Velichka Miteva; 33b Ashley Whitworth; 40l robin foto; 43tl Ines Behrens-Kunkel; 45t Glynsimages2013; 46bl Bo Valentino; 47 Manfred Ruckszio; 48t guentermanaus; 52b De Visu; 54 Pongwisa Dechapun; 56 siambizkit; 58t zebra0209; 63b Becca D; 65t Elis Blanca; 67t Olya Detry; 68t Goncharov_Artem; 68b jeep2499; 70b Richard Griffin; 71t Fotos593; 71b Benny Marty; 72t ronstik; 72b Jubal Harshaw; 73b Frank Fennema; 74b panphai; 76b Ethan Daniels; 78t ckchiu; 78b Lee Prince; 79t AAMLERY; 79b Jubal Harshaw; 80r Dr Morley Read; 81b Alexander Mazurkevich; 84t Worachat Tokaew; 86t Iurii Kruglikov; 86b Morphart Creation; 89 Sundry Photography; 94b DmitryKomarov; 97m Iryna Loginova; 101br kpboonjit; 102t Mike Rosecope;

104t Claudio Divizia; 104m grapher_golf; 106t Elena Elisseeva; 107t koliw; 108t Eric Buermeyer; 117t Dean Pennala; 120l Ines Behrens-Kunkel; 121tm Mike Rosecope; 125tr Chris Murer; 126 Igor Normann; 128t lcrms; 130r abdusselam fersatoglu; 131ml hareluya; 131mr JFFotografie; 131br WeihrauchWelt; 132br Kyle T Perry; 134 LutsenkoLarissa; 136tl Bildagentur Zoonar GmbH; 137t ANGHI; 138r SAJE; 139bl NuiGetSetGo; 139bmr Jojoo64; 139mr Puttiporn; 139br Jeff Holcombe; 140t george photo cm; 140bl bonchan; 141bl Rattiya Thongdumhyu; 141br Bihrmann; 143br Unkas Photo; 144t Mr3d; 144bl PJ photography; 144br LeStudio; 147r Gallinago_media; 148tl PFMphotostock; 148tr Brzostowska; 149tr Anton Foltin; 149b Iva Vagnerova; 150l Nomad_Soul; 150r r. classen; 151t AlessandroZocc; 151m Rattiya Thongdumhyu; 151b Vitalii_Mamchuk; 153tl&tr Stephen Farhall; 153m Sina Jasteh; 154r Jaboticaba Fotos; 155b simona pavan; 156l Imladris; 156r Pixeldom; 157t nevodka; 157m Matsuo Sato; 159b Rattiya Thongdumhyu; 160t Korotkov Oleg; 161tl svf74; 161tr Aniroot; 161b Nik Merkulov; 162t smileimage9; 163t pisitpong2017; 163br Pascale Gueret; 166b Jeff Holcombe; 169m marako85; 169br pittaya; 171tl photolike; 172b frank60; 174t&b cpaulfell; 177b alybaba; 178 Brian Maudsley; 180bl Anest; 180br Kenneth Dedeu; 181t Andrew M Allport; 181b Noppharat888; 182t Carlos Rondon; 183t Lebendkulturen.de; 183b Daniel Poloha; 185m Chad Zuber; 185b ChWeiss; 187t IanRedding; 187b Starover Sibiriak; 190tr Kichigin; 190bl Henri Koskinen; 190br Anest; 192 Sibiriak; 193t Tatjana Romanova; 193ml,mr&br Jubal Harshaw; 193bl Rattiya Thongdumhyu; 196t Konstantin Bratsikhin; 197m Elisa Manzati; 201b Antonio Gravante; 202r bevz tetiana; 203t IreneuszB; 203m Bildagentur Zoonar GmbH; 204tr Fanfo; 205b Sanit Fuangnakhon; 206t aleksandr shepitko; 206m rwkc; 206b Ihor Bondarenko; 212t Manfred Ruckszio; 213b Tim Zurowski; 214l Dobryanska Olga; 215t Suman_Ghosh; 215bl Stephen B Goodwin; 215br Isabelle

OHara; 218t Ivaschenko Roman; 218b Nataliia Zhekova; 219t Alter-ego; 220 Blue Rose photos; 222tl&tr arka38; 222b Heiti Paves; 223tl Gurcharan Singh; 223tr Gavin Budd; 224l ingamiv, 224r Dionisvera; 225tl Doug Armand; 225tr Sherjaca; 225b Suttipon Yakham; 226 elementals; 227t Worraket; 227b Protasov; 228t basel101658; 228b Oleksandr Kostiuchenko; 230b Lano Lan; 231b Snow At Night; 232t vilax; 232bl Nick Pecker; 232br tr3gin; 233b Dr Morley Read; 235t Kerrie W; 236tr Nella; 236bl Oleksandr Kostiuchenko; 238b Bihrmann; 239b Foto2rich; 240t Kateryna Larina; 240b teekayu; 241b Byron Ortiz; 242 Oleksandr Kostiuchenko; 243tl Swetlana Wall; 243tr Oleksandr Kostiuchenko; 244t Michael Richardson; 244b Dory F; 245tl Wagner Campelo; 245tr IamTK; 247tl Xuanlu Wang; 248r revenaif; 249t Ole Schoener; 249b SeDmi; 250b Joe McDonald; 251b mhgstan; 253m&r kc_film; 254r Karel Gallas; 255t Galina Savina; 255bl Digital Media Pro; 255br EpicStockMedia; 257mr Florian Andronache; 258bl zaferkizilkaya; 258br Vespa; 259b catus; 261tr Laurens Hoddenbagh; 261b Rudchenko Liliia; 262br Armando Frazao; 263t yevgeniy11; 266b Ivaschenko Roman; 267b divedog; 268 puchan; 269l Volcko Mar; 269r Pakhnyushchy; 270m Alexey Wraith; 271l Mit Kapevski; 274 John_T; 275b Henri Koskinen; 279t Abdecoral; 279bl Fernando Tatay; 281l Sigur; 282l songsak; 282r

ChristinaDi; 283b Tony BKK; 284tl Nuttapong Wongcheronkit; 284tr Nataliia K; 284b wisawa222; 285l Tukaram. Karve; 285r Taisiia_89; 287m Mirelle; 290 Daimond Shutter; 291t Artistas; 292l Andrew M Allport; 292r Pavel Kovacs; 293t Dennis W Donohue; 294l JayPierstorff; 294m Jarun Tedjaem; 294b Sarah2; 295t D Kucharski K Kucharska; 295m Kazakov Maksim; 295b wasanajai; 296t Manfred Ruckszio; 297m prapann; 298b viks74; 299l Zhukerman; 300t Christian Roberts-Olsen; 300b Vlad Siaber; 301t KarenHBlack; 302t Leonid Ikan; 303m elod pali; 304t emmor; 304m brackish_nz; 304bl Ole Schoener; 304br watcher fox; 305t kalmukanin; 305m Microgen; 306tm Aleks Kvintet; 306tr ueuaphoto; 306b ElenVD; 307tl Alter-ego; 307tr Avigator Thailand; 307bl Anna Kucherova; 307bml Tim UR; 307bmr Bekshon; 307br Atwood; 308b Maksim Aan; 309tl Peter Hermes Furian; 309tr Tim UR; 309bm krungchingpixs; 309br inewsfoto; 310t Sergei25; 311tl HelloRF Zcool; 311tr Martin Fowler; 311bl Filipe B Varela; 312l Fotos593; 312r Jeff Grabert; 313br Elise Lefran; 314 Jakob Fischer; 320t Yellow Cat; 320b BasPhoto; 321t LAURA_VN; 322t nednapa; 322b homydesign; 323t SOMMAI; 323b margouillat photo; 325bl FatManPhoto; 325br Kovaleva_Ka; 326t El Greco; 326b John Copland; 327b Photology1971; 328tl Reinhold Leitner; 328tml vardy0; 328tm Africa Studio; 328tmr

IMG Imagery; 328tr Ragnarock; 328b Mark Winfrey; 329m Dale Stagg; 329b LUNNA TOWNSHIP; 330t Swapan Photography; 331t Manfred Ruckszio; 331bl oksana2010; 331bml NIPAPORN PANYACHAROEN; 331bmr Binh Thanh Bui; 331br Kyselova Inna; 332 Sara Winter; 333t LiliGraphie; 333b Rawpixel.com; 334t Le Do; 335tl cooperr; 335tr Pierre-Yves Babelon; 335b Veniamin Kraskov; 337b KULISH VIKTORIIA; 338t kzww; 338bl Maryan Melnyk; 338bm UMB-O; 339b Giorgio Rossi; 340t Matteo Ceruti; 341tl Orest lyzhechka; 341tr Vasilius; 341b barbajones; 342tl Dionisvera; 342tm Tommy Atthi; 342tr Nattika; 343t Dionisvera; 343m Swapan Photography; 343b Muellek Josef; 344tl Rich Carey; 344tr dolphfyn; 345b Olga Vasilyeva; 346t Yasonya; 347t Grandpa; 347b Andrei Rybachuk; 351t Claudine Van Massenhove; 354 Anna Nikonorova.

Ria Tan: 23b.

Irene Terry: 201m.

Vida van der Walt: 239t.

Jill Walker: 136b, 137l&br, 139bml, 140mr, 143t, 143bl, 147l, 148bl&br, 149tr, 152b, 153b, 160b, 163bl, 167t&b, 169t, 169bl, 176t, 190tl, 223b, 241t, 247r, 248l, 257t, 265tl&tr, 280b, 283t, 301b, 306tl, 309m, 309bl, 311br.

Daniel Wallick: 103bl, 130l.

Wellcome Collection, London: 307m, 316–17, 318b, 319t&b.

Wikimedia Commons: 8 Des Callaghan; 16r Mike Bayly; 19b Christian Fischer; 21t kvd; 21b Hermann Schachner; 26 Jason Hollinger; 28t Plantsurfer; 28bl Peter Coxhead; 28br Matteo De Stefano/MUSE; 31t James St John; 31b Rodney E Gould; 35b ANE; 37r Luis Fernández García; 39b Stefan.lefnaer; 41t LiquidGhoul; 45b Anatoly Mikhaltsov; 48tr Tim Bartel; 51m Melburnian; 58b Verisimilus T; 61b Bernd Haynold; 85b Louisa Howard; 87t Joachim Schmid; 87b William Wergin & Richard Sayre (colorized by Stephen Ausmus); 96t Canley; 96b Falconaumanni, 117b David McSpadden; 124 Stan Shebs; 131tl Harry Rose; 133b Casliber; 138l Christian Fischer; 138t Muriel Bendel; 140ml Kevmin; 162b Schurdl; 164t H Zell; 164bl Robb Hannawacker; 165bl&br Sten Porse; 166t Noah Elhardt; 168t Júlio Reis; 172t C T Johansson; 173tl Scott Zona; 173tr H-U Küenle; 175tl JoshNess; 175tr Qwert1234; 175b Mark Marathon; 176b Robert Kerton, CSIRO; 194t Jeffdelonge; 199m tanetahi; 200t Alex Lomas; 204tl BotBln; 205t gbohne; 205m Hans Hillewaert; 207b Heiti Paves; 208b Abbieeturner; 210l Scott Zona; 213t El Grafo; 228m John Game; 230t Frank Vincentz; 231m Christian Fischer; 233t Frank Vincentz; 234l Beentree; 234r Stan Shebs; 235b (all 4) Frank Vincentz; 243b Tino Ehrhart; 245b

Fritzflohrreynolds; 246t Wilferd Duckitt; 247ml Dcrjsr; 249m Hildesvini; 251t Frank Vincentz; 254l Haneeshkm; 256t&m Gideon Pisanty; 256b Manfred Werner; 257ml Gnangarra; 257b joanvicent; 260t Orchi; 260b Dalton Holland Baptista; 261tl Bob Peterson; 262t Orchi; 264b Alex Jones; 275t Dartmouth College Electron Microscope Facility; 276l James St John; 276r Matteo De Stefano/MUSE; 277l James St John; 278b John Tann; 279br Didier Descouens; 280t Gerhard Elsner; 286t Orchi; 287t Wouter Hagens; 287b Melburnian; 288l Alessandro Wagner Coelho Ferreira; 289b Tomas Figura; 297t Walter Siegmund; 299r JonRichfield; 305b Paul Henjum; 338br LepoRello; 348t Martin Stübler; 350t Miksu; 353b MurielBendel.

Adam Wilson: 191b.

Joseph S Wilson: 238t, 253b.

Emanuele Ziaco: 129.

Vojtěch Zavadil: 201t.

All other images in this book are in the public domain.

Every effort has been made to credit the copyright holders of the images used in this book. We apologize for any unintentional omissions or errors, and will insert the appropriate acknowledgement to any companies or individuals in subsequent editions of the work.

Acknowledgements

Andrew Drinnan would like to acknowledge Susi Bailey for her excellent copy-editing and headlines.

Jarmila Pittermann would like to acknowledge Drs A. Groover, E. Ziaco, F. Biondi, M. Pace, W. Anderegg, A. Jacobsen, S. Mayr, C. Leitinger, C. Brodersen, J. C. Domec, F. Lens and J. Rosell, along with L. Pittermann, A. Baer, C. Martinez, J. Wilson, and K. Cary, for sharing images and/or assisting with content and editing.